Methods in Cell Biology

VOLUME 68

Atomic Force Microscopy in Cell Biology

Series Editors

Leslie Wilson
Department of Biological Sciences
University of California, Santa Barbara
Santa Barbara, California

Paul Matsudaira
Whitehead Institute for Biomedical Research and
Department of Biology
Massachusetts Institute of Technology
Cambridge, Massachusetts

Methods in Cell Biology

Prepared under the Auspices of the American Society for Cell Biology

VOLUME 68
Atomic Force Microscopy in Cell Biology

Edited by

Bhanu P. Jena
Department of Physiology and Pharmacology
Wayne State University School of Medicine
Detroit, Michigan

J. K. Heinrich Hörber
Cell Biology and Biophysics Program
European Molecular Biology Laboratory
Heidelberg, Germany

ACADEMIC PRESS
An imprint of Elsevier Science

Amsterdam   Boston   London   New York   Oxford   Paris
San Diego   San Francisco   Singapore   Sydney   Tokyo

Academic Press
An imprint of Elsevier Science.
525 B Street, Suite 1900, San Diego, California 92101-4495, USA
http://www.academicpress.com

Academic Press
84 Theobalds Road, London WC1X 8RR, UK
http://www.academicpress.com

International Standard Book Number: 0-12-544171-1 (hb)
International Standard Book Number: 0-12-383851-7 (pb)

PRINTED IN THE UNITED STATES OF AMERICA
03 04 05 06 07 MM 9 8 7 6 5 4 3 2

CONTENTS

CONTRIBUTORS

Numbers in parentheses indicate the pages on which authors' contributions begin.

S. M. Altmann (311), Cell Biology and Biophysics Program, European Molecular Biology Laboratory, D-69117 Heidelberg, Germany

Nathan Becker (213), Department of Physics, University of California, Santa Barbara, Santa Barbara, California 93106

Martin Benoit (91), Center for Nanoscience, Ludwig-Maximilians-Universität München, D-80799 Munchen, Germany

Guillaume Charras (171), Bone and Mineral Center, Department of Medicine, The Rayne Institute, University College London, London WC1E 6JJ, United Kingdom

Aileen Chen (301), Department of Physiology and Biophysics, University of Miami School of Medicine, Miami, Florida 33136

Sang-Joon Cho (33), Department of Physiology and Pharmacology, Wayne State University School of Medicine, Detroit, Michigan 48201

Daniel M. Czajkowsky (231), Department of Molecular Physiology and Biological Physics, University of Virginia School of Medicine, Charlottesville, Virginia 22908

Andreas Engel (257), M. E. Müller Institute, Biocenter, University of Basel, CH-4056 Basel, Switzerland

Ernst-Ludwig Florin (193), Cell Biology and Biophysics Program, European Molecular Biology Laboratory, D-69117 Heidelberg, Germany

Marie-Cécile Giocondi (51), Center of Structural Biochemistry, French National Institute for Health and Medical Research U414, 34090 Montpellier Cedex, France

Christian Le Grimellec (51), Center of Structural Biochemistry, French National Institute for Health and Medical Research U414, 34090 Montpellier Cedex, France

Helen G. Hansma (213), Department of Physics, University of California, Santa Barbara, Santa Barbara, California 93106

Peter Hinterdorfer (115), Institute for Biophysics, University of Linz, A-4040 Linz, Austria

J. K. Heinrich Hörber (1), Cell Biology and Biophysics Program, European Molecular Biology Laboratory, D-69117 Heidelberg, Germany

Mike Horton (171), Bone and Mineral Center, Department of Medicine, The Rayne Institute, University College London, London WC1E 6JJ, United Kingdom

A. D. L. Humphris (337), H. H. Wills Physics Laboratory, University of Bristol, Bristol BS8 1TL, United Kingdom

Bhanu P. Jena (33), Department of Physiology and Pharmacology, Wayne State University School of Medicine, Detroit, Michigan 48201

Johannes H. Kindt (213), Department of Physics, University of California, Santa Barbara, Santa Barbara, California 93106

Assen Koitschev (141), Department of Otorhinolaryngology, Universität Tübingen, D-72076 Tübingen, Germany

Matthias G. Langer (141), Division of Sensory Biophysics, Universität Tübingen, D-72076 Tübingen, Germany[*]

Petri Lehenkari (171), Departments of Surgery and Anatomy, University of Oulu, FIN-90014 Oulu, Finland

P.-F. Lenne (311), Cell Biology and Biophysics Program, European Molecular Biology Laboratory, D-69117 Heidelberg, Germany

Eric Lesniewska (51), Laboratory of Physics, National Center for Scientific Research URA 5027, UFR Sciences et Techniques, 21078 Dijon Cedex, France

M. J. Miles (337), H. H. Wills Physics Laboratory, University of Bristol, Bristol BS8 1TL, United Kingdom

Pierre Emmanuel Milhiet (51), Center of Structural Biochemistry, French National Institute for Health and Medical Research U414, 34090 Montpellier Cedex, France

Vincent T. Moy (301), Department of Physiology and Biophysics, University of Miami School of Medicine, Miami, Florida 33136

Daniel J. Müller (257), Max-Planck-Institute of Molecular Cell Biology and Genetics, D-01097 Dresden, Germany

Emin Oroudjev (213), Department of Physics, University of California, Santa Barbara, Santa Barbara, California 93106

Lia I. Pietrasanta (213), Department of Physics, University of California, Santa Barbara, Santa Barbara, California 93106

Arnd Pralle (193), Cell Biology and Biophysics Program, European Molecular Biology Laboratory, D-69117 Heidelberg, Germany[†]

Manfred Radmacher (67), Drittes Physics Institute, Georg-August Universität, 37073 Göttingen, Germany[‡]

Bruno Samorì (357), Department of Biochemistry, University of Bologna, 40126 Bologna, Italy

Zhifeng Shao (231, 243), Department of Molecular Physiology and Biological Physics, University of Virginia School of Medicine, Charlottesville, Virginia 22908

Sitong Sheng (243), Department of Molecular Physiology and Biological Physics, University of Virginia School of Medicine, Charlottesville, Virginia 22908

John C. Sitko (213), Department of Physics, University of California, Santa Barbara, Santa Barbara, California 93106

Mario B. Viani (213), Department of Physics, University of California, Santa Barbara, Santa Barbara, California 93106

Giampaolo Zuccheri (357), Department of Biochemistry, University of Bologna, 40126 Bologna, Italy

[*]Present address: HNO-Klinik, D-72076 Tübingen, Germany

[†]Present address: Department of Molecular Cell Biology, University of California, Berkeley, Berkeley, California 94720

[‡]Present address: Department 1, Universit t Bremen, D-28359 Bremen, Germany

PREFACE

In the last decade, the atomic force microscope (AFM) has emerged as a powerful tool for cell biology research giving ultrahigh resolution in real time under near physiological conditions. Studies revealing nanometer-scale details of the living cell, subcellular organelles, and biomolecules, previously impossible due to the resolution limits of light microscopes, are now accessible using the AFM. Pioneering work and instrumental development were carried out by the groups of Paul Hansma, University of California, Santa Barbara; and Gerd Binnig, IBM Physics, Munich. For the first time, in 1989, Binnig visualized the process of pox virus release on living cells. Meanwhile, many other groups contributed exciting new insights at the cellular and molecular levels using the AFM. This book contains examples of more recent studies done with instruments that have reached a stage of development in which the biological question and the preparation procedures become the major objectives. The first section focuses on the application on cells and their membrane structures. The contribution by Benoit deals with cell adhesion, whereas Radmacher demonstrates how the elastic properties of cells can be determined using the AFM. The elastic properties of single stereocilia of haircells are studied by Langer and Koitschev, who with a combined AFM/patch-clamp setup simultaneously measure membrane potentials. The last two contributions of this section deal with membrane structures. Lesniewska *et al.* investigate special lipid structures, and Jena and Cho identify new cellular structures involved in exocytosis combining, for the first time, biochemical and AFM techniques. The second section focuses on extracted molecular structures. The contribution by Müller and Engel demonstrates the resolution possibilities of the instrument on two-dimensional protein crystals. Czajkowsky and Shao explain how supported lipid bilayers can be used as substrates for AFM investigations of various molecular structures. The contribution by Sheng and Shao introduces cryo preparation for the AFM, a procedure developed for electron microscopy. Zuccheri and Samorì describe DNA studies with the AFM, and in the last contribution of this second section, Kindt *et al.* provide a more general overview on AFM studies on molecular structures and introduce a force-measuring technique, which is the main theme of the third section. The last section focuses on actual instrumental developments and new methods. The contribution by Hinterdorfer describes how, at the AFM tip, ligands can be used to measure specific interactions even on cell surfaces. Humphris and Miles developed a new type of AFM which is able to measure forces in a dynamic way, whereas Chen and Moy describe static force measurements with a conventional AFM. Altmann and Lenne invented a new type of active stabilization for the AFM making force-clamp measurements, used for protein unfolding studies, more accessible.

Recently, the photonic force microscope (PFM) was developed (see first chapter by Hörber) by combining the principles of AFM, confocal microscopes, and optical tweezers into a new nanotechnological tool. The advantage of the PFM is its capability of entering the force range from 50 pN down 1/10 pN. This allows imaging of very soft membrane structures. Furthermore, the instrument provided new methods to study molecular structures with the observation of the thermal movement of the small particles used, e.g., the tip in an AFM. This became possible by using a new optical technique to detect the three-dimensional position of the particle with respect to the trapping laser focus, which allows imaging of three-dimensional networks as formed by the cytoskeleton with the position resolution determined by the instrument, which is actually about 1 nm. Thermal fluctuations of a particle also reflect all the influences of its environment. In this way, the technique can be used to map surface potentials, to study mechanical properties at the molecular level, and to measure viscosity. Pralle and Florin demonstrate in the last chapter how the PFM can be used to examine the biophysical properties of the plasma membrane in live cells.

In general, the book is designed to provide a working knowledge of the AFM and its potential for use in cell biology studies. The strengths and limitations of the AFM technique are discussed from a practical perspective. The book provides a wide range of applications in cell biology, which by no means are exhaustive. The examples described in the book will enable the reader to appreciate the power and scope of the AFM to study various aspects of cellular structure and function. Additionally, sample preparation and use of various approaches to study cells with the AFM provide practical guidelines to the reader. Since nothing can replace hands-on experience, once investigators make the determination that AFM or PFM could substantially contribute to their studies, collaboration with experienced people is advisable to determine feasibility and to gain hands-on experience prior to investing on equipment and personnel.

Bhanu P. Jena, Ph.D.
Heinrich Hörber, Ph.D.

CHAPTER 1

Local Probe Techniques

J. K. Heinrich Hörber

EMBL Meyerhofstrasse 1
69117 Heidelberg, Germany

I. Introduction

About 400 years ago, the invention of telescopes and microscopes not only extended our sense of seeing but also revolutionized our perception of the world. Extending this perception further and further has since been the driving force for major scientific developments. Local probe techniques extend our sense of touching into the micro- and nanoworld and in this way provide complementary new insight into these worlds with microscopic techniques. Furthermore, touching things is an essential prerequisite to manipulating things, and the ability to feel and to manipulate single molecules and atoms certainly marks another of these revolutionizing steps in our relation to the world we live in.

Local probes are small objects, e.g., the very end of sharp tips, whose interactions with a sample, or better, the surface of a sample, can be sensed at selected positions.

Proximity to or contact with the sample is required for high spatial resolution. This, in principle, is an old idea that appeared in literature from time to time, in context with bringing a source of electromagnetic radiation into close contact with a sample (Synge, 1928; O'Keefe, 1956; Ash and Nicolls, 1972), yet found no resonance and therefore was not pursued until recently. Nanoscale local probes require atomically stable tips and high-precision manipulation devices. The latter, based on mechanical deformations of spring-like structures by given forces—piezoelectric, mechanical, electrostatic, or magnetic—to ensure continuous and reproducible displacements with precision down to the picometer level, also require very good vibration isolation. The resolution that can be achieved with local probes is mainly determined by the effective probe size, its distance from the sample, and the distance dependence on the interaction between the probes and the samples measured. The latter can be considered to create an effective aperture by selecting a small feature of the overall geometry of the probe tip, which then corresponds to the effective probe.

The first of these local probe instruments was the scanning tunneling microscope (STM), which emerged during the early 1980s as a response to an issue in semiconductor technology (Binnig *et al.*, 1982). Inhomogeneities on the nanometer scale had become increasingly important as miniaturization of electronic devices progressed. The STM is an electronic–mechanical hybrid. The probe positioning is mechanics, whereas the interaction sensed by the tunneling current between probe and sample is of quantum mechanical origin. The physical effect of electron tunneling describes the strongly distant-dependent probability of electrons to cross a gap between two conducting solids before they really form a contact. The STM for the first time showed the atomic structure at the crystalline surface of silicon in real space and demonstrated that it was even possible to manipulate single atoms. The importance of this development was recognized when the Nobel Prize in Physics was awarded to Binnig and Rohrer in 1986.

In 1986, Binnig together with Quate and Gerber demonstrated that the short-range van der Waals interaction can also be used to build a scanning probe microscope (Binnig *et al.*, 1986). This new device was called the atomic force microscope (AFM). With no electron transport involved, even insulators could be studied down to atomic resolution. The essential part of an AFM, as for all scanning probe microscopes, is the tip that determines by its structure the type of interaction with a surface; and by its geometry, the area of interaction. The original idea for the AFM was to measure the van der Waals interaction of an atom at the very end of the tip with atoms at a surface of a solid substrate. To bring a single atom at a tip close to within angstrom distance toward a surface is only possible if the surface is atomically flat (Fig. 1c), such as, for example, the crystalline surface of mica. If the surface is rough on a nanometer scale (Fig. 1b), groups of atoms can interact and determine, according to their size, the possible resolution. With a roughness at the micrometer scale (Fig. 1a) the macroscopic level is reached where instruments like the surface profiler are able to measure surface roughness. A similarly important part of the scanning probe microscope is the mechanism which moves the tip closer to the surface and scans it across with precision fitting to the highest resolution. What enables such precise manipulation is the property of some materials to change size proportional

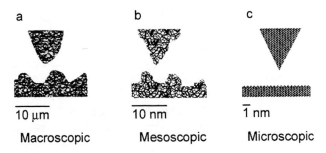

a   b   c

10 µm   10 nm   1 nm

Macroscopic   Mesoscopic   Microscopic

Fig. 1 Scanning probe tip structures shown at different scales.

to an applied electric field. These materials can also generate an electric field if a force is applied, an effect first described by Pierre and Jacques Curie in 1880 for quartz. The piezo-tube scanner is widely used to produce movements in all three directions easily and consists of a thin-walled hard piezo-electric ceramic that is radially polarized. Electrodes are attached to the internal and external faces of the tube. The external electrode is split into quarters parallel to the axis as shown in Fig. 2. By applying a voltage between the inner and all the outer electrodes, the tube expands or contracts and in this way either moves a tip closer to a surface or retracts it from a surface, respectively. If the voltage is applied just between the inner and one outer electrode, the tube will bend, i.e., moving the tip along the surface, with a precision determined by both the noise of the voltage source used and the overall mechanical stability. The disadvantage of these piezo-tubes is that the tip is not scanned exactly parallel to the surface but is moved on an arc, leading to an effect known as "eyeballing" when large scans are carried out. Another problem of piezo-materials is the hysteresis, which like the arc motion must be corrected by the electronic equipment controlling the movement by providing the necessary voltage.

In the meantime, many other types of scanning probe microscopes using various types of interactions have been developed and are too numerous to mention in this short

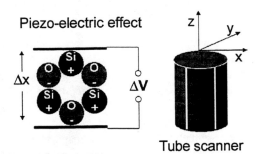

Piezo-electric effect

Tube scanner

Fig. 2 Piezo-electric effect of quartz and the piezo-ceramic tube scanner with inner and segmented outer electrodes used in scanning probe microscopes.

introduction. I prefer, therefore, to name only one other: the scanning nearfield optical microscope (SNOM), developed by Pohl *et al.* (1988), which is, as the name implies, the near-field equivalent to the conventional optical microscope working in the farfield of the radiation. The STM, on the other hand, can be seen as the nearfield equivalent to the electron microscope. The optical microscope, like other types of microscopes using radiation in the farfield range, is limited in its resolution by the wavelength of the radiation. This limit, reported by Abbe in 1873, restricts the optimal resolution to several hundred nanometers for using visible light. The only way of overcoming this limit is by using nearfield effects observed within a wavelength from a radiation source. In high resolution, the very small tip can be used again. The tip of a SNOM is, at least in many instruments, a specially prepared end of an optical fiber, which acts as a light source. The interaction of the electromagnetic nearfield at the tip with the surface determines how much light is radiated from the source and how much is reflected back into the optical fiber. In this way the aspects of the surface structure correlated to the interaction with electromagnetic fields can be studied.

Many types of scanning probe microscopes have been developed and can be used not only for measuring surface topologies but also for measuring various material properties at or close to surfaces. This can be done in vacuum, in gas, or in liquids in a broad temperature range with a resolution down to either the atomic or the molecular level. In this way, it is the only type of microscopy that can complement optical microscopy in biology on a smaller scale. Additionally, these instruments allow manipulations at either the single-atomic or the molecular level, making experiments which no one ever dreamed of 20 years ago possible. Experiments at the nanometer scale provide a complete new insight into processes which, before the development of these instruments, were accessible only by ensemble-average processes, where all of the elements can never be identical, and all of the information concerning the behavior of individuals is lost. With the available information on single components using scanning probe techniques we can now learn how processes, which we were previously unaware of, are determined by the properties of the single elements of such ensembles.

II. Scanning Tunneling Microscopy

It is of particular interest to understand the images of biological structures obtained by the STM, as this technique allows imaging with a signal-to-noise ratio unequalled by other techniques and under near-physiological conditions (Hörber *et al.*, 1988; Hörber, Schuler, Witzemann, Schröter *et al.*, 1991; Hörber, Schuler, Witzemann, Müller *et al.*, 1991; Heckl *et al.*, 1989; Ruppersberg *et al.*, 1989; Göbel *et al.*, 1992; Maaloum *et al.*, 1994). This is an advantage that can only be exploited by having a deeper knowledge of both the "tunneling" or electron transport mechanism and the environmental conditions under which it takes place. Furthermore, a means to understand the nature of the images produced, namely, a model that can be used as reference, is necessary. For this purpose, a sample can, for instance, be imaged by scanning tunneling and electron microscopy, and the results can then be compared to investigate the physical mechanism of image contrast formation in the STM for biological samples. For example, the electron microscope can produce a three-dimensional image of a helical structure, e.g., a bacteriophage tail. For such experiments performed by my group, T5 tails were purified and adsorbed to glow discharged indium tin oxide (ITO) surfaces in solution (Guénebaut *et al.*, 1997). The surface was washed with distilled water, which was removed partially by blotting, leaving only a thin layer of aqueous solution. The STM used was a noncommercial "pocket-size" type (Smith and Binnig, 1986), equipped not only with tungsten tips etched in KOH by alternating currents but also with a patch-clamp amplifier allowing measurements down to 0.5 pA with an equivalent noise current of 200 fA. Importantly, all the current measurements were carried out in the picoampere range. The feedback circuit controlling the movement of the tip in the z direction, which is the distance to the sample, was equipped with a logarithmic amplifier to correct for the exponential behavior of the current. However, the direct measurements of the current variation giving the constant height images usually are not corrected for this exponential behavior. Therefore, the logarithms of these images were extracted, before combining the left-to-right and right-to-left scans, to produce a real-space representation of the specimen. Both scans can be normalized by histogram equalization and, after combining the different scan directions, they could be compared to the transmission electron microscope (TEM) reconstruction of the phage tail structure. The time constant of the feedback used was the limiting factor in the tip movement, and adding both right- and left-scan images significantly suppressed the z feedback effect. With this setup, recording the feedback signal simultaneously with the current signal is also possible. In principle, this combined constant height and constant current imaging mode increases the height resolution of the instrument, showing the fine structure on the top of the phage tails as a constant height image (Fig. 3).

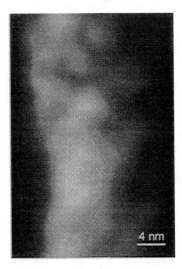

4 nm

Fig. 3 STM image of the tail of the bacteriophage T5 prepared on an ITO surface. The scan size is 18×18 nm^2. The picture was taken with a 30-pA current at a 120-mV tip voltage within a thin layer of water.

The general characteristics of the bacteriophage T5 tail make it an excellent test specimen for comparing TEM and STM results. Bacteriophage T5 is a member of the T-odd phage family having an icosahedral head with a diameter of 80 nm. Its noncontractile flexible tail is 160 nm long and is composed of 120 copies of a 58 kDa protein. The proteins are arranged as trimers, each trimer forming a ring with an external diameter of 11 nm. The superposition of 40 of these rings, with a 40-degree angular shift between each stack, confers a helical symmetry to the tail. The tail model was calculated from cryo-TEM images using helical reconstruction methods. The general dimensions of the tail allowed for its easy identification in the STM images. These Bacteriophage T5 tail images exhibit size features approaching 3 nm, which were used in comparison to the reference obtained from electron microscopy data.

As for other biological materials observed using STM, the tail appears with a positive contrast and exhibits complex features that prevent trivial interpretation of the images. It is difficult to correlate these two observations with the classical concept of the electron "tunneling" mechanism between two conductors through an energetically forbidden region. Nevertheless, it is clear from the many experiments performed thus far that it is possible to image nonconductive molecular structures using the STM. The imaging of cyanobiphenyl monomolecular layers of liquid crystals, where near-atomic details were observed, confirmed the transfer of electrons through thin, nonconductive, and organic materials (Smith *et al.*, 1989, 1990). However, the mechanism by which this phenomenon occurs through thicker nonconductive layers of organic material, either a multilayered arrangement of small molecules or larger molecular structures, is still not understood. The role of water, which is always present under ambient conditions (Freund *et al.*, 1999), while keeping biological samples under physiological conditions, remains unknown.

By comparing TEM results to those of STM on these bacteriophage tails it became clear that, although the STM images did not show the surface of tail structures, they, however, could be directly compared to contrast-inverted TEM images. The actual situation for STM imaging such samples, i.e., the position of the tip with respect to the sample, can be studied using current/distance measurements. It was found that phage tails freshly adsorbed on ITO-coated glass retained a thin (50- to 100-nm) film of water. While imaging, the tip was immersed several tens of nanometers into this film; at these distances, currents of 5–50 pA were observed. In this situation (Fig. 4), the electrons had to cross a water layer of up to several tens of nanometers in addition to the molecular structure, but still could provide a resolution of 3 nm. The exponential distance dependence of the measured current decays faster in water than through the macromolecules, leading to a positive contrast. Without hypothesizing on the nature of the electron transfer mechanism across biological material and through water, the observation that the protein structure has less resistance to the current than to the surrounding aqueous solution is very interesting. This produces a positive image of the specimen, while cryo-TEM, based on high-energy electron scattering by the specimen, produces a negative image. A possible explanation might be that as denser protein structures are more ordered low-energy electrons do not scatter as frequently.

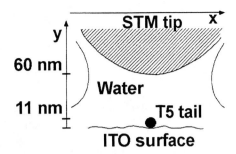

Fig. 4 Scaled schematic drawing of the imaging situation as determined by current/distance measurements. The diameter of the tail is 11 nm and the tip surface distance while imaging is about 60–70 nm above the surface. The water is kept by cooling the sample as a thin layer of 100–200 nm on top of the sample.

If the physical basis for the use of the STM on biological structures can be identified, then the STM can become an important complement to TEM in structural studies, as completely different preparation methods are used and the samples remain hydrated under close to physiological conditions.

III. Atomic Force Microscopy

A. Combination with Optical Microscopy

It has been shown in many experiments that the AFM can be used to study biological structures under physiological conditions. It is even possible for the AFM to both image living cells (Häberle *et al.*, 1991) and study dynamic processes at the plasma membrane, although such experiments are quite difficult, as the AFM cantilever is by far much more rigid than cellular membrane structures (Schneider *et al.*, 1997; Jena and Cho in this book). The preparation of cells and the parallel optical observation, which are necessary for having standard biological controls for cell activities available, present other problems. To address these problems, in 1988 we initiated an IBM Physics project in Munich to develop a special AFM built into an inverted optical microscope. This instrument could make the first reproducible images of the outer membrane of a living cell, fixed only by a pipette in its normal growth medium (Hörber *et al.*, 1992; Ohnesorge *et al.*, 1997). This pipette was moved by a conventional piezo-tube scanner. The detection system, in principle, was a normal optical detection scheme using a glass fiber as a light source and placed very close to the cantilever (Fig. 5). This configuration allowed a very fast scanning speed for imaging cells in the variable deflection mode, as the parts moving in the liquid are very small. Therefore, in contrast to the standard procedure of imaging cells attached to a flat substrate on the scanning stage, neither significant excitation of disturbing waves nor convection in the liquid occurs. Additionally, the severe deformation of the cells was avoided, which normally occurs when they are squeezed between a solid substrate and a cantilever. Since it was possible to keep the

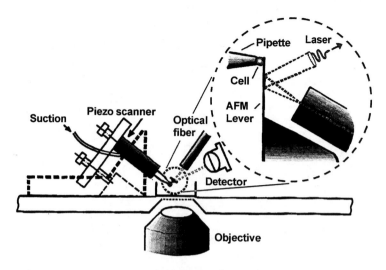

Fig. 5 A schematic drawing of the AFM built onto an inverted optical microscope with a patch-clamp pipette as a sample holder. An optical fiber as a light source very close to the cantilever is used for the optical detection of the cantilever deflection. The detection of the reflected light is done by a quadrant photo-diode above the sample chamber.

cell alive and well for days while imaging, this made studies of live activities and kinematics in addition to the application of other measuring techniques possible. With this step in the development of scanning probe instruments, the capability of optical microscopy to investigate the dynamics of biological processes of cell membranes under physiological conditions could be extended into the nanometer range with the help of the AFM.

In the initial experiments with the AFM, we observed the reaction of cultured monkey kidney cells infected by orthopox viruses. We usually saw no reaction during the first few minutes after adding the virus suspension to the fluid chamber where the cells were kept in buffer solution. Yet in one case we observed a decaying protrusion after about 1 h. The size of the protrusion was comparable to that of a virus (200–300 nm), but we observed an effect like this only once. The fact that we usually did not observe the endocytosis of the virus might have been due to a shadowing by the lever and the imaging tip, which prevented the penetration of viruses into this area. On the other hand, at about the time when the virus would be expected to enter the cell (a few minutes after adding viruses according to estimates of diffusion times in the surrounding liquid) we noticed a strong softening of the cells, which was always accompanied by the danger of the tip easily penetrating the membrane and the images losing considerable contrast. One might imagine that a virus only locally modifies the membrane to enable its entry into the cell. However, from the fact that the dramatic softening of the cell membrane is always observed when viruses are added, we conclude that the cell membrane as a whole is affected by the penetration or adhesion of the viruses. It is known that 4 to

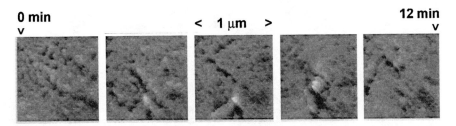

Fig. 6 Exocytotic process imaged by AFM 3 h after monkey kidney cells were infected by pox viruses. The size of the structure seen is about 200 nm and similar to the size of viral particles.

6 h after infection the first viruses reproduced inside the cell and emerged from the cell through the cell membrane. However, approximately 2.5 h after infection we observed a series of processes occurring in our SFM images. Single clear protrusions became visible and grew in size. The objects quickly disappeared and the original structures on the cell surface were more or less restored. Such processes can occur several times in the same area and last about 90 s for a small protrusion (about 20-nm lateral extent) and up to 10 min for a larger one (cross section of about 100 nm). Each process proceeds distinctly, apparently independently of the others, and is never observed with uninfected cells and never prior to 2 h after infection.

The fact that the growing protrusions abruptly disappeared after a certain time led us to believe that we observed an exocytodic process but not the virus release. First-progeny viruses are known to appear 5–8 h after infection and they are clearly bigger than the structures observed. It is also known, however, that after 2–3 h only the early stage of virus reproduction is finished and the final virus assembly has just begun. Since the protrusions are observed after this characteristic time span, we believe that they are related to the exocytotic processes connected to the virus assembly. Significantly more than 6 h after infection even more dramatic changes are seen in the cell membrane (Fig. 6). Large protrusions, with cross sections of 200–300 nm, grow out of the membrane near deep folds. These events occur much less frequently than those which occur after only 2 h. These protrusions also abruptly disappear, leaving behind small scars on the cell surface. Considering the timing and their size, we believe these protrusions are progeny viruses exiting the cell. Assuming that approximately 20–100 viruses exit the living cell and that roughly 1/40 of the cell surface is accessible to our SFM, one should be able to observe one or two of these events for each infected cell. We actually observed two processes exhibiting the correct size and timing during one 46-h experiment on a single infected cell: one after 19 h and the other after 35 h. It is known from electron microscopy that individual viruses exit the cell at the end of finger-like microvilli that are formed at the cell membrane. Figure 7 actually shows a finger-like protrusion at whose end an exocytotic process is observed. The release of the particle observed also occurs in a region where the cell membrane is dominated by finger-like structures. This striking similarity to results from electron microscopy made us believe that we indeed had imaged the exocytosis of a progeny virus through the membrane of an infected live cell.

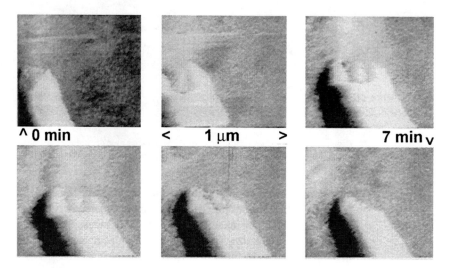

Fig. 7 Sequence of images showing the escape of a viral particle at the end of a microvillus 19 h after infection of the cells.

With the setup developed, it was finally possible to observe structures as small as 10–20 nm at high-imaging rates of up to one frame per second. This, in principle, gives one access to processes besides endo- and exocytosis such as the binding of labeled antibodies, pore formation, and the dynamics of surface structures in general. Nevertheless, still after more than 10 years much work must be done to control the interaction between tip and plasma membrane structures, which can be influenced quite strongly by the so-called extracellular matrix of cells containing a broad variety of sugars and other polymer structures.

As with the integrated tip of the cantilever, forces in the range of some 10 to 100 pN are applied to the investigated cell membrane, and the mechanical properties of cell surface structures dominate the imaging process. On the one hand, topographic and elastic properties of the sample in the images are combined; on the other hand, additional information is provided regarding cell membranes and their dynamics in various situations during the life of the cell. To separate the elastic and topographic properties, additional information is needed, which can be provided either by topographic data from electron microscopy or by the use of AFM modulation techniques. The pipette–AFM concept is very well suited for such modulation measurements, because, as mentioned earlier, perturbation by the excitation of convection or waves in the solution are extremely small compared to the normal situation in AFM measurements. Furthermore, the cells held by a pipette are supposedly in a state much more comparable to the natural situation than a cell adhering to a substrate. For a thorough analysis of a cell membrane elasticity map, one would have to record pixel by pixel a complete frequency spectrum of the cantilever response and derive image data from various frequency regimes. This would require too much time for a highly dynamic system like a living cell. Nevertheless, we performed experiments on live cells which showed in some cases a certain weak mechanical resonance in the regime of several kilohertz. Such resonance might be used to both characterize cells and provide a kind of spectroscopic fingerprint for either cellular processes or even drug effects, which may lead to new medical diagnosic methods at the cellular level.

B. Combination with Patch–Clamp Technique

In principle a force microscope, where the cell is fixed on top of a patch-clamp pipette, makes the combination of patch-clamp measurements on ion channels in the membrane of whole cells with force microscope studies already possible. Therefore, the next logical step along this line is the development of a combined patch-clamp/AFM setup that can be used to investigate excised membrane patches (Hörber et al., 1995). The motivation to develop such a setup was to study specialized ion channels in the membrane, which become activated by mechanical stress. These channels are quite important to our sense of touching and hearing. In 1991, we initiated the development of this instrument at the Max-Planck Institute for medical research in Heidelberg. A new much more stable patch-clamp setup had to be developed to satisfy the needs of AFM applications (Fig. 8). The chamber, where a constant flow of buffer solution guarantees the proper conditions for the experiments, consists of two glass plates, one on top of the other, and the water between is kept in place just by its surface tension. In this way, this "flow cell" is freely accessible from two sides. The chamber, along with the optical detection of the AFM lever movement and a double-barrel application pipette, is mounted on an *xyz* stage.

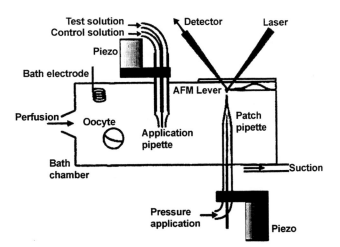

Fig. 8 Schematic top view of the patch-clamp/AFM setup within a bath chamber. The patch pipette is shown in front of the cantilever. Bath and pipette electrodes are used for electrophysiological recordings. A glass plate is positioned behind the cantilever to reduce fluctuations of laser beam direction by movement of the water–air interface. Different solutions can be applied to the excised patch by a two-barrelled application pipette. If necessary, continuous perfusion of the path chamber can be stopped during AFM measurements.

The pipette is integrated so that the setup can also be used for standard patch-clamp measurements in either the presence or absence of various chemicals. The patch-clamp pipette itself is mounted on a piezo-tube scanner fixed with respect to the objective of an inverse optical microscope necessary to control the approach to both the cell and the AFM lever. In such experiments, small membrane pieces are excised from the cell containing none, one, or only a few ion channels. This enables the study of currents through a single ion channel opening on a micro- to millisecond time scale resulting in currents in the picoampere range.

In initial experiments, the structure of the patch-clamp pipettes made of different types of glass was studied in standard solutions. There was no indication that thermal polishing could improve the structure at the end of the pipettes to make a so-called giga-seal more likely. Such an extremely good contact between glass and membrane is necessary to prevent leak currents that would make measurements of currents at the picoampere range impossible if the normal membrane resistivity were to get below 1 GΩ.

In such a setup with a membrane patch at the end of the pipette, the AFM can image the tip of the pipette with the patch on top (Fig. 9). Furthermore, structural changes in the membrane patch according to the changing pressure in the patch pipette can be monitored, along with the reaction of the patch to the change of the electric potential across the membrane patch. Important information was obtained from the first images demonstrating that excised membranes still contain cytoskeleton structures. These stabilizing structures of the membrane are stiff enough to obtain a 10- to 20-nm resolution in the images, showing reproducible structures with dynamics that can be activated by the application of forces.

Voltage-sensitive ion channels change shape in electrical fields leading eventually to the opening of the ion-permeable pore. To investigate the size of this electromechanical transduction, we examined the relevant movement with the AFM on membrane patches

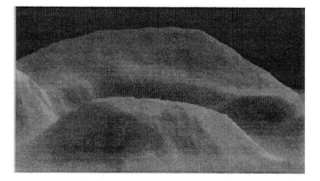

Fig. 9   AFM image of an excised membrane patch at the end of a pipette pulled to a 2-μm opening. The image is obtained in constant-force mode with a scan frequency of 40 Hz. Height differences seen are about 1 μm. The pipette is seen in the background, and the smoother structure of the inside-out patch is seen in the front.

obtained from cells of a cancer cell line (HEK 293), which are kept at a certain membrane potential (voltage-clamped) (Mosbacher *et al.,* 1998). We used either normal cells as controls or cells transfected with Shaker K+ ion channels. In control cells, we found movements of 0.5 to 5 nm normal to the plane of the membrane. These movements tracked a ±10 mV peak–peak AC carrier stimulus to frequencies >1 kHz with a 90- to 120-degree phase shift, from a displacement current. The movement was outward with depolarization, and the holding potential was only weakly influenced by the amplitude of the movement. In contrast, cells transfected with a noninactivating mutant of Shaker K^+ channels showed movements that were sensitive to the holding potential, decreasing with depolarization at between −80 and 0 mV.

Further control experiments used open or sealed pipettes and cantilever placements just above the cells. The results suggested that the observed movement is produced either by the cell membrane rather than by the artificial movement of the patch pipette or by the acoustic or electrical interaction of the membrane and the AFM tip. The large amplitude of the movements and the fact that they also occur in untransfected cells with a low density of voltage-sensitive ion channels imply the presence of multiple electromechanical motors. These experiments suggest that the AFM may be able to exploit the voltage-dependent movements as a source of contrast for imaging membrane proteins. Due to this newfound motility, the consequences for cell physiology remain to be determined.

IV. Force Spectroscopy

The approach to molecular structures in cells and to their interaction in molecular biology is traditionally a chemical one. Therefore, molecular interactions are characterized by binding constants, on- and off-rates, and corresponding binding energies. The relevant energies for single binding events range from thermal energy up to some one hundred times that of the thermal energy if covalent bonds are involved. In addition to covalent bonds, especially hydrogen bonds with energies between 4 and 16 kT play an important role. At interfaces of macromolecular structures, coulomb and dipole forces determine the interaction with the aqueous environment and in this way also between these molecules. An important question arises regarding these interactions: is their distance dependent on the actual environment? This question can be addressed at the single-molecule level by the AFM, which presents a new view of molecular interactions in terms of interaction forces and their distance dependence. At the molecular level, an important aspect in the measurement of forces is the dependence of the force measured on the time scale it was applied. With respect to molecular interaction potentials, an applied force simply deforms the potential. Due to the thermal fluctuations of molecular structures, it becomes more likely that the bond will break during a certain observation time. For instance, the off-rate for biotin/avidin binding at room temperature is on the order of 6 months. If a small force of about 80 pN is applied, the binding potential is deformed such that it becomes lower, reducing the off-rate to about 9 s. Doubling this force decreases the lifetime of the binding further by three orders of magnitude. With the available computers, the normal time scales of molecular computer simulations are restricted to from picosecond to nanosecond time scales. Therefore, at these time scales, simulations of rupture forces of biotin/avidin lead to forces of 600 pN. In AFM measurements done at the 100-ms time scale, the actual measured forces are between 100 and 200 pN (Florin *et al.*, 1994).

A. Molecular Adhesion

Protein adsorption is a very important aspect in many biomedical and biotechnological applications. For instance, many chromatographic separations, such as hydrophobic, displacement, and ion-exchange chromatography, are based on differences in the binding affinities of proteins to surfaces. In addition, *in vitro* cell cultures require cell-surface adhesion, which is mediated by a sublayer of adsorbed proteins.

Protein adsorption is a net result of various complex interactions between and within all components including the solid surface, the protein, the solvent, and any other solutes present. These interaction forces include dipole and induced dipole moments, hydrogen bond forming, and electrostatic forces. All these inter- and intramolecular forces contribute to a decrease in the Gibbs energy during adsorption.

An important question regarding the protein adsorption process is its reversibility. One approach to this problem is an analysis of the time course of adsorption. As adsorption is a multistep process, an important question is, at which stage does the process become irreversible? The most common way to quantify adsorption is by means of the adsorption isotherm, where at constant temperature, the amount of molecules adsorbed is plotted against the steady-state concentration of the same molecules in bulk solution. Adsorption isotherms provide a convenient method for determining whether an adsorption process is reversible. Reversibility is commonly observed in the adsorption of small molecules on solids, but only rarely in the case of more complex and randomly coiled polymers.

Proteins are polymers; however, globular proteins are not true random coils. The native state of these proteins in aqueous solution is highly ordered. Most polypeptide backbones have little or no rotational freedom. Therefore, significant denaturing processes must occur to form numerous contacts with any surface. Structural rearrangements may occur in such a way that the internal stability of globular proteins prevent them from completely unfolding on a surface into loose "loop-and-tail"-like structures. Thus, the number of protein-to-surface contacts formed at the steady state is determined by a subtle balance between intermolecular and intramolecular forces. Therefore, the thermodynamic description of the protein adsorption process in general should be based on the laws of irreversible thermodynamics (Norde, 1986; Haynes and Norde, 1994). The process is strongly time dependent, and some of the involved steps of molecular rearrangement are remarkably slow and probably lead to the significant binding of proteins only on a time scale of seconds (Hemerlé *et al.*, 1999). With different time scales for the various interactions taking place, the adsorption process can be divided into fast steps, which can be reversible, and slow steps, where protein structure rearrangements that are determined by the surface environment can occur. The latter processes in many cases may become irreversible.

The adhesion forces established by single proteins, e.g., protein A and tubulin molecules, within contact times from milliseconds to seconds, can be measured by the AFM. Protein A and tubulin, both globular proteins, can be seen as examples for different types of protein binding. The molecules can be attached to the cantilever tip and then brought into contact with different surfaces, which also can be covered with various molecular structures. These surfaces can be metal surfaces, for example, gold, titanium, and indium-tinoxide (ITO). Such investigations are relevant for many biomedical applications, such as the optical transparent ITO, for the development of interfaces between biological molecules and electro-optic devices.

By approaching gold, indium-tinoxide, and titanium surfaces with protein-coated tips and then retracting, the adhesion forces between proteins and these metal surfaces can be measured. As a result, for the first contact, there is a specific interaction characteristic for different molecules and metals (Eckert *et al.,* 1998). With the high reproducibility of one certain value of adhesion forces for a series of measurements, it must be assumed that in these experiments a certain type of interaction between a certain amino acid group within a protein and the metal determines the first contact. The exact nature of the measured adhesion forces still must be determined. Nevertheless, it could be demonstrated that, with adequate preparation, such a technique can be used not only to measure interactions at the single-molecule level but also to study the dependence of these interactions under various environmental conditions.

B. Intramolecular Forces

The modular structure of proteins seems to be a general strategy for resistance against mechanical stress not only in natural fibers but also in the cytoskeleton. One of the most abundant modular proteins in the cytoskeleton is spectrin. In erythrocytes, spectrin molecules are part of a two-dimensional network that is assumed to provide red blood cells with special elastic features. The basic constituent of spectrin subunits is the repeat, which has about 106 amino acids and is made of three antiparallel α-helices, folded into a left-handed coiled-coil. The repeats are connected by helical linkers. The mechanical properties of several modular proteins, for example, titin, have already been investigated by the AFM (Rief *et al.,* 1997). Such experiments have demonstrated that the elongation events observed during stretching of single proteins may be attributed to the unfolding of individual domains, and experiments with optical tweezers have corroborated these results (Kellermayer *et al.,* 1997; Tskhovrebova *et al.,* 1997). These studies suggest that single domains unfold one at a time in an all-or-none fashion when subjected to directional mechanical stress.

A major technological breakthrough in protein folding studies using the AFM was our development, during the last few years, of the multiple detection system (Fig. 10). This concept allows long-term stabilization of the distance between the cantilever tip

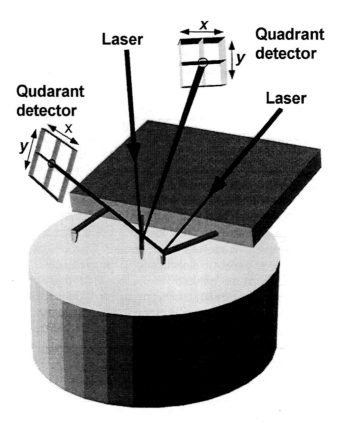

Fig. 10 Schematic drawing of the double-detection system using two cantilevers to separate force and distance measurements. One cantilever is always in contact with the surface and controls, with the help of a feedback circuit, the distance between the second cantilever and the surface.

and the surface with a fraction of angstrom precision and, in this way, makes so-called force-clamp measurements possible. By using this technique, unfolding and refolding experiments on small protein structures, such as spectrin, with only four repeats, became possible. In these experiments, proteins are attached to clean, freshly prepared gold surfaces (Lenne *et al.*, 2000). While approaching this surface with the tip of an AFM cantilever, a single molecule can become attached to the tip, and forces in the range of piconewtons to nanonewtons can be applied to the molecule (Fig. 11). Folded proteins can be stretched to more than 10 times their length, reaching almost their total contour length. The force–extension curves show a characteristic sawtooth-like pattern. The reaction coordinate of unfolding is imposed by the direction of pulling, and in this way unfolding events occurring in a single protein can be examined. Each peak is attributed to a breakage of a main stabilizing connection of a folded protein structure. Recombinant DNA techniques were also used to construct tandem repeats from one single spectrin domain for such experiments. The method extends the monomer (R16) at both ends so

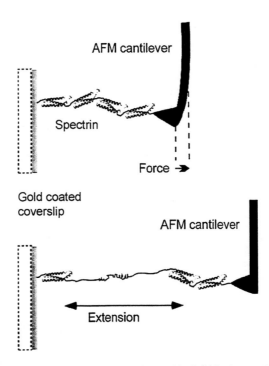

Fig. 11   Unfolding by AFM of an artificial spectrin polypeptide chain fixed on a gold surface by cysteines.

that the polymeric protein product contains a 13-residue linker between the consecutive R16 units with two cysteine residues at the C-terminal end. The force–extension relation could be measured on a four-repeat construct fixed with the cysteine residues introduced at the C-terminal end to the gold surface. In the sawtooth-like pattern of the force–extension curves, each peak represents the breaking of the major stabilizing connection. After each peak, the force drops back as the additional length of the protein chain becomes available. The maximal extension in these experiments is variable as the polymer either is not always picked up at the end or has attached at more than one site to the surface. Some of the measured force–extension curves depending on the actual unfolding process can be described by the so-called worm-like chain model (WLC), which predicts the isothermal restoring forces of a flexible polymer, acts as an entropic spring during extension, and defines a persistence length to the polymer chain, which turns out to be 0.6 nm for the spectrin chain. Two consecutive force peaks are spaced by a distance that represents the gain in length produced by a single unfolding event. These distances were measured for several hundred unfolding events. The histogram of the measured distances reveals a length with two statistically relevant peaks. When fitted with two Gaussian distributions the maxima are at 15.5 and 31 nm. The force distribution associated with the short and long elongation events is clearly visible. The mean forces for the 15.5- and 31-nm elongation events are 60 and 80 pN, respectively, at a pulling speed of 3 nm/ms. With these experiments it was demonstrated for the first time at the single protein level that structural domains might unfold in several discrete steps and in a much more complex manner than that expected before such experiments became possible. The appearance of different elongation events suggests that at least one intermediate state for the spectrin repeat exists between the folded and the completely unfolded state. Thermodynamic studies could not show such an intermediate. Statistical analyses of the force curves demonstrate that well-defined short elongation events are half the length of the total unfolded length of one domain. It may thus correspond to either the gain of length between a folded and partially unfolded domain or between an already partially unfolded and a completely unfolded domain. The molecular pathway of unfolding is unknown for the spectrin domain, but it may well involve a partial opening of the bundle and/or the loss of the secondary structure at the ends of the helices, which may be elongated before the most stable part finally breaks.

The complexity of protein folding is high. However, by applying a force the possible unfolding pathways obviously become very strongly restricted. We can explain our results with only two possible unfolding pathways. One leads directly from the folded to the unfolded state; the other, via an intermediate state. The intermediate state is conceptually available along both pathways, but each pathway by itself leads from the native folded state 0 either to the partially unfolded state 1 or to the completely unfolded state 2. The relative difference of the height of the free energy barrier along either of the two pathways determines whether state 1 or state 2 is attained. The advantage of such strongly simplified modeling is that only one free parameter is needed for differentiating between the two pathways.

In the native folded state, the protein is in state 0 at the bottom of the potential well. The mechanical stress applied by the AFM tip not only decreases the barrier height for thermally activated unfolding but also reduces the options of the protein to follow either path 1 or path 2 during unfolding. A certain protein will follow only one path leading to the observed bimodal probability distribution with 35 and 65% probability for paths 1 and 2, respectively. By including this scenario in a simple Monte Carlo simulation, the reaction kinetics can be tested simultaneously for the short and long elongation events. This means that along paths 1 and 2 the kinetics can be characterized by both the width of the first barrier and an effective so-called attempting frequency, which includes the barrier height as an exponential multiplication factor, normalized by the thermal energy.

These experiments not only shed new light on the mechanical behavior of spectrin, which is an essential cytoskeletal protein with unique elastic properties but also demonstrate for the first time that the unfolding of spectrin repeats can occur in a stepwise fashion during stretching. The force–extention patterns exhibit features that are compatible with the existence of at least one intermediate state. These new details regarding the unfolding of single domains revealed by precise AFM measurements show that force spectroscopy can be used not only to determine forces that stabilize protein structures but also to analyze the energy landscape and the transition probabilities between different conformational states. For more information about such experiments, see the chapter by Lenne *et al.* (2000).

C. Combination with Optical and Patch–Clamp Techniques

For studies on mechanosensitive ion channels, the AFM can be used to apply calibrated forces in the piconewton range to the membrane while measuring the ion currents in parallel with the patch-clamp technique. Still, the structure of such mechanosensitive ion channels is not known and no genetic information is available. In this situation, a possible approach is to find a natural system where a certain type of channel is important for the function. Such a highly specialized system is the mechanoelectrical transduction system of the hair cells of the inner ear.

Cochlear hair cells of the inner ear are responsible for both detecting sound and encoding the magnitude and time course of an acoustic stimulus such as an electric receptor potential, which is generated by a still unknown interaction of cellular components. Specialized hair bundles called stereocillia are present at the apical end of these cells. In the literature, different models for the mechanoelectrical transduction in hair cells are discussed (Corey and Hudspeth, 1983; Furness *et al.*, 1996; Howard *et al.*, 1988; Hudspeth, 1983). All hypotheses suggest that a force applied to the hair bundle in the positive direction opens transduction channels, whereas negative deflection closes them. For a better understanding of the transduction process, it is important not only to know what elements of the hair bundle contribute to the opening of transduction channels but also to study their mechanical properties. The morphology of hair cells is precisely described by scanning and transmission electron microscopy. Unfortunately, this method is restricted to a fixed and dehydrated specimen. Therefore, we developed an optimized AFM setup which is built onto an upright differential interference contrast (DIC) light microscope (Langer *et al.*, 1997), to study cochlear hair cells in physiological solution. A water immersion objective ($40 \times /0.75$) provides a high resolution of about 0.5 μm even on organotypic cell cultures with a thickness of about 300 μm. Using this setup (Fig. 12), it is possible not only to see ciliary bundles of inner and outer cochlear hair cells extracted from 6- and 8-day-old rats but also to approach these structures in a controlled manner with the AFM cantilever tip (Langer *et al.*, 2000). The question of whether morphological artifacts occurred at the hair bundles during the AFM investigation can be clarified by preparing the cell cultures for the electron microscope directly after the AFM measurements. Forces of up to 1.5 nN applied in the direction of the stereocilia axis did not change the structure of the hair bundles. The AFM was also used to image both the tips of individual stereocilia and the typical shape of the ciliar bundle of inner and outer hair cells.

These studies were performed in a collaboration between my group and the clinical group of Peter Ruppersberg at the University of Tübingen. Our experiments demonstrate that the AFM can be used to image the stereocillia of these haircells and to apply controlled forces to these cellular structures. Our measurements also could determine the mechanical stiffness of the stereocillia, which is in the range of some tens of millinewtons per meter, clearly reflecting the stiffness of the cytoskeleton filaments. Finally, the first single-channel recordings of mechano-sensitive ion channels became possible by using such an approach, helping to compare such ion channels to others previously examined. For further information, see the chapter by Langer *et al.* (2000).

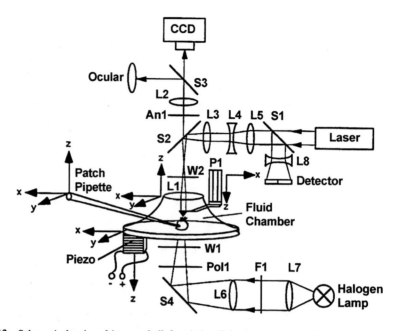

Fig. 12 Schematic drawing of the setup built for whole cell tissue patch-clamp/AFM applications. A piezo-tube scanner P1 moves the cantilever mounted on a titanium arm in xy and z directions. An xz translation stage allows the lateral and vertical coarse positioning of the cantilever. Additionally, an xy translation device moves the optical microscope. Thus, in combination with the vertical objective translation stage of the optical microscope, the laser beam is adjustable in all three orthogonal directions relative to the cantilever. The cantilever deflection is detected by the laser deflection method. A parallel beam of a laser-diode propagates through the beam-splitter cube S1 and is reduced to a diameter of 1 mm by the convex lens L5 and the concave lens L4. The convex lens L3 focuses the beam into the back-focal plane of the water immersion objective. Notch filter S2 reflects the laser light to the objective lens L1, transmits light below 600 nm and above 700 nm, and reflects with 98% in between. A Zeiss Achroplan $40 \times /0.75W$ water immersion objective lens creates a parallel laser beam, which is positioned on the cantilever. The reflected laser light propagates the same way back and is reflected by the beam-splitter cube S1 through the concave lens L8 onto a quadrant photo-diode. With the help of this optical setup the shift of the reflected laser spot caused by a cantilever deflection or torsion is detected. For fine adjustment of the tip–sample distance the specimen chamber is moved vertically by a piezo-stack. A hydraulic micromanipulator, directly mounted on the specimen stage, is used to move the patch-clamp head stage together with the glass pipette in three orthogonal directions. The whole setup is built into an upright optical DIC-Microscope. For infrared DIC imaging the beam of a halogen light source propagates through an edge filter F1 cutting below 700 nm. The light is focused by lenses L6 and L7 and reflected by mirror S4 on the specimen chamber. The polarizer Pol1, the Wollaston prisms W1 and W2, and the Analyzer An1 produce the DIC contrast. The imaging tubus lens L2 creates an image which is detected by a CCD camera simultaneously viewed through the $10 \times$ eyepiece. It is possible to use all three units (optical microscope, AFM, and patch-clamp setup) simultaneously.

V. Photonic Force Microscopy

The AFM, which is used to examine nonconducting materials under ambient conditions or in solution, was considered from the beginning to be an ideal tool for physical studies of live biological specimens. However, cells have a rough surface, which often

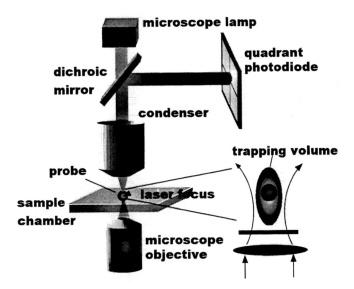

Fig. 13 Schematic drawing of the photonic force microscope setup on top of an inverted optical microscope. The probe, a nanometer-sized spherical particle, is trapped by optical forces, and its position is measured with respect to laser focus using the interference of the laser light with the light scattered by the particle, which is detected by a quadrant photo-diode in the back-focal plan of the condenser.

prevents the tip of a mechanical cantilever from following fine topographic details. Furthermore, forces of about 10 pN are necessary for stable imaging and are often strong enough to cause deformation in soft cellular structures such as the plasma membrane. Only stiff structures connected to the cytoskeleton are clearly visible in the pictures. Imaging inside cells is impossible due to the mechanical connection to the imaging tip. Therefore, a scanning probe microscope without a mechanical connection to the tip and working with extremely small loading forces is desirable. During the past few years, we developed such an instrument, the photonic force microscope (PFM), at the European Molecular Biology Laboratory (EMBL) in Heidelberg (Florin *et al.*, 1996, 1997, 1998; Pralle *et al.*, 1998, 1999).

In the case of the PFM, the mechanical cantilever of the AFM is replaced by the three-dimensional trapping potential of a laser focus (Fig. 13). Ashkin (1986) first described the possibility of trapping small particles with high-stability in the focus of a laser. In the PFM a nanometer- to micrometer-sized particle, e.g., a latex, glass, or metal bead, is used as a tip. The differences in the refractive index of the medium and the bead, the diameter of the bead, the laser intensity and the intensity profile in the focal volume determine the strength of the trapping potential. Depending on the application, the potential is adjusted by changing the laser power. Usually the spring constant of the potential is two to three orders of magnitude softer than that of the softest commercially available AFM cantilevers.

The PFM contains two position-sensing systems to determine the position of the trapped sphere relative to that of the potential minimum. The first sensor records the fluorescence intensity emitted inside the trapped sphere by fluorophores, which are excited by the trapping laser via a two-photon process. The fluorescence intensity provides an axially sensitive position signal with millisecond time and nanometer spatial resolution. The second sensor, which is based on the interference of the forward scattered light from the trapped particle with the unscattered laser light at a quadrant photodiode, provides a fast three-dimensional recording of the particle's position and is most sensitive perpendicular to the optical axis. The position of the bead can be measured with a spatial resolution of better than 1 nm in a range determined by the wavelength at a temporal resolution of 1 μs. Depending on the application, one or both of these detection systems are used.

At room temperature, the thermal position fluctuations of the trapped bead in weak trapping potentials reach several hundred nanometers. At first glance, the fluctuation seems to be disturbing noise that limits the resolution. However, due to the speed and resolution of the position sensor based on the forward-scattered light detection, the fluctuations of the bead can be tracked directly, presenting new methods of analyzing the interaction of the bead with its environment. The PFM developed is based on an inverted optical microscope with a high numerical aperture objective lens providing good optical control of the investigated structures. The laser light is coupled into the microscope using techniques known from laser scanning microscopy and is focused by the objective lens into the specimen plane. A 1064-nm Nd:YVO4 laser is used since neither water nor biological material has significant absorption at this wavelength. A dichroic mirror behind the condenser deflects the laser light onto the quadrant photo-diode. The difference between the left and the right half of the diode provides the x position; the difference between upper and lower half, the y position; and the sum signal, the z position. The signals change linearly with the position of the bead in the trap for displacements which are small relative to the focal dimensions. The trapped particle acts as a Brownian particle in a potential well, and its position distribution is therefore described by the Boltzmann distribution. Such distributions $p(r)$, which are readily measured with the available high spatial and temporal resolutions, allow the calibration of the three-dimensional trapping potential $E(r)$, for instance, just by knowing the temperature, using

$$E(r) = -kT^* \ln p(r) + E_0.$$

A precision of one-tenth the thermal energy kT is achieved with a sufficient number of statistically independent position readings.

Scanning either the laser focus or the sample, the PFM can be used very similar to a conventional AFM to image surface topographies (Fig. 14). The applied forces range from a few piconewtons down to fractions of a piconewton. The resolution is determined by both the interaction area of the spherical bead used with the sample and the thermal fluctuations of the bead. The PFM, like an AFM, can be used in different imaging modes. Beside the constant height mode the constant force mode also can be used by implementing a feedback loop for force control. This mode provides not only the force and the error signal but also the measurement of lateral friction forces. Furthermore, due to large thermal fluctuations of the bead, which avoid stationary objects, a new imaging

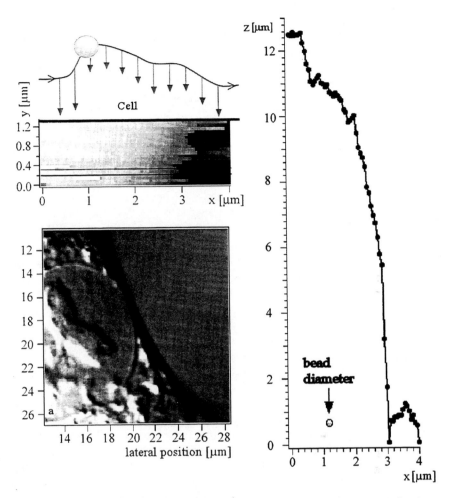

Fig. 14 Tapping mode imaging with the PFM. In this mode, the position of the imaged surface in space is not important for the resolution, unlike that with an AFM. The deep slope seen in the optical DIC image in the upper left corner of a fibroblast cell if the nucleus is close to the edge it is shown in the PFM image in the lower left corner. As an 8-bit grey scale cannot reproduce details of such a vertical structure, a line scan is shown on the right.

mode becomes feasible with the PFM by observing the spatial distribution of the thermal fluctuations of the bead. In this way, it is possible to image, e.g., the three-dimensional structure of polymer networks, with a resolution now limited only by the precision of the detection system. Additionally, detailed information about the interaction potential between the bead and the surface of the objects imaged can be obtained. At present, PFM imaging of biological material is limited by nonspecific adhesion events between the bead and the sample, which can lead to the loss of the probe due to the weak forces of the laser focus. Although there seems to be no general strategy to prevent non-specific adhesion to complex biological material, it is possible to achieve specific binding, for instance, to membrane and other molecular structures for a certain time and to track their motion.

A. Mechanics of Molecular Motors

Kinesins and kinesin-like proteins are of wide interest in biology because of their fundamental functions in the cell. They are responsible, e.g., for both targeting organelles through the cells and setting up the mitotic spindle during cell division. The directed transport along the cytoskeletal filaments of microtubules is powered in an ATP-dependent way by these molecular motors. Conventional kinesin is a so-called plus-end-directed motor, which is able to transport organelles in a defined direction over a distance of up to several microns as a single molecule. Structural and biochemical data of kinesin reveal a heavy chain folded as a globular N-terminal motor domain containing an ATP and a separate microtubule binding site. The catalytic domain is followed by a neck region that consists of two short β-sheets and a coiled-coil helical structure which is probably responsible for dimerization. Behind this region, a hinge of variable length that is not predicted to form a coiled-coil is followed by the kinesin stalk that apparently is an α-helical coiled-coil interrupted by another hinge. The stalk finally connects to a poorly conserved tail structure that interacts with its light chain, binding to the cargo. Several studies on kinesins have shown that the neck and the first hinge region of the motor play important roles in kinesin directionality, velocity, and ATPase activity (Yang et al., 1989; Hackney, 1995; Hirose et al., 1995; Crevel et al., 1996; Henningsen and Schliwa, 1997; Arnal and Wade, 1998). Most of the molecules dimerize in vivo and therefore have two enzymatic head domains. The mechanism of movement and force generation required for intracellular transports is not yet fully understood. Several authors suggest a hand-over-hand model where both heads would alternate to "walk" along the filaments. The kinesin motor, having all its enzymatic and binding machinery in its heavy chain, has the unique property of operating completely on its own either as a dimer or even as a monomer. Previous studies on single molecules have shown that the maximum force generated by the kinesin molecule to transport a microsphere along a microtubule is in the range of 5 pN (Svoboda and Block, 1994). Obviously, characterizing mechanical properties of molecular motors is essential for a better understanding of how nanomachines, like kinesin, convert chemical energy into mechanical movement. So far, attempts to satisfactorily explain the function of kinesin using the knowledge gained by dynamic and structural studies have not been very successful. A major problem is that the high-resolution structural data are obtained on an ensemble of molecules in a rigid state and cannot resolve the dynamic of intermediate states required for the kinesin motility. Furthermore, the molecules used for structural analysis are truncated constructs missing the stalk. On the other hand, for in vitro motility assays, full-length constructs of the motors are used in an environment which is rather different from that of a protein crystal.

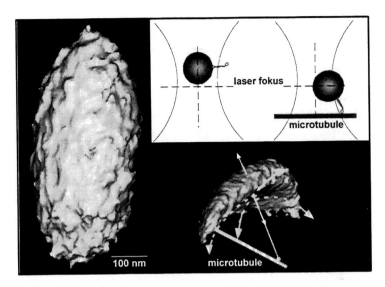

Fig. 15 Isoenergy surfaces of a particle (a) trapped in the laser focus compared to (b) the particle bound by the molecular motor protein kinesin to a microtubule structure fixed on the glass coverslip.

We focused our PFM studies on the intrinsic mechanical properties of kinesin, e.g., elasticity, comparing the mechanical behavior of two different full-length kinesin constructs by adsorbing them onto glass microspheres and letting them interact with a microtubule in the presence of different nucleotides. We studied two wild-type motors: *Drosophila* with two head groups (Jeney *et al.*, 2001) and a chimera which was missing one of the heads. To determine the mechanical properties of the kinesin molecule bound to both a bead and a microtubule in thermal equilibrium, the thermal position fluctuations of the bead were measured by the PFM. Using the Boltzmann distribution to analyze these fluctuations in three dimensions, the energy landscape defined by the kinesin as a molecular linker between bead and microtubule structure fixed to the surface can be determined (Fig. 15). Finally, from the energy landscape measured the mechanical properties of single kinesin molecules for the two different types of motors studied and their different nucleotide binding states can be extracted in three dimensions.

The first characteristic features one can observe (see Table I) are the overall differences in the rotational stiffness along and perpendicular to the microtubule. Thus, the restoring forces the molecules develop against lateral bending are dependent on the microtubule orientation and are influenced by the microtubule's presence. A comparison between the one- and two-headed molecules shows that the wild-type motor behaves stiffer in all three directions. This points to the influences of the second head, which restricts the bead movement due to steric interactions with the microtubule and perhaps in cases of high stiffness even binds to it. Furthermore, the stiffness of the molecules bound with the

Table I

Experimental conditions	Rotational stiffness		Stalk stiffness k_x (mN/m)
	k_x (pN nm/rad)	k_y (pN nm/rad)	
Two-headed kinesin with 2 mM AMPPNP	6	2.5	120
Two-headed kinesin without nucleotide	3	1.3	80
Single-headed kinesin with 2 mM AMPPNP	1	0.2	70
Single-headed kinesin without nucleotide	0.5	0.1	50

nonhydrolyzable ATP analogue, AMPPNP, is much higher than that of the molecules with ADP for both motor types. This leads to the conclusion that the binding strength between kinesin and tubulin influences the elasticity of the entire kinesin/bead/microtubule system. Nevertheless, the binding states with ADP are not always comparable for both motor constructs. In the experiments with the two-headed kinesin, due to low ATP concentration, the measured binding states are probably a mixture of weak ADP-binding events and strong ones without nucleotides. In the experiments with the one-headed chimera in the presence of 2 mM ATP, the weakly bound state of kinesin was measured (Crevel *et al.*, 1996). This finding confirms the fact that different nucleotides influence the position of the free domain relative to the bound state of kinesin (Hirose *et al.*, 1995; Arnal and Wade, 1998). Several arguments can be made demonstrating that the measured stiffness comes from a functional part of the protein, for example, a part that is involved in its intrinsic properties for the dynamical interaction between microtubules and motors. First, we can exclude that the stable coiled–coiled stalk contributed to the stiffness measured, since this was found to be more than two orders of magnitude higher in computer simulation studies (H. Grubmüller, priv. com.). Second, the influences of the microtubule orientation and the nucleotide state on molecular-restoring forces demonstrate that these forces come from flexible parts of the molecule that are dependent on these parameters. Such a functional part is the neck region that seems heavily involved in kinesin directionality and dynamics (Henningsen and Schliwa, 1997). The neck also determines the relative orientation of the two heads, which is influenced by the nucleotide state of the bound motor domain.

Our results show that, strikingly, the kinesin molecule comprises no free molecular joints since the laser stiffness was never observed to be predominant over the motor stiffness. In Fig. 16, we propose a model for the mechanical behavior of kinesin. The stable coiled–coiled parts of the stalk are represented as rods S1 and S2. The flexible parts in the bead–kinesin system are assumed to be the attachment between the bead and the kinesin tail H3, the predicted hinge (Henningsen and Schliwa, 1997) in the stalk H2, and the hinge next to the motor domain H1, which includes the flexibility of the kinesin head itself. We describe these parts by three-dimensional springs placed in a series. These springs can be stretched, squeezed, and bent back and forth in all directions. The spring H1 of the neck domain is the one dominating the PFM measurements and thus the weakest. The other possible soft connections of the kinesin–bead attachment and the hinge in the middle of the molecule are obviously stiffer.

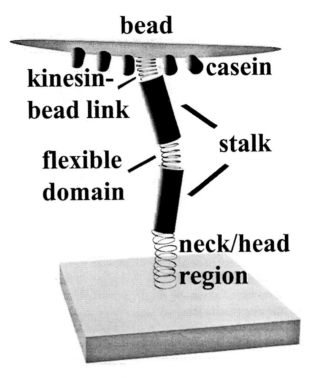

Fig. 16 Mechanical model of the kinesin molecular motor protein showing a possibly flexible region as springs and the stiff stalk parts as rods. Measured by the PFM, the thermal fluctuations of a particle coupled by the molecule to the surface demonstrate that the overall flexibility is dominated by the neck region H1. In this way the technique allows one to examine the function of the essential part.

With these measurements, it is also possible to estimate the total length of kinesin by calculating the tilt angle of the three-dimensional position distribution of the bead probe. This angle never exceeded the possible angle allowed by a bead–kinesin configuration where the bead radius was either 450 or 215 nm and the tether was 70–85 nm in length, which fits nicely with the kinesin's full length. This also supports the assumption that the region close to the head is responsible for the bending stiffness we measured. In the experiments the position fluctuations of the bead linked by a single kinesin molecule to a microtuble fixed to a solid surface can be directly connected to the structural and mechanical properties of the linker molecule. A crucial step in this procedure is to adjust the trap stiffness to be at least one order of magnitude smaller than the contribution of the molecule to ensure a negligible influence of the laser focus on the molecules. The values used to describe the molecule's mechanical behavior when it was exposed to thermal energy are its rotational stiffness in piconewton nanometers per rad along and perpendicular to the microtubule and its stalk stiffness in micronewtons per meter. The advantage that the units have is that measured values become independent of the used bead size and more characteristic for the molecule.

In the future systematic studies on three-dimensional potential profiles of different molecular structures will lead to a better understanding of the correlation between molecular structure, mechanical properties, and function. Studies on the kinesin molecular motor, which analyze the mechanical properties of the neck and hinge regions of different chimera, e.g., mutated in neck and hinge regions, could provide further insight into the influence of these regions on kinesin directionality, velocity, and ATPase activity.

B. Local Viscosity Measurements

The motion of a thermally fluctuating particle in a harmonic potential like that of the laser trap can be characterized to a certain level by an exponentially decaying position autocorrelation function. This function has a characteristic autocorrelation time of $\tau = \gamma/\kappa$, where γ denotes the viscous drag on the sphere and κ denotes the force constant of the optical trap. Thus, the local viscous drag and the diffusion coefficient, $D = kT/\gamma$, of a sphere in a harmonic potential can be calculated from both the measured autocorrelation time of the motion and the stiffness of the trapping potential. The stiffness of the trapping potential is determined as described in the previous section for the calibration of the laser trap. To measure the diffusion coefficient with a statistical error smaller than 10%, the observation interval must be about 1000 times longer than the autocorrelation time. Hence, the motion of the bead limits the temporal resolution of the viscosity measurement and not the bandwidth of the detection system. The diffusion is reduced near surfaces, e.g., glass surfaces or cell membranes, because of the spatial confinement (Happel and Brenner, 1965). Diffusion within an obstacle-free lipid bilayer was described by Saffman and Delbrück (1975) using a hydrodynamic model treating the bilayer as a continuum and assuming weak coupling to the surrounding liquid medium. The viscous drag γ_m on a cylindrical particle with radius r in a homogeneous lipid bilayer of thickness h is $\gamma_m = 4\pi \eta_m h/(\ln(\eta_m h/\eta_w r) - e)$, where η_w denotes the viscosity of the surrounding fluid; η_m, the viscosity of the lipid bilayer; and e, the Euler constant. This approximation

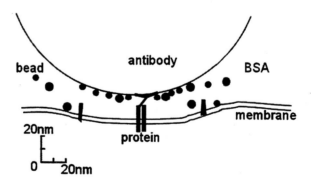

Fig. 17 Scaled model of the experimental situation when a sphere ($r = 108\,\mathrm{nm}$) is bound via an adsorbed antibody to a transmembrane protein on a cell.

is valid for proteins with radii, large compared to the size of the lipid molecules, and for $\eta_m \gg \eta_w$, which is true for cellular membranes. Thus, the membrane viscosity can be determined both by binding a bead to a membrane structure of known diameter and by measuring the total viscous drag on the bead as described earlier (Fig. 17). Using our PFM, such measurements can be performed with a temporal resolution of 0.3 s and within areas of 100 nm in diameter.

The measured viscous drag γ for the bead connected to a membrane component is the sum of the Stokes drag of the sphere $\gamma_s = 6\pi \eta_w r$ and the viscous drag γ_m of, e.g., a single protein in the lipid bilayer. The Stokes drag of the sphere near the cell membrane increases due to the confinement and this must be taken into account for the correct calculation of γ_m. The observation of the strength of the lateral potential ensures that the observed membrane component diffuses freely. By using the PFM technique on a single transmembrane protein of known size, it is possible to determine the membrane viscosity in living cells in areas about 100 nm in diameter. Once the membrane viscosity has been obtained, the technique allows one to determine the diameter of other membrane structures of unknown size and to continuously monitor their diffusion characteristics. We performed such experiments to determine the size of membrane structures called "rafts" (Pralle et al., 2000). Lipid rafts are involved in the polarized sorting of proteins and cellular signaling (Simons and Ikonen, 1997; Keller and Simons, 1998). According to our study of the plasma membrane of fibroblast-like cells, these structures are indeed quite stable on the time scale of minutes, justifying the model of "rafts" as lipid–protein complexes floating raft-like in the membrane, whose diameter was determined to be about 50 nm. In these experiments, the PFM proved to be a powerful tool in the study of not only the biophysical properties of the plasma membranes but also the interaction of single molecules or complexes of molecules within a membrane. Future applications of the PFM will cover a wide range from a three-dimensional in vivo imaging of biological samples to a detailed investigation of single-molecule properties. For further information, see the chapter by Pralle et al. (2000).

References

Arnal, I., and Wade, R. H. (1998). *Structure* **6**, 33.

Ash, E. A., and Nicolls, G. (1972). *Nature* **237**, 510.

Ashkin, A. (1986). *Opt. Lett.* **11**, 288.

Binnig, G., Rohrer, H., Gerber, C., and Weibel, E. (1982). *Appl. Phys. Lett.* **40**, 178.

Binnig, G., Quate, C. F., and Gerber, Ch. (1986). *Phys. Rev. Lett.* **56**, 930.

Corey, D. P., and Hudspeth, A. J. (1983). *J. Neurosci.* **3**, 962.

Crevel, I. M.-T. C., Lockhart, A., and Cross, R. A. (1996). *J. Mol. Biol.* **257**, 66.

Eckert, R., Jeney, S., and Hörber, J. K. H. (1998). *Cell Biol. Intern.* **21**, 707.

Florin, E.-L., Hörber, J. K. H., and Stelzer, E. H. K. (1996). *Appl. Phys. Lett.* **69**, 446.

Florin, E.-L., Moy, V. T., and Gaub, H. E. (1994). *Science* **264**, 415.

Florin, E.-L, Pralle, A., Hörber, J. K. H., and Stelzer, E. H. K. (1997). *J. Struct. Biol.* **119**, 202.

Florin, E.-L., Pralle, A., Stelzer, E. H. K., and Hörber, J. K. H. (1998). *Appl. Phys. A* **66**, S75.

Freund, J., Halbritter, J., and Hörber, J. K. H. (1999). *Microsc. Res. Tech.* **44**, 327.

Furness, D. N., Hackney, C. M., and Benos, D. J. (1996). *Hearing Res.* **93**, 136.

Göbel, H. D., Hörber, J. K. H., Gerber, Ch., Leitner, A., and Hänsch, T. W. (1992). *Ultramicroscopy* **42–44,** 1260.

Guénebaut, V., Maaloum, M., Bonhivers, M., Wepf, R., Leonard, K., and Hörber, J. K. H. (1997). *Ultramicroscopy* **69,** 129.

Häberle, W., Hörber, J. K. H., and Binnig, G. (1991). *J. Vac. Sci. Technol.* **B9,** 1210.

Hackney, D. D. (1995). *Nature* **377,** 448.

Happel, J., and Brenner, H. (1965). "Low Reynolds Number Hydrodynamics," Prentice Hall, Englewood Cliffs, NJ.

Haynes, C. A., and Norde, W. (1994). *Colloids Surf. B, Biointerfaces* **2,** 517–566.

Heckl, W. M., Kallury, K. M. R., Thompson, M., Gerber, Ch., Hörber, J. K. H., and Binnig, G. (1989). *Langmuir* **5,** 1433.

Hemerlé, J., Altmann, S. M., Maaloum, M., Hörber, J. K. H., Heinrich, L., Voegel, J.-C., and Schaaf, P. (1999). *Proc. Natl. Acad. Sci. USA* **96,** 6705.

Henningsen, U., and Schliwa, M. (1997). *Nature* **389,** 93.

Hirose, K., Lockhart, A., Cross, R. A., and Amos, L. A. (1995). *Nature* **376,** 277.

Hörber, J. K. H., Häberle, W., Ohnesorge, F., Binnig, G., Liebich, H. G., Czerny, C. P., Mahnel, H., and Mayr, A. (1992). *Scan. Microsc.* **6,** 919.

Hörber, J. K. H., Lang, C. A., Hänsch, T. W., Heckl, W. M., and Möhwald, H. (1988). *Chem. Phys. Lett.* **145,** 151.

Hörber, J. K. H., Mosbacher, J., Häberle, W., Ruppersberg, P., and Sakmann, B. (1995). *Biophys. J.* **68,** 1687.

Hörber, J. K. H., Schuler, F. M., Witzemann, V., Müller, H., and Ruppersberg, J. P. (1991). *In* "Scanned Probe Microscopies: STM and Beyond" (H. Kumar Wickramasinghe, ed.), p. 241. AIP Conference Proceedings.

Hörber, J. K. H., Schuler, F. M., Witzemann, V., Schröter, K. H., Müller, H., and Ruppersberg, J. P. (1991). *J. Vac. Sci. Technol.* **B9,** 1214.

Howard, J., and Hudspeth, A. J. (1988). *Neuron* **1,** 189.

Hudspeth, A. J. (1983). *Annu. Rev. Neurosci.* **6,** 187.

Jeney, S., Florin, Ernst-Ludwig, and Hörber, J. K. H. (2001). Series: "Methods in Molecular Biology; Kinesin Protocols" (Isabelle Vernos, ed.). Humana Press Inc. Totowa, NJ.

Keller, P., and Simons, K. (1998). *J. Cell Biol.* **140,** 1357.

Kellermayer, M. S. Z., Smith, S. B., Granzier, H. L., and Bustamante, C. (1997). *Science* **276,** 1112.

Langer, M. G., Öffner, W., Wittmann, H., Flösser, H., Schaar, H., Häberle, W., Pralle, A., Ruppersberg, J. P., and Hörber, J. K. H. (1997). *Rev. Sci. Instrum.* **68,** 2583.

Langer, M. G., Koitschev, A., Haase, H., Rexhausen, U., Hörber, J. K. H., and Ruppersberg, J. P. (2000). *Ultramicroscopy* **82,** 269.

Lenne, P.-F., Raae, A. J., Altmann, S., Saraste, M., and Hörber, J. K. H. (2000). *FEBS Lett.* **476,** 124.

Maaloum, M., Chretien, D., Karsenti, E., and Hörber, J. K. H. (1994). *J. Cell Sci.* **107,** 3127.

Mosbacher, J., Langer, M., Hörber, J. K. H., and Sachs, F. (1998). *J. Gen. Physiol.* **111,** 65.

Norde, W. (1986). *Adv. Colloid Interface Sci.* **25,** 267.

Ohnesorge, F. M., Hörber, J. K. H., Häberle, W., Czerny, C.-P., Smith, D. P. E., and Binnig, G. (1997). *Biophys. J.* **73,** 2183.

O'Keefe, J. A. (1956). *J. Opt. Soc.* **46,** 359.

Pohl, D. W., Fischer, U. Ch., and Dürig, U. T. (1988). *J. Microsc.* **152,** 853.

Pralle, A., Florin, E.-L., Stelzer, E. H. K., and Hörber, J. K. H. (1998). *Appl. Phys. A* **66,** S71.

Pralle, A., Keller, P., Florin, E.-L., Simons, K., and Hörber, J. K. H. (2000). *J. Cell Biol.* **148,** 997.

Pralle, A., Prummer, M., Florin, E.-L., Stelzer, E. H. K., and Hörber, J. K. H. (1999). *Microsc. Res. Tech.* **44,** 378.

Rief, M., Gautel, M., Oesterhelt, F., Fernandez, J. M., and Gaub, H. E. (1997). *Science* **276,** 1109.

Ruppersberg, J. P., Hörber, J. K. H., Gerber, Ch., and Binnig, G. (1989). *FEBS Lett.* **257,** 460.

Saffman, P. G., and Delbrück, M. (1975). *Proc. Natl. Acad. Sci. USA* **72,** 3111.

Schneider, S. W., Kumudesh, C. S., Geibel, J. P., Oberleithner, H., and Jena, B. P. (1997). *PNAS* **94,** 316.

Simons, K., and Ikonen, E. (1997). *Nature* **387,** 569.

Smith, D. P. E., and Binnig, G. (1986). *Rev. Sci. Instrum.* **57,** 2630.

CHAPTER 2

The Atomic Force Microscope in the Study of Membrane Fusion and Exocytosis

Bhanu P. Jena and Sang-Joon Cho

Departments of Physiology & Pharmacology
Wayne State University School of Medicine
Detroit, Michigan 48201

I. Introduction

The atomic force microscope (AFM), which was discovered just over a decade ago, has emerged in recent years as a powerful tool, especially for the study of cellular structure and function at ultrahigh resolution. Morphological studies revealing nanometer-scale details of living cells, which were previously impossible due to the resolution limits of the light microscope, are now possible using the AFM. The AFM has been used to examine the cellular surface, intracellular structures, and isolated organelle and biomolecules, revealing previously unknown cellular and subcellular structures at nanometer resolution. The AFM's major contribution to biology has been its ability to study the dynamics of live cells in physiological medium at ultrahigh resolution and in real time. By using the AFM, structures unidentifiable by electron microscopy on fixed or freeze-fractured tissues, probably due to perturbation during processing, can now be imaged at nanometer resolution in living cells. Recently, the AFM has made both the identification of new structures at the plasma membrane of live pancreatic acinar cells and the study of their dynamics in real time during exocytosis possible. These AFM studies determine the structures to be fusion pores, which we wish to call "porosomes," where secretory vesicles transiently dock and fuse to release their contents.

Conventional microscopes use light or electron beams to image objects. The optical resolution of microscopes using light is limited by its wavelength. Although angstrom resolution of biological samples can be achieved by using the electron microscope, these samples must be frozen, fixed, dried, and/or processed to gain contrast prior to imaging. Morphological changes due to freezing, tissue fixation, and processing for electron microscopy have always been a major concern. As a result, even nanometer resolution of live cells had been impossible until the discovery of the atomic force microscope (AFM) (Binnig *et al.,* 1986) and its development for imaging biological samples (Alexander *et al.,* 1989; Rugard and Hansma, 1990; Albrecht *et al.,* 1990). In AFM, a probe tip microfabricated from silicon or silicon nitride and mounted on a cantilever spring is used to scan the surface of the sample at a constant force (Albrecht *et al.,* 1990). Either the probe or the sample can be precisely moved in a raster pattern using an *xyz* piezo-tube to scan the surface of the sample (Fig. 1) (Binnig and Smith, 1986). The deflection of the cantilever measured optically is used to generate an isoforce relief of the sample (Alexander *et al.,* 1989). Force is thus used to image surface profiles of objects by the AFM, allowing imaging of live cells and subcellular structures submerged in physiological buffer solutions. Since force is used to survey the topology of the material under investigation, the mechanical properties of the sample critically influence image formation. In examining soft biological samples, the elastic properties of the sample significantly contribute to

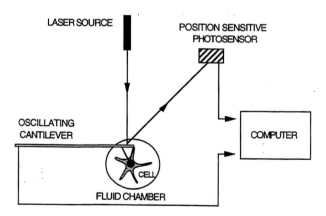

Fig. 1 Schematic diagram depicting key components of an AFM.

the image formed. To minimize elastic effects, biological samples should be scanned by the AFM at the lowest possible imaging force (approximately 1 nN or less) without compromising resolution. Even imaging forces of 1 nN or less were found to influence the image (Radmacher *et al.*, 1995, 1997). The precaution necessary to obtain an image with negligible influence on the elastic properties of a cell would require loading forces of only a few piconewtons, which is not possible at the current state of the technology. Although it was recently discovered that photonic force microscopy (Hörber; Pralle; in this book; and Pralle *et al.*, 1999), overcomes the requirement of large loading forces for imaging biological samples, it does so at the expense of image resolution. Hence, imaging soft biological samples by the AFM is challenging, and achieving subangstrom resolution images of the topology of live cells is virtually impossible at the present time. The other major problem in the study of biological samples using the AFM is the ability of proteins and cellular debris to adhere to the AFM tip, resulting in loss of resolution, generation of image artifacts, or the total inability to generate an isoforce relief of the sample using such contaminated tips. Even with all these limitations, the AFM, alone (Dufrene *et al.*, 1999; Hörber *et al.*, 1992; Henderson *et al.*, 1992; Fritz *et al.*, 1994; Hoh and Schoenenberger, 1994; Le Grimellec *et al.*, 1998; Spudich and Braunstein, 1995; Schneider *et al.*, 1997) and in combination with optical and electrophysiological devices (Danker and Oberleithner, 2000; Langer *et al.*, 1997), has been successfully used to study cellular structure and function. In this article, cellular structure–function studies using the AFM and the promise and potential of the instrument for future use in understanding the biology of living cells are discussed. Use of the AFM in understanding membrane fusion and exocytosis, has been explored in this chapter. Several AFM studies using fixed cell or subcellular organelle preparations and studies measuring inter- and intramolecular interactions in biomolecules, have therefore not been discussed.

II. Methods

A. AFM Operation Modes

To examine the topology of living cells, the scanning probe of the AFM must operate in fluid (physiological or near physiological buffers) and may do so under two modes: *contact* or *tapping*. In the contact mode, the probe is in direct contact with the sample surface as it scans at a constant vertical force. Although high-resolution AFM images can be obtained in this mode of AFM operation, the sample height information generated may not be accurate because the vertical scanning force may depress the soft cell. However, even though AFM imaging forces may depress cells in the contact mode, information on the viscoelastic properties of the cell and the actual spring constant of the cantilever would enable the measurement of the cell height. On the other hand, in the tapping mode of operation, the cantilever resonates and the tip of the probe makes brief contacts with the sample, too brief to allow adhesive forces between the probe tip and the sample surface. In the tapping mode in fluid, the lateral forces are virtually negligible. It is therefore important that information on the topology of living cells be obtained using both these modes of AFM operation in physiological buffers maintained at a stable temperature. The scanning rate of the tip over the sample also plays a critical role in the quality of the image. Because cells are soft samples, a high scanning rate would influence their shape. Therefore, a slow movement of the AFM tip at the cell surface would be ideal and result in minimal distortion and better resolved images. Rapid cellular events could nonetheless be monitored using section analysis, obtained by a rapid scan line passing through the sites of interest at the cell surface.

B. AFM Probes

As mentioned earlier, AFM scanning tips are microfabricated from silicon or silicon nitride and mounted on a cantilever spring. Cantilevers range from 85 to 320 μm in length and vary from 0.3 to 2 μm in thickness. Several types of cantilevers graded in their ability to flex are available. To image soft biological samples in the contact mode, more flexible cantilevers are ideal because they would deflect without deforming the sample surface. Scanning tips are available in various shapes and sizes; however, pyramidal silicon nitride tips, approximately 3 μm in length with their fine end measuring 50–200 Å in diameter, are commonly used to scan biological samples in physiological buffers. Tip geometry and composition play a critical role in generating AFM images of the sample (Taatjes *et al.*, 1999). Tip-imaging artifacts or convolutions may occur if the scanning tip is less sharp than the profile of the object at that point (Fig. 2) (Hoh and Hansma, 1992). When the scanning tip is sharper than the feature on the sample, the true profile of the feature is obtained. On the contrary, the use of very sharp tips would more likely result in cell damage. Therefore, depending on the cell type, appropriate tips (different sharpness or aspect ratios) should be used without compromising the viability of cells.

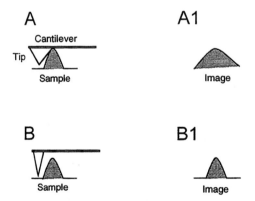

Fig. 2 Schematic diagram demonstrating how the shape of the scanning tip may influence the image of the object scanned by the AFM. When the scanning tip is wide and inaccessible to every part of the surface of the object (A), an erroneous profile of the object is seen in the AFM isoforce relief (A1). On the contrary, when a fine tip is used to scan the same object (B), the AFM micrograph depicts a near identical image of the object (B1). Similarly, when the AFM tip is scanning a depressed area, the tip angle limits the maximum depth to which the tip could access the depression.

C. The Sample

A major problem in imaging live whole cells using the AFM has been to achieve immobilized live cells with a clean but intact cell surface. The sensitivity of cells to the cantilevers' disruptive effect is dependent on the cell type (You *et al.*, 2000). Hardy cells such as fibroblasts and cells that adhere well to the substrate are ideal for study using the AFM. Cell cultures grown in a monolayer on collagen-coated glass coverslips, e.g., laminin, poly-L-lysine, or Cel-Tak, and immersed in physiological buffer are ideal candidates for examination using the AFM (Henderson *et al.*, 1992; Nagayama *et al.*, 1996). Freshly cleaved Cel-Tak-coated mica were used with great success in the study of isolated cells with the AFM (Schneider *et al.*, 1997). The contents of the bathing medium and the cell surface to be scanned need to be devoid of any large-molecular-weight proteins or cellular debris. Prior to the study, the careful selection of filtered physiological medium and the thorough washing of plated cells with debris-free medium would alleviate these major problems. As discussed earlier, a careful selection of the cantilever and tip to not only maintain cell viability but also obtain optimal resolution of the scan surface is critical for different types of cells.

III. AFM Studies on Live Cells

In the last decade, the AFM was successfully utilized to study the morphology of a number of different living cells. As mentioned earlier, immobilization of the live cell is critical in its examination using the AFM. In 1992, Häberle, Hörber, and colleagues were able to successfully immobilize cells on a suction pipette and scan them with an AFM (Ohnesorge *et al.*, 1997). In these studies, they observed and reported the exit dynamics of the pox virus from live monkey cells. Using the AFM, live hippocampal neurons and glia cells cultured on collagen-coated glass coverslips were imaged, revealing ongoing intracellular filament dynamics underneath the cell membrane (Henderson *et al.*, 1992; Parpura *et al.*, 1993). These studies demonstrated that besides being able to examine the surface topology of living cells, the AFM, using a higher imaging force of >100 nN, is capable of both imaging cytoskeletal structures and studying their arrangement and dynamics in live cells (Henderson *et al.*, 1992). The fibrillar network of medial midbrain astrocytes, however, which prevents neuritic growth, could only be displayed by the AFM using 4% paraformaldehyde-fixed astrocytes (Weissmuller *et al.*, 2000). This study indicated the difficulties for imaging soft samples using the AFM (Braet *et al.*, 1998), and fixation to harden the sample is sometimes required.

By using the AFM it was possible to observe wave-like movements, possibly organized rearrangements of cytoplasm, at the plasma membrane in live lung carcinoma cells and MDCK cells (Kasas *et al.*, 1993; Schoenenberger and Hoh, 1994). Similarly, the lamellipodia of migrating live MDCK-F cells examined under the AFM suggests the possible molecular mechanisms involved in cellular movement (Oberleithner *et al.*, 1993; Schneider *et al.*, 2000). These studies implicate membrane turnover during cell migration.

Active endocytosis of distinct membrane patches (one patch/minute) at the lamellipodia is suggested to regulate locomotion in these MDCK-F cells.

Using the AFM, granula motion and membrane spreading in activated human platelets were also imaged and examined (Fritz *et al.*, 1994).

Fine structural changes in live neurons obtained from the central ganglia of the pond snail *Lymnaea stagnalis* were seen at nanometer resolution using the AFM (Nagayama *et al.*, 1996). In this *Lymnaea stagnalis* ganglia study, time-dependent changes of the lamellipodia and growth-cone terminals in primary neuron cultures were observed in the AFM micrographs, revealing dynamic structural details that were important in understanding the development of neural networks.

Individual stereocilia of living sensory cells of the organ of Corti were successfully imaged for the first time by the AFM; this provided new insight into studying structural and mechanical properties of live auditory hair cells under physiological conditions (Langer *et al.*, 2000). In this study, the tips of individual stereocilia and the typical V-shaped ciliary bundles were displayed.

Changes in morphology and growth dynamics of NIH 3T3 and PC12 cells in culture were examined using the AFM (Rotsch *et al.*, 1999; Lal *et al.*, 1995). In the Lal *et al.* study (1995), differentiating PC12 cells with a neuronal phenotype containing abundant cytoskeletal elements, neuritic processes, and growth cones was examined by the AFM over long periods of time. Fifty- to 700-nm-wide growth cones were identified and found to reorganize in a time-dependent manner. The fibrous organization of the cell and the bead-like growth cones appeared and disappeared over time, revealing cellular structural dynamics that may drive cell behavior. The Rotsch *et al.* study (1999) demonstrates that the AFM force-mapping mode allows the dynamics, thickness, and mechanical properties of the protrusion at active cell edges to be examined.

With the AFM, Gad and Ikai (1995), by using 3% agar to immobilize yeast cells (*Saccharomyces cerevisiae*), were not only able to successfully scan these cells, but also study their growth and budding dynamics. Permanent bud scars at the yeast cell wall, never previously reported, were found to be present. Furthermore, these AFM studies reveal that buds remain for a long time without dissociating from the mother cell. Understanding these real-time morphological events at the cell surface is important in our understanding of cellular growth and development.

Scanning force microscopy performed on the rat basophilic leukemia cell surface demonstrates the appearance of pits, approximately 1.5 μm in diameter, at the plasma membrane following stimulation (Spudich and Braunstein, 1995). The authors suggest that these structures may be involved in membrane retrieval following intense stimulation of these cells. The AFM has thus enabled the identification of pits at the cell surface and the study of their dynamics over time following stimulation of secretion in live rat basophilic leukemia cells.

Using the AFM, the elastic properties and the morphology of living osteoblasts on the metallic substrates were compared with those cultured on glass and polystyrene (Domke *et al.*, 2000). This study showed the capability of the AFM to measure adhesion of live cells on different substrates and provided new information on medical implant research.

IV. Identification of New Plasma Membrane Structures Involved in Exocytosis

A. Apical Plasma Membrane Structures: Pits and Depressions

Our studies using the AFM reveal a new group of plasma membrane structures in live pancreatic acinar cells involved in exocytosis (Schneider *et al.*, 1997). Exocytosis is a fundamental cellular process responsible for neurotransmission, enzyme secretion or hormone release. The final step in the exocytotic process involves the docking and fusion of membrane-bound secretory vesicles at the cell plasma membrane. Although several proteins have been implicated in exocytosis, the molecular mechanism of secretory vesicle docking and fusion at the cell plasma membrane is unclear. Isolated acinar cells of the exocrine pancreas were used to understand the exocytotic process. Pancreatic acinar cells are polarized, slow secretory cells that secret digestive enzymes following stimulation of secretion.

Membrane-bound secretory vesicles called zymogen granules, ranging from 0.2 to 1.2 μm in diameter (Jena *et al.*, 1997), dock and fuse at the apical plasma membrane of acinar cells to release vesicular contents. Using isolated pancreatic acinar cells plated on Cel-Tak-coated mica, the surface topology of the plasma membrane before, during, and after stimulation of secretion was obtained using the AFM. In resting acinar cells, pits measuring 0.5–2 μm and containing 3–20 depressions measuring 100–180 nm in diameter were identified only at the apical region of these cells where membrane-bound secretory vesicles are known to dock and fuse (Fig. 3). Following stimulation of secretion, dynamic size changes occur only in depressions. No change in either pits or topology of the plasma membrane surface at the basolateral end of the cell was noted (Fig. 4). Following stimulation of secretion, a 35% increase in depression diameter was observed, which correlated with an increase in measured enzyme release. Thirty minutes following stimulation of secretion, a 20% decrease in depression size and no further increase in enzyme secretion were observed (Fig. 5).

Exposure of acinar cells to cytochalasin B, a fungal toxin that inhibits actin polymerization, results in a 50–60% loss of stimulation of amylase secretion. A significant decrease in the depression diameter was observed following the treatment of acinar cells with cytochalasin B. Since the AFM uses force to image objects, it was unclear whether pits and depressions were actual surface structures at the plasma membrane or represented an arrangement of underlying cytoskeletal elements at the apical end of pancreatic acinar cells. Additionally, apically localized microvilli which are observed in electron micrographs of pancreatic acinar cells were absent from our AFM images (Fig. 3). The possibility that the absence of microvilli in our previous study reflects the early onset in loss of cell polarity due to loss of tight junction was investigated. The preparation of large acinar cell clusters (10–15 cells), having intact tight junction, demonstrated the presence of microvilli with interspersed pits and depressions (Fig. 6). Analogous to a cornfield with potholes, the microvilli and pits could be observed at the apical end of live pancreatic acinar cells. When transmission electron microscopy was performed on these cells, the presence of microvilli was confirmed

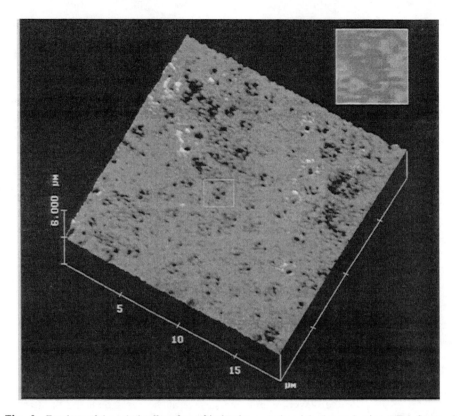

Fig. 3 Topology of the apical cell surface of isolated pancreatic acini, observed using AFM. Scattered pits are seen at the apical plasma membrane. One pit (inset) with four depressions is shown. Reproduced with permission from Schneider, S. W., Sritharin, K. C., Geibel, J. P., Oberleithner, H., and Jena, B. P. (1997). Surface dynamics in living acinar cells imaged by atomic force microscopy: Identification of plasma membrane structures involved in exocytosis. *Proc. Nat Acad. Sci. U.S.A.* **94,** 316–321. Copyright (1997) *National Academy of Sciences, U.S.A.* (See Color Plate.)

(Fig. 7A). Cells without intact tight junction were devoid of microvilli (Fig. 7B) (Jena *et al.,* unpublished). In the studies outlined previously, the AFM has not only enabled the identification of a new group of plasma membrane structures at ultrahigh resolution in live pancreatic acinar cells but also helped to determine their dynamic involvement in exocytosis (Fig. 8). In a recent study, Tojima *et al.* (2000) also demonstrated the presence of pits at the surface of fixed NG108-15 cells. Their studies suggest that these pits are exocytotic apertures where secretory vesicles may dock and fuse.

B. Depressions are Fusion Pores

To determine whether pits and depressions in pancreatiaccinar cells are sites where vesicles dock and fuse to release vesicular contents, the detection of release of secretory products at these sites was investigated. A major content of the secretory vesicle in

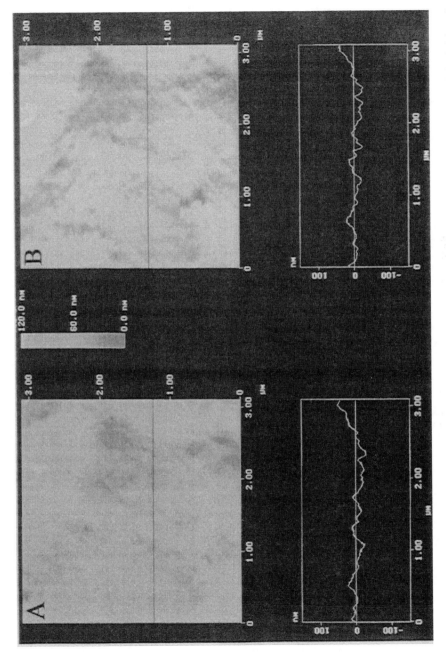

Fig. 4 Topology of the basolateral surface of isolated pancreatic acini, observed using the AFM (A). The scan line depicts the relative height and depth of the surface imaged. Following stimulation of secretion, no major change in the basolateral surface profile is observed (B). (See Color Plate.)

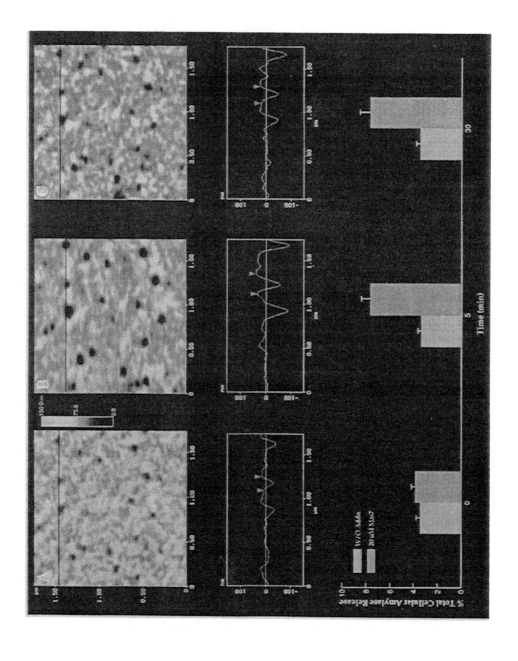

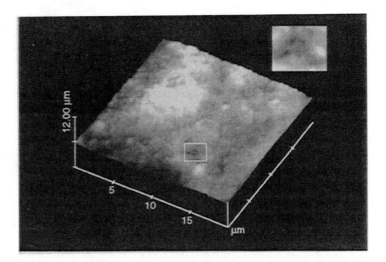

Fig. 6   AFM-generated image of the apical surface of several freshly isolated pancreatic acinar cells demonstrating the presence of microvilli with interspersed pits and depressions. One pit with four to five depressions is shown (Jena *et al.,* unpublished).

pancreatic acinar cells is amylase. Since all attempts to image amylase being released from pits and depressions in live acinar cells failed, acini were stimulated to secrete in the presence of a gold-tagged amylase antibody. Cells were then immediately fixed using a fixative that retained pit and depression morphology. Following fixation, cells were thoroughly washed prior to imaging in buffer using the AFM. Our studies demonstrate the presence of gold clusters decorating primarily pits and depressions, confirming these structures to be sites where vesicles dock and fuse transiently to release vesicular contents (Fig. 9) (Quinn *et al.,* unpublished).

C. Transient Fusion Occurs at Depressions

It is commonly accepted that the final step in exocytosis is the total incorporation of secretory vesicle membrane with the cell plasma membrane (Fig. 10A). The

Fig. 5   Dynamics of depressions following stimulation of secretion. (A) Several depressions within a pit are shown. The scan line across three depressions in the top panel is represented graphically in the middle panel and defines the diameter and relative depth of the depressions. The middle depression is represented by red arrowheads. The bottom panel represents the percentage of total cellular amylase release in the presence and absence of the secretagogue Mas 7. (B) Notice an increase in the diameter and depth of depressions, correlating with an increase in total cellular amylase release at 5 min after stimulation of secretion. (C) At 30 min after stimulation of secretion, there is a decrease in diameter and depth of depressions, with no further increase in amylase release over the 5-min time point. No significant increases in amylase secretion or depression diameter were observed in either resting acini or those exposed to the nonstimulatory mastoparan analog Mas 17. Reproduced with permission from Schneider, S. W., Sritharin, K. C., Geibel, J. P., Oberleithner, H., and Jena, B. P. (1997). Surface dynamics in living acinar cells imaged by atomic force microscopy: Identification of plasma membrane structures involved in exocytosis. *Proc. Nat Acad. Sci. U.S.A.* **94,** 316–321. Copyright (1997) *National Academy of Sciences, U.S.A.* (See Color Plate.)

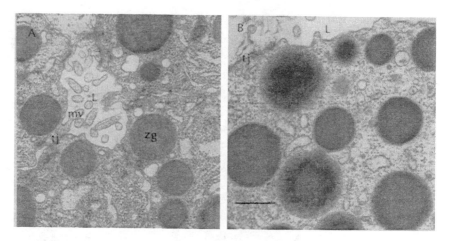

Fig. 7 Electron micrographs of the apical end of pancreatic acinar cells freshly prepared and exhibiting tight junction (tj) and microvilli (mv) at the apical lumen (L) (A). Isolated cells exhibiting no tight junction and stabilized for 1 h prior to fixation and processing for transmission electron microscopy (B). Notice the near absence of microvilli in (B).

compensatory retrieval of excess membrane by endocytosis occurs at a later time. The results from our AFM study (Schneider *et al.*, 1997) and the fine-tuned electrophysiological measurements by other investigators (Alvarez de Toledo *et al.*, 1993; Maruyama and Peterson, 1994) demand a reconsideration of the mechanism of this vital cellular process. Would it be beneficial for cells to release vesicular contents by total incorporation of the vesicle membrane at the cell plasma membrane? Such a mechanism of total

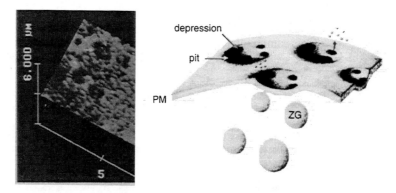

Fig. 8 Atomic force micrograph (left) of the apical plasma membrane (PM) in live acinar cells. Large pits containing small depressions (inset) are seen, where secretory vesicles may dock and fuse transiently to release their contents to the extracellular space. A diagram of PM pits containing small depressions, where vesicles fuse to release their contents to the outside (right). (See Color Plate.)

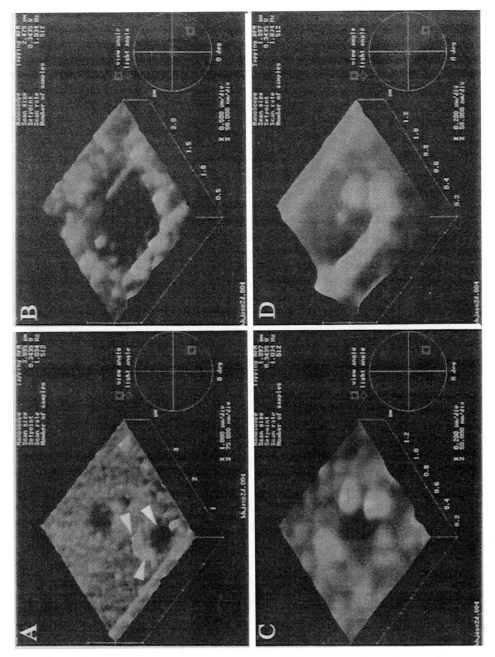

Fig. 9 Localization of amylase at the "pit." Atomic force micrography of the apical plasma membrane of fixed pancreatic acinar cells following exposure to amylase-gold during stimulation of secretion. Note gold clusters decorating primarily the edges of "pits." (See Color Plate.)

EXOCYTOSIS

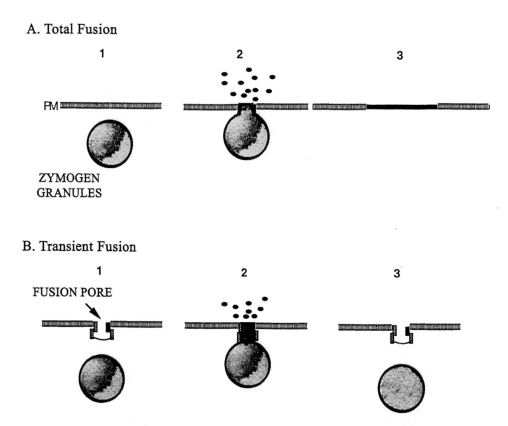

A. Total Fusion

B. Transient Fusion

Fig. 10 Schematic representation of total fusion (A) where secretory vesicle membranes are completely incorporated with the plasma membrane and transient fusion (B) of secretory vesicles at specific sites (depressions) of the cell plasma membrane to release vesicular contents to the cell exterior during exocytosis. Reproduced with permission from Jena, B. P. (1997). Exocytic fusion: Total or transient? *Cell. Biol. Int.* **21**(5), 257–259.

fusion would require the cell to expend much more energy than transient fusion (Fig. 10B). In fact, there is little evidence to support total fusion. Studies using transmission electron microscopy (TEM) rarely show the total incorporation of secretory vesicle membrane at the cell plasma membrane. On the contrary, the majority of the TEM studies demonstrate that following stimulation of secretion, there is an enriched presence of several intact but empty or partially empty secretory vesicles within cells (Lawson *et al.*, 1975). Similar changes are also identifiable in the pancreas and parotid exocrine cell of rats (Jamieson, 1972). Fast freeze-fracture EM studies on mast cells (Chandler and Heuser, 1980) show that, even after fusion of the vesicle membrane at the cell plasma membrane, much of the granule membrane is present and well separated from the plasma membrane (Chandler and Heuser, 1980). Electrophysiological studies in mast cells (Alvarez de Toledo *et al.*, 1993; Monck *et al.*, 1995) as well as in adrenal chromaffin cells (Chow *et al.*, 1992) suggest that the fusion pores either irreversibly expand (total fusion) or close following stimulation of secretion (transient fusion). Quantitative electron microscopy on stimulated and resting bovine chromaffin cells demonstrates no significant change following stimulation of secretion, in the number of peripheral densecore vesicles (Plattner *et al.*, 1997). An increase followed by a decrease in plasma membrane capacitance suggests that vesicles transiently fuse and dissociate. Alternately, the step increase in membrane capacitance is interpreted as evidence of total vesicle fusion. The step increase observed could result from secretory vesicles undergoing transient fusion at the plasma membrane, and before they dissociate, others fuse, causing a step increase in membrane capacitance. The capacitance measurements in a slow secretory cell such as the pancreatic acinar cell in either mice (Maruyama and Petersen, 1994), or rats (personal observation) demonstrate the occurrence of only transient fusions following stimulation of secretion. In fast secretory cells like neurons or mast cells, the number of secretory vesicles fusing at the plasma membrane at one time is far greater than that in pancreatic acinar cells. Therefore, the possibility of encountering a step increase in membrane capacitance is greater in mast cells or neurons. Due to the rapid fusion of secretory vesicles at the presynaptic membrane of a neuron, the rapid and selective retrieval of vesicle membrane would be a requirement. Since time and energy are critical factors, it would be efficient for such fast secretory cells (neurons or neuroendocrine cells) to exocytose via the transient mechanism of vesicle fusion. If total fusion at the presynaptic membrane were the case, what use do neurotransmitter transporters have in the synaptic vesicle membrane? Similarly, none of the secretory vesicle-associated proteins implicated in exocytosis (Rothman, 1992), have been confirmed to incorporate at the cell plasma membrane following stimulation of secretion. In our AFM studies (Schneider *et al.*, 1997) a 35% increase in the diameter and a 25–50% increase in the depth of depressions are observed during exocytosis. If secretory vesicles were to fuse completely at depressions, these structures would have dilated much more than what was observed. Since zymogen granules measure 0.2–1.2 μm in diameter, their total fusion at the cell plasma membrane would obliterate depressions. Based on these findings and supporting evidence, the transient fusion of secretory vesicles at the plasma membrane occurs at depressions in pancreatic acinar cells during exocytosis. Total fusion may occur when plasma membrane receptors, signal transducing molecules, transporters, or ion channels are required to be incorporated at the cell plasma membrane or when the plasma membrane undergoes recycling.

V. Future of AFM in the Study of Live Cells

Although much progress in AFM research has been achieved due to the optimization of sample preparation (Shao *et al.*, 2000; Linder *et al.*, 1999), delicate image acquisition software (Möller *et al.*, 1999; Müller *et al.*, 1999; Vie *et al.*, 2000) and continuous developments in instrumentation (Lehenkari *et al.*, 2000) are required. Soft and elastic properties of a live cell surface still remain as major hurdles in obtaining atomic or even angstrom resolution images. The development of highly flexible cantilever springs, extremely sensitive in the contact mode of AFM operation, combined with fine yet less damaging and nonsticky probes will greatly alleviate major problems in AFM studies on live cells. Although current technology precludes such high-resolution imaging of living cells by the AFM, the AFM in combination with excellent optics and electro-physiological measurements has and will greatly enhance our understanding of cellular structure–function. Additionally, functionalized scanning probes have recently enabled us for the first time to understand the secretion, interaction, and the biophysical and biochemical properties of molecules at the surface of live cells. These multicapabilities of the AFM will certainly be exploited further in the studies of living cells, bringing our understanding of cellular structure and function to a new dimension. The full potential, however, of the AFM on examining the structure–function of live cells has yet to be realized.

Acknowledgments

This study was supported by grants from the National Institute of Health (BPJ).

References

Albrecht, T. H., Akamine, S., Carver, T. E., and Quate, C. F. (1990). Microfabrication of cantilever styli for the atomic force microscope. *J. Vac. Sci. Technol.* **A8**, 3386–3396.

Alexander, S., Hellemans, L., Marti, O., Schneir, J., Elings, V., and Hansma, P. K. (1989). An atomic resolution atomic force microscope implemented using an optical lever. *J. Appl. Phys.* **65**, 164–167.

Alvarez de Toledo, G., Fernández-Chacón, R., and Fernandez, J. M. (1993). Release of secretory products during transient vesicle fusion. *Nature* **363**, 554–558.

Binnig, G., Quate, C. F., and Gerber, Ch. (1986). Atomic force microscope. *Phys. Rev. Lett.* **56**, 930–933.

Binnig, G., and Smith, D. P. F. (1986). Single-tube three-dimensional scanner for scanning tunneling microscopy. *Rev. Sci. Instrum.* **57**(8), 1688–1689.

Braet, F., Seynaeve, C., De Zanger, R., and Wisse, E. (1998). Imaging surface and submembranous structures with the atomic force microscope: a study on living cancer cells, fibroblasts and macrophages. *J. Microsc.* **190**(Pt 3), 328–38.

Chandler, D. E., and Heuser, J. E. (1980). Arrest of membrane fusion events in mast cells by quick-freezing. *J. Cell Biol.* **86**, 666–674.

Chow, R. H., von Rüden, R., and Neher, E. (1992). Delay in vesicle fusion revealed by electrochemical monitoring of single secretory events in adrenal chromaffin cells. *Nature* **356**, 60–63.

Danker, T., and Oberleithner, H. (2000). Nuclear pore function viewed with atomic force microscopy. *Pfluegers Arch.* **439**(6), 671–81.

Domke, J., Dannohl, S., Parak, W. J., Muller, O., Aicher, W. K., and Radmacher, M. (2000). Substrate dependent differences in morphology and elasticity of living osteoblasts investigated by atomic force microscopy. *Colloids Surf. B, Biointerfaces* **19**(4), 367–379.

Dufrene, Y. F., Boonaert, C. J., Gerin, P. A., Asther, M., and Rouxhet, P. G. (1999). Direct probing of the surface ultrastructure and molecular interactions of dormant and germinating spores of Phanerochaete chrysosporium. *J. Bacteriol.* **181**(17), 5350–5254.

Fritz, M., Radmacher, M., and Gaub, H. E. (1994). Granula motion and membrane spreading during activation of human platelets imaged by atomic force microscopy. *Biophys. J.* **66**, 1328–1334.

Gad, A., and Ikai, A. (1995). Method for immobilizing microbial cells on gel surface for dynamic AFM studies. *Biophys. J.* **69**, 2226–2233.

Häberle, W., Hörber, J. K. H., Ohnesorge, F., Smith, D. P. E., and Binnig, G. (1992). In situ investigation of single living cell infected by viruses. *Ultramicroscopy* **42–44,** 1161–1167.

Henderson, E., Haydon, P. G., and Sakaguchi, D. S. (1992). Actin filament dynamics in living glial cells imaged by atomic force microscopy. *Science* **257,** 1944–1946.

Hoh, J. H., and Hansma, P. K. (1992). Atomic force microscopy for high-resolution imaging in cell biology. *Trends Cell Biol.* **2,** 208–213.

Hoh, J. H., and Schoenenberger, C. A. (1994). Slow cellular dynamics in MDCK and R5 cells monitored by time-lapse atomic force microscopy. *Biophys. J.* **67,** 929–936.

Hörber, J. K. H., Häberle, W., Ohnesorge, F., Binnig, G., Liebich, H. G., Czerny, C. P., Mahnel, H., and Mayr, A. (1992). Investigation of living cells in the nanometer regime with the scanning force microscope. *Scanning Microsc.* **6,** 919–930.

Jamieson, J. D. (1972). Transport and discharge of exportable proteins in pancreatic exocrine cells: *In vitro* studies. *In* "Current Topics in Membranes and Transport" (F. Bronner and A. Kleinzeller, eds.), pp. 273–338. Academic Press, New York/London.

Jena, B. P. (1997). Exocytotic fusion: Total or transient? *Cell. Biol. Int.* **21**(5), 257–259.

Jena, B. P., Schneider, S. W., Geibel, J. P., and Sritharan, K. C. (unpublished).

Jena, B. P., Schneider, S. W., Geibel, J. P., Webster, P., Oberleithner, H., and Sritharan, K. C. (1997). Gi regulation of secretory vesicle swelling examined by atomic force microscopy. *Proc. Natl. Acad. Sci. U.S.A.* **94**(24), 13,317–13,322.

Kasas, S., Gotzos, V., and Celio, M. R. (1993). Observation of living cells using the atomic force microscope. *Biophys. J.* **64,** 539–544.

Lal, R., Drake, B., Blumberg, D., Saner, D. R., Hansma, P. K., and Feinstein, S. C. (1995). Imaging real-time neurite outgrowth and cytoskeletal reorganization with an atomic force microscope. *Cell Physiol.* **38,** C275–C285.

Langer, M. G., Koitschev, A., Haase, H., Rexhausen, U., Hörber, J. K. H., and Ruppersberg, J. P. (2000). Mechanical stimulation of individual stereocilia of living cochlear hair cells by atomic force microscopy. *Ultramicroscopy* **82**(1–4), 269–78.

Langer, M. G., Öffner, W., Wittmann, H., Flösser, H., Schaar, H., Häberle, W., Pralle, A., Ruppersberg, J. P., and Hörber, J. K. H. (1997). A scanning force microscope for simultaneous force and patch-clamp measurements on living cell tissues. *Rev. Sci. Instr.* **68**(6), 2583–2590.

Lawson, D., Fewtrell, C., Gomperts, B., and Raff, M. C. (1975). Anti-immunoglobulin-induced histamine secretion by rat peritoneal mast cells studied by immunoferritin electron microscopy. *J. Exp. Med.* **142,** 391–401.

Le Grimellec, C., Lesniewska, E., Giocondi, M. C., Finot, E., Vie, V., and Goudonnet, J. P. (1998). Imaging of the surface of living cells by low-force contact-mode atomic force microscopy. *Biophys. J.* **75**(2), 695–703.

Lehenkari, P. P., Charras, G. T., Nykanen, A., and Horton, M. A. (2000). Adapting atomic force microscopy for cell biology. *Ultramicroscopy* **82**(1–4), 289–295.

Linder, A., Weiland, U., and Apell, H. J. (1999). Novel polymer substrates for SFM investigations of living cells, biological membranes, and proteins. *J. Struct. Biol.* **126**(1), 16–26.

Maruyama, Y., and Petersen, O. H. (1994). Delay in granular fusion evoked by repetitive cytosolic Ca^{2+} spikes in mouse pancreatic acinar cells. *Cell Calcium* **16,** 419–430.

Meyer, E. (1992). Atomic force microscopy. *Prog. Surf. Sci. (UK)* **41,** 3–49.

Möller, C., Allen, M., Elings, V., Engel, A., and Müller, D. J. (1999). Tapping-mode atomic force microscopy produces faithful high-resolution images of protein surfaces. *Biophys. J.* **77**(2), 1150–1158.

Monck, J. R., Oberhauser, A. F., and Fernandez, J. M. (1995). The exocytotic fusion pore interface: a model of the site of neurotransmitter release. *Mol. Membr. Biol.* **12,** 151–156.

Müller, D. J., Fotiadis, D., Scheuring, S., Müller, S. A., and Engel, A. (1999). Electrostatically balanced subnanometer imaging of biological specimens by atomic force microscope. *Biophys. J.* **76**(2), 1101–1111.

Nagayama, S., Morimoto, M., Kawabata, K., Fujito, Y., Ogura, S., Abe, K., Ushiki, T., and Ito, E. (1996). AFM observation of three-dimensional fine structural changes in living neurons. *Bioimages* **4,** 111–116.

Oberleithner, H., Giebisch, G., and Geibel, J. (1993). Imaging the lamellipodium of migrating epithelial cells in vivo by atomic force microscope. *Pfluegers Arch.* **425,** 506–510.

Ohnesorge, F. M., Horber, J. K., Haberle, W., Czerny, C. P., Smith, D. P., and Binnig, G. (1997). AFM review study on pox viruses and living cells. *Biophys. J.* **73**(4), 2183–2194.

Parpura, V., Haydon, P. G., and Henderson, E. (1993). Three-dimensional imaging of living neurons and glia with the atomic force microscope. *J. Cell Sci.* **104,** 427–432.

Plattner, H., Artalejo, A. R., and Neher, E. (1997). Ultrastructural organization of bovine chromaffin cell cortex-analysis by cryofixation and morphometry of aspects pertinent to exocytosis. *J. Cell Biol.* **139**(7), 1709–1717.

Pralle, A., Prummer, M., Florin, E.-L., Stelzer, E. H. K., and Hörber, J. K. H. (1999). Three-dimensional high-resolution particle tracking for optical tweezers by forward scattered light. *Microsc. Res. Technol.* **44**(5), 378–386.

Quinn, A. S., Taatjes, D. J., and Jena, B. P. (unpublished).

Radmacher, M. (1997). Measuring the elastic properties of biological samples with the atomic force microscope. *IEEE Eng. Med. Biol.* **16,** 47–53.

Radmacher, M., Fritz, M., and Hansma, P. K. (1995). Imaging soft samples with the atomic force microscope: Gelatin in water and proponal. *Biophys. J.* **69,** 264–270.

Rothman, J. E. (1992). Mechanism of intracellular transport. *Nature* **355,** 409–415.

Rotsch, C., Jacobson, K., and Radmacher, M. (1999). Dimensional and mechanical dynamics of active and stable edges in motile fibroblasts investigated by using atomic force microscopy. *Proc. Natl. Acad. Sci. U.S.A.* **96**(3), 921–926.

Rugard, D., and Hansma, P. (1990). Atomic force microscopy. *Physics Today* **43,** 23–30.

Schneider, S. W., Pagel, P., Rotsch, C., Danker, T., Oberleithner, H., Radmacher, M., and Schwab, A. (2000). Volume dynamics in migrating epithelial cells measured with atomic force microscopy. *Pfluegers Arch.* **439**(3), 297–303.

Schneider, S. W., Sritharan, K. C., Geibel, J. P., Oberleithner, H., and Jena, B. P. (1997). Surface dynamics in living acinar cells imaged by atomic force microscopy: Identification of plasma membrane structures involved in exocytosis. *Proc. Natl. Acad. Sci. U.S.A.* **94,** 316–321.

Schoenenberger, C. A., and Hoh, J. H. (1994). Slow cellular dynamics in MDCK and R5 cells monitored by time-lapse atomic force microscopy. *Biophys. J.* **67**(2), 929–936.

Shao, Z., Shi, D., and Somlyo, A. V. (2000). Cryoatomic force microscopy of filamentous actin. *Biophys. J.* **78**(2), 950–958.

Spudich, A., and Braunstein, D. (1995). Large secretory structures at the cell surface imaged with scanning force microscopy. *Proc. Natl. Acad. Sci. U.S.A.* **92,** 6976–6980.

Taatjes, D. J., Quinn, A. S., Lewis, M. R., and Bovill, E. G. (1999). Quality assessment of atomic force microscopy probes by scanning electron microscopy: Correlation of tip structure with rendered images. *Micro. Res. Tech.* **44**(5), 312–326.

Tojima, T., Yamane, Y., Takagi, H., Takeshita, T., Sugiyama, T., Haga, H., Kawabata, K., Ushiki, T., Abe, K., Yoshioka, T., and Ito, E. (2000). Three-dimensional characterization of interior structures of exocytotic apertures of nerve cells using atomic force microscopy. *Neuroscience* **101**(2), 471–481.

Vie, V., Giocondi, M. C., Lesniewska, E., Finot, E., Goudonnet, J. P., and Le Grimellec, C. (2000). Tapping-mode atomic force microscopy on intact cells: optimal adjustment of tapping conditions by using the deflection signal. *Ultramicroscopy* **82**(1–4), 279–288.

Weissmuller, G., Garcia-Abreu, J., Mascarello Bisch, P., Moura Neto, V., and Cavalcante, L. A. (2000). Glial cells with differential neurite growth-modulating properties probed by atomic force microscopy. *Neurosci. Res.* **38**(2), 217–20.

You, H. X., Lau, J. M., Zhang, S., and Yu, L. (2000). Atomic force microscopy imaging of living cells: a preliminary study of the disruptive effect of the cantilever tip on cell morphology. *Ultramicroscopy* **82**(1–4), 297–305.

CHAPTER 3

Atomic Force Microscope Imaging of Cells and Membranes

Eric Lesniewska,* Pierre Emmanuel Milhiet,[†,‡]
Marie-Cécile Giocondi,[‡] and Christian Le Grimellec[‡]

*Laboratory of Physics, National Center for Scientific Research, URA 5027
UFR Sciences et Techniques
21078 Dijon Cedex, France

†Laboratory CRRET
Université Paris 12
94000 Créteil Cedex, France

‡Center of Structural Biochemistry
French National Institute for Health and Medical Research, U414
34090 Montpellier Cedex, France

I. Introduction

Because of the plasma membrane's fundamental role in nature, i.e., no membrane no cell, and in the relationships between a living cell and its environment, including the supply of nutrients and the necessary transmission of signals needed for the various adaptation processes, the knowledge of both cell membrane and cell-surface organizations remains a major goal in biology. Numerous techniques and methods, from physiopathological studies to X-rays and neutron diffraction, have been developed or used to better define the structural dynamic arrangement of molecules in membranes. Biological membranes are now modeled as two-dimensional structures essentially composed of lipids and proteins organized in domains (Jacobson *et al.*, 1995; Simons and Ikonen, 1997). Sugar residues, most often covalently linked to the lipid polar headgroups and to peptidic chains exposed at the cell/liquid medium external interface, and cytoskeleton elements are the other partners involved in the plasma membrane organization.

The atomic force microscope (AFM), which can obtain the highest resolution images of surfaces in aqueous medium, has recently attracted the interest of cell and membrane biologists (Radmacher *et al.*, 1992). As reported 10 years ago, AFM imaging of RBC examined in liquid medium strongly suggested that this new approach could provide topological information on the imaging of both cell surface and membrane structures at a significantly higher resolution than the optical microscope (Butt *et al.*, 1990). This approach is now established, although the highest lateral resolution that can be obtained remains a matter of debate. Topographical images of eukaryotic, prokaryotic, plant cell surfaces, and isolated membranes were published and revealed structural details that could not be detected by other approaches, thus presenting the AFM as a potent additionnal tool for the biologist.

The principle of the AFM, which consists in raster scanning a surface with a tip of finite size, imposes constraints on the type of material that can be examined. For instance, the AFM does not perform imaging of cells in suspension. Due to the very soft nature of biological material, particular attention must also be paid to not only the sample preparation but also the adjustment of experimental conditions. This chapter aims to provide a practical basis for the imaging of cells and membranes by the AFM.

II. AFM Equipment

Most of the commercial equipment currently used in physics and chemistry departments is suited to image cells and membranes and must be equipped with a liquid cell for the work in aqueous medium and, if possible, with the accessories necessary for scanning in an oscillating mode. Stand-alone AFMs coupled to an inverted optical microscope are also available. For heterogeneous or dispersed samples, such as nonconfluent cell cultures or membrane preparations, stand-alone AFMs offer the advantage of allowing the selection, by either morphological criteria or by the use of fluorescent markers, of the zone to be probed rather than probing at random. A generally slightly decreased stability and resolution are the consequences. So far, practically all the biological applications of the AFM have been conducted at room temperature, but temperature-controlled stages are presently being proposed by different companies.

III. AFM Operating Modes

The essential difference between the physicochemical, and the biological applications of the AFM is the constraint imposed by the softness and the fragility of the samples. The modes of operation are identical, i.e., contact and oscillating modes. These modes have been described in detail in other chapters of this book and will be only briefly discussed here. The principle of the AFM is based on the measurement of the repulsive (hard sphere) or attractive (van der Waals) interaction forces between the atom(s) at the extremity of a fine tip and the atom(s) at the sample surface (Binnig et al., 1986). In the contact mode, the user fixes the value of the repulsive force between the tip and the sample which will be maintained constant during the raster scan of the sample, providing an isoforce image of the surface. Theoretically, the tip extremity and the sample remain in contact during scanning. In the oscillating mode, the tip oscillates at a high frequency, determined by the cantilever spring constant, and interacts with the surface only at the lower end of each cycle. When the interaction involves atomic repulsion (Putman et al., 1992) the mode is usually called the tapping mode (Digital Instruments, Santa Barbara, CA). The main advantage of this mode, as compared to the contact mode, resides in a marked reduction of the friction forces during scanning. Biological applications of the oscillating noncontact mode, which is based on the use of attractive interactions, remain to be established.

IV. Requirements for the Imaging of Intact Cells

By imaging intact cells we mean imaging cells, living or fixed, in their natural environment, which is essentially aqueous for eukaryotes. The requirements for imaging under liquid are the same for cells and isolated membranes.

A. Cell Immobilization

The first requirement for cell imaging is general and applies to all categories of samples: by principle, the AFM can only image material which adheres or is fixed to a support. Thus, cells or membranes in suspension would be pushed away by the scanning tip and could not be imaged by this technique. Trapping of cells like red blood cells or bacteria in pores of filters can allow their immobilization and provide access to the AFM imaging of their upper exposed surface (Kasas and Ikai, 1995). Upon settling, cells in suspension can also spontaneously adhere to the surface of glass coverslips or mica. The well-known recipe among cell biologists which consists of coating a glass coverslip with positively charged material like polylysine can also help in attaching cells to a flat surface. A very elegant way to immobilize cells for AFM examination is by using micropipettes. An adapted, home-built AFM must, however, be used in conjunction with

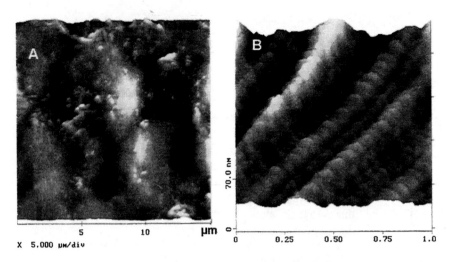

Fig. 1 Contact mode AFM imaging of corneal tissue in liquid medium. (A) Corneal epithelial surface of adult albino rabbit. (B) Collagen fibrils of corneal stroma arranged in bundles after mechanical ablation.

the micropipette holder (Hörber *et al.*, 1995). Fortunately, a large number of cell types can be grown directly either on glass coverslips or on plasticware treated by a Bunsen burner flame which markedly reduces the plastic surface corrugations (Thimonier *et al.*, 1997). Before placing a cell preparation under the AFM, a good test to run consists of rinsing the preparation several times with medium or buffer and then, by using an optical microscope, checking whether the cells are still attached to their substrate. Recently, flat pieces of fresh tissues were examined by the AFM (Fig. 1) (Tsilimbaris *et al.*, 2000). Such experiments were made possible by gluing the tissue pieces directly to the AFM magnetic disks. Firm fixation of the glass coverslips or plastic or filter supports used for cell immobilization onto the magnetic disks is also of crucial importance. For cells or membranes adsorbed on glass coverslips, after drying the bottom of the support, we use cyanoacrylate glue which resists to long exposure to aqueous medium. The stability of the imaging on stand-alone equipment can also be improved by screwing a teflon ring to the round glass coverslips used as a support on the inverted microscope stage.

B. Scanning Forces

The second requirement is using imaging forces, as low as possible, to avoid damaging the soft cell structures. Thermal fluctuation experienced by a free cantilever provides a first indication about the lower limit of the imaging forces. Considering the cantilever as an uncompressed spring, thermal fluctuation results in spontaneous tip movements whose amplitude Δ can be estimated using (Shao *et al.*, 1996)

$$\Delta = (k_B T / k)^{1/2},$$

where k_B is the Boltzmann constant (1.38×10^{-23} J/K), T is the temperature in Kelvin, and k is the cantilever spring constant (N/m). Fluctuation amplitudes and the

Table I
Thermal Fluctuation, Uncertainties, and Signal-to-Noise Ratio of
Soft Cantilevers[a]

Spring constant (N/m)	Thermal fluctuation Å (equivalent force, pN)	Force determination uncertainty pN (for 50-pN force)	Signal/noise (for 50-pN force)
0.01	6.4 (6.4)	0.4	126
0.03	3.7 (11.1)	1.2	42
0.06	2.6 (15.6)	2.4	21
0.10	2.0 (20.0)	4.0	12.5
0.36	1.1 (39.6)	14.4	3.5

[a]Calculated from Shao *et al.* (1996).

corresponding «force noise» at room temperature (20°C) for the cantilevers commonly used in contact mode imaging under aqueous medium are given in Table I. These amplitudes vary from 6.4 to 1.1 Å for cantilevers with spring constants between 0.01 and 0.36 N/m. The equivalent forces, F, are 6.4 and 39 pN, respectively ($F = k\Delta$). Thus, the use of softer cantilevers markedly reduces the limiting value of the scanning forces, which must be larger than that of the thermal fluctuations. Fluctuations are reduced, as a function of the force applied, once the tip is in contact with the surface, according to

$$\Delta_c = (k_B T)/2F,$$

where Δ_c is the amplitude of the fluctuation of the tip position in contact, and F is the force applied. For instance, applying a force of 20 pN to a 0.01 N/m cantilever, corresponding to a 2-nm deflection of the tip position, results in a decrease in fluctuation from 6.4 to 1.0 Å; i.e., the spontaneous fluctuation is 1/20 of the imposed deflection value. This value of 20 pN is close to the limiting value for imaging reported in the literature (Le Grimellec *et al.*, 1998). It must be mentionned that, due to the drift of piezo-electric elements, maintaining a constant imaging value below 100 pN requires the readjustment of the tension applied to the z element during imaging. The use of radiation pressure brought by a second laser acting on the cantilever, associated with very low spring constant homemade cantilevers, has been reported to decrease the limiting imaging force to values below 1 pN (Tokunaga *et al.*, 1997).

Adjustment of the scanning force via the force versus distance curves to the lowest accessible value, prior to cell imaging, is the first step to avoiding cell damage and implies that both the spring constant of the cantilever and the sensitivity of the detection of the AFM tip response to a known vertical displacement must be, at least approximately, known before imaging. Sensitivity to vertical displacement is simply obtained by force versus distance curves, under the liquid medium, on regions of the support (glass, mica, HOPG) devoid of biological material. Both the spring constant and the sensitivity of the detection system transform the voltage reading from the photo-diode into an absolute displacement (d) leading to the force experienced (F) by the cantilever during the scan ($F = k \cdot d$). In the spring constant of the cantilever (k), the values provided by manufacturers are in general a reasonable estimate of the mean value obtained for a population of cantilevers of comparable length and thickness. A better evaluation of the spring constant of the cantilever used for the experiment can be obtained through the determination of its resonance frequency or of its thermal noise (Cleveland *et al.*, 1993; Butt and Jaschke, 1995). The laser light reflected by the cantilever dissipates heat around the tip, resulting in instabilities in the position of the cantilever as long as the thermal equilibrium with the surrounding medium is not achieved. For the softest commercial cantilevers ($k = 0.01$ N/m) the equilibrium period can take more than 1 h. For low-force AFM in the contact mode, this equilibrium must be obtained before starting the imaging. The oscillating mode is less sensitive to temperature equilibration.

C. Cell Viability

Until very recently, there was no temperature control unit commercially available for the AFM. Accordingly, with few exceptions, experiments on cells were performed at about 3°C above room temperature, because of the heating of the solution by the energy dissipated from the laser diode. Most of the mammalian cell lines which grow at 37°C are still active, albeit at a reduced rate, at this temperature. It is important, however, to check under an optical microscope whether the shape of the cells is not altered by the change in temperature. In the 5% CO_2 atmosphere of incubators, growth media are buffered by bicarbonate. In an air atmosphere, maintaining the pH constant requires the replacement, at least partial, of the culture medium. Hank's medium, phosphate saline buffer (PBS) complemented with glutamine and calcium, or mixtures of the growth medium (without serum) with PBS (1/1 or 1/2, vol/vol) generally offer satisfying solutions. For long-duration experiments on cells, care must be taken to prevent changes in the buffer composition resulting from water evaporation. Dye exclusion tests, like the trypan blue test, allow the establishment of the viability of the preparation at the end of the experiment (Le Grimellec *et al.*, 1994). Hoh and Schoenenberger (1994) demonstrated that the viability of cells under the AFM can be maintained for periods as long as 44 h.

V. Imaging of Cells

Using stand-alone AFM coupled with an inverted optical microscope makes the positioning of the tip above a cell body easy, even when using either nonconfluent cell cultures or isolated plated cells. For traditional AFMs, with the piezo-system located below the sample, finding dispersed cells often requires patience, and working under liquid also imposes to protect the piezo from leaks. Some manufacturers propose special chambers with O-ring seals. Very often these seals are not perfectly efficient. Most investigators work without the O rings and prefer to protect the piezo with pieces of thin aluminum foil, stretch n' seal, or equivalent films.

A. Contact Mode

Once the sensitivity of the optical lever is determined and the thermal equilibrium is reached, the system is ready for cell imaging. For reasons cited previously, cantilevers with the lowest spring constants (0.01 to 0.06 N/m) are preferred. The usual silicon nitride tips are also preferred for intact cell imaging because sharpened tips penetrate the cell membrane (Haydon *et al.*, 1996). The first step consists of the obtention on the cell surface of a force versus distance (f–d) curve, where the approaching and retracting curves coincide as much as possible. To limit the damage, the force curve is performed with the x,y scan size set to 0 nm. The vertical (z) scan range is fixed to a value between 500 and 1500 nm, with a scan rate of 1 Hz. Generally, using the default engagement conditions, force versus distance curves obtained from the cells have a poor aspect (Fig. 2A): a marked hysteresis between approaching and retracting curves is observed, and it is very difficult, if not impossible, to determine the onset of the tip–cell surface interaction and thus the scanning force. Such curves reflect the fact that, for the whole vertical scanning distance z, the tip is in contact with the surface which is progressively pushed down. This is most likely a consequence of the low spring constant of the cell surface. Imaging under conditions where a large hysteresis between approaching and retracting f–d curves occurs results, for cells as for any material, in very poor AFM height images. Relatively good deflection images of the cell surface can, however, be obtained under these conditions, but they need to be critically analyzed due to lack of control on the scanning force. It is generally possible to circumvent this problem by raising, step by step, the position of the tip (tip-up command on Digital Instruments AFMs) while

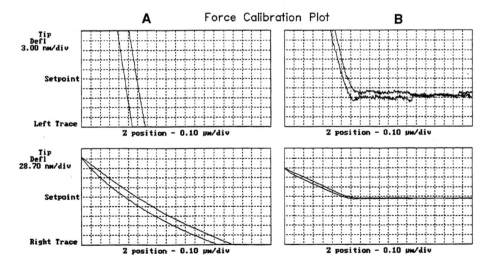

Fig. 2 Contact mode force versus distance curves at the surface of living CHO cells. (A) Force versus distance curves obtained using the default engagement conditions. (B) Corresponding curves obtained after raising the AFM tip according to the procedure described in the text. Cantilever spring constant: 0.03 N/m, z scan rate 1 Hz.

examining the *f–d* curves. A progressive decrease in the hysteresis accompanied by a flattening of curves is observed, which ultimately leads to the detection of the onset of the tip–cell surface interaction (Fig. 2B). For living cells, the tip often has to be raised by distances greater than 1 or 2 μm above the engagement position. The *z* piezo is accordingly working in an extended configuration. Obtention of the flat portion of *f–d* curves, which corresponds to the position of the free cantilever, allows the precise setting of the imaging force, using the predetermined sensitivity in nanometers, of the cantilever response. The spring constant of the cell surface is significantly increased by fixation using either glutaraldehyde or paraformaldehyde (Hoh and Schoenenberger, 1994). Practice in adjusting the tip position in *f–d* plots on fixed cells helps in subsequent experiments on living cells.

Once the scanning force has been adjusted to low values, i.e., below 100 pN, a *x–y* scanning range of, for example, 10 μm is chosen. Imaging conditions must be refined using the oscilloscope traces accessible to the user via the AFM equipment. As for force curves, good imaging requires that the ≪trace≫ and ≪retrace≫ traces coincide with each other as much as possible. Again, a poor coincidence means poor quality imaging. For large scans (>5 μm), the adjustments begin by using a scan rate of 1 Hz, integral and proportional gains around 2 (Nanoscope III, D.I), and a scanning angle of 0°. The effect of varying the scanning force, through changes in the set-point voltage, is first checked in the oscilloscope traces. Then gains and scanning rate are readjusted and, finally, the best scanning angle is selected. This procedure is applied each time the imaging field is changed. Once this is achieved, a new *f–d* curve is recorded which enables the determination of the precise value of the force used for starting the imaging. A second *f–d* plot is recorded once the image is acquired, leading to the force value at the end of the imaging. If the forces prior to and after imaging are too different, which is often the case for imaging forces below 200 pN, the set point must be readjusted manually during the image acquisition. Following this procedure, height images of the surface of living CV-1 cells and MDCK cells were obtained with forces as low as 20 pN (Le Grimellec *et al.*, 1998). On these cell types, imaging below 100 pN was routinely achieved. It is important to mention that for such low values it is difficult to estimate the tip force experienced by the cell surface during scanning. What is measured is a deflection of the cantilever, which does not mean that the tip is in real contact with the cell surface. Electrostatic repulsion linked to the fact that Si_3N_4 tips are negatively charged, as are most of the cell surfaces, is likely to play a role in the bending of the cantilever (Butt, 1992; Müller *et al.*, 1999). Unfortunately, working with living cells limits the possibilities of changing the medium composition to precisely determine to what extent the electrostatic repulsion participates in the recorded cantilever deflection.

B. Oscillating Mode

Basically, for cell imaging in the oscillating mode, one must follow the same procedure as that for the contact mode. The same triangular cantilevers with low spring constants are used while working in aqueous medium, and the first step before imaging consists of selecting the resonance peak (Putman *et al.*, 1994). Using an oscillating piezo-drive which excites mechanical resonances in both the cantilever holder and the bathing fluid generates numerous peaks in the frequency sweep. Experimentally, we found that the peak centered around 8 kHz gives satisfactory results. The imaging frequency is not at the peak value but is slightly shifted toward the lower frequency end. Establishing the sensitivity, in nanometers, of the response of the detection system for the cantilever chosen constitutes the second step. This is obtained through amplitude versus distance (a–d) plots performed on a clean region of the support, taking care to limit the z scanning range to the point where the oscillation is completely damped. Alternatively, to limit the risk of tip damage, an approximate value corresponding to the mean value, determined in previous experiments from a series of cantilevers of identical nominal spring constant, can be fixed. The true sensitivity is established at the end of the experiments, and the amplitude values used for imaging are corrected accordingly. Before engagement, the peak amplitude on the frequency sweep is adjusted to ∼50 nm, with a set point corresponding to a damping of ∼10 nm. The automatic tip approach command is then activated, setting a 0-nm xy scan size. Upon engagement, a new a–d plot is performed on the living cell surface. Most often, as in the contact mode, the plot shows a high curvature which renders the detection of the onset of the change in amplitude slope practically impossible (Fig. 3A). This precludes the precise adjustment of the feedback damping (set point) value. By manually readjusting the vertical position of the photodetector it is possible to acquire a deflection versus distance (de–d) plot simultaneously with the a–d plot (Vié et al., 2000). Such plots, which compare with the f–d curves obtained in

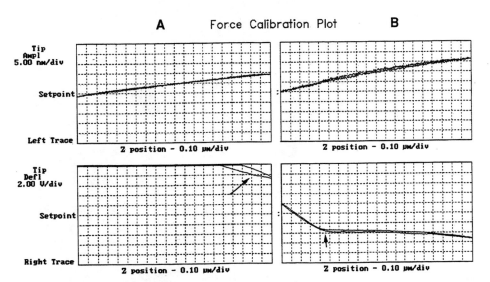

Fig. 3 Tapping mode amplitude and deflection versus distance curves at the surface of living CHO cells. The plots were performed with the same cells and the same cantilever as those used in Fig. 2. (A) and (B) correspond to the amplitude (upper panels) and deflection (lower panels) versus distance curves obtained using the default engagement condition and after the tip position adjustment, respectively. The arrow in (A) indicates that the tip interacts with the cell surface during the entire z scan. On the other hand, in (B), the change in slope in the deflection curve (arrow) allows the precise determination of the onset of the tip-cell surface interaction and the adjustment of the amplitude damping for imaging.

contact mode, result from the change in the cantilever median position upon interaction of the oscillating tip with the sample surface and they are obtained only when working under liquid. As previously described for low-force imaging in the contact mode, raising the position of the tip to work with the *z* piezo in an extended configuration markedly reduces the hysteresis in both the *a–d* and the de–*d* curves (Fig. 3B). Once this is achieved, the damping amplitude set point can be adjusted to the chosen value at the position corresponding to the change in slope in the de–*d* plot. The end of the procedure for cell imaging is the same as that for the contact mode. The exact determination of the loading force applied to the cells during scanning in the oscillating mode is actually not possible, but a rough estimate can be made from the amplitude and the spring constant of the cantilever. Images of cells have been obtained using 0.01 N/m cantilevers with a set point corresponding to a reduction in amplitude of less than 2 nm, i.e., with estimated imaging forces below 20 pN. Interestingly, the deflection signal obtained in the oscillating mode is sensitive to local stiffness (Vié *et al.*, 2000) and can provide images of the organization of the submembranous cytoskeleton simultaneously with the acquisition of the cell-surface topography via the height and amplitude signals. Deflection images are close to but not identical to phase images which are reporting, at least partly (Chen *et al.*, 1998), on the local viscoelastic properties of the surfaces (Fig. 4).

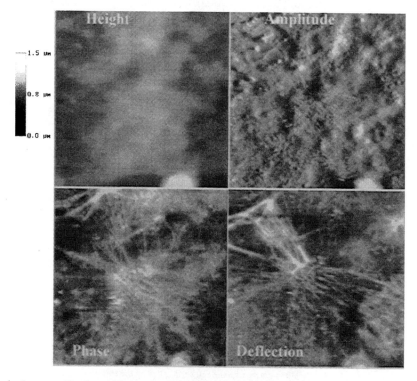

Fig. 4 Low-magnification tapping mode images of living CV-1 cells. Scan size, 25 μm; scan rate, 1 Hz; approximate scanning force; 50 pN (cantilever spring constant: 0.01 N/m).

The Oscillating mode in liquid can also be obtained via an oscillating magnetic field driving a magnetized cantilever (Han *et al.*, 1996; MAC mode, Molecular Imaging, Phoenix, AZ). This new mode offers the advantage of driving the tip directly, thus leading to a unique resonance peak, with a reduced perturbation of the liquid medium. The performances of this mode in cell imaging mostly remain to be established.

C. Actual Possibilities and Limits

Topographical images, under aqueous media, of the surface of various cell types either in culture or even in tissue slices have now been reported with a resolution much better than that of an optical microscope (see, for example, Fig. 5). Because it reduces the friction forces, and therefore reduces the accumulation of membrane components on the tip (Schaus and Henderson, 1997), the oscillating mode seems to be actually the best choice for imaging cells. In terms of topography, several groups have reported lateral resolutions close to 10 nm for the surface of various unfixed cells (Butt *et al.*, 1990; Hörber *et al.*, 1992; Le Grimellec *et al.*, 1994, 1998). Indeed, this can only be achieved using very low scanning forces and most often after an enzymatic treatment for disrupting the glycocalix organization. Clearly not all cells are suitable for high-resolution imaging. Besides the abundancy of the cell surface glycocalix, cells moving or growing fast at room temperature can only provide fuzzy images when examined at a sufficiently high magnification. Another type of problem that can be encountered is the presence of tightly packed microvilli covering the surface, a situation often found in epithelial

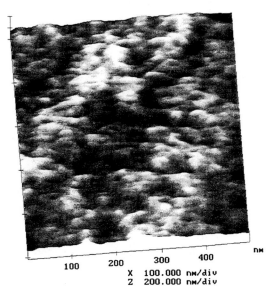

Fig. 5 High-magnification imaging of outer hair cells from the inner ear. Cells were fixed with 0.5% glutaraldehyde in buffer, and rinsed and examined in the same buffer. Height image, mode: tapping; scan size, 500 nm; scan rate, 1.5 Hz; approximate force, 20 pN; z range, 200 nm/division.

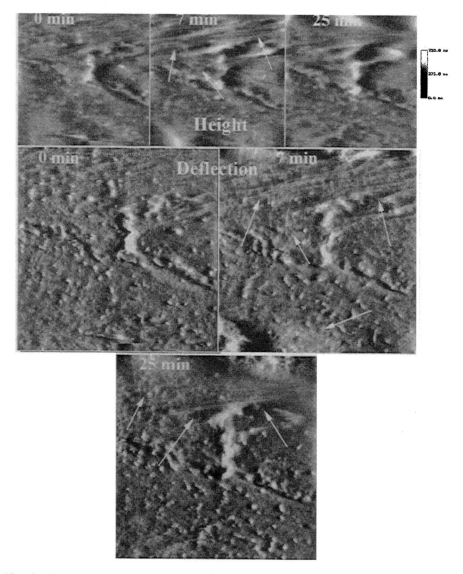

Fig. 6 Time-dependent reorganization of the surface of living cells after addition of 2.5 m*M* db-cAMP. Upper part, height images; lower part, deflection images. Scan size, 25 μm; scan rate, 1.5 Hz; force, <100 pN (contact mode); cantilever spring constant, 0.01 N/m.

cells. In that case, even working at low forces, the tip can only access the upper part of the flexible microvilli, precluding the obtention of high-resolution images. Lateral diffusion of membrane constituents, in the order of 1 μm^2/s for membranes in the fluid state, constitutes the ultimate limitation to high-resolution cell-surface imaging by AFM. Accordingly, fixation is an absolute requirement for a precise localization of a particular protein (unless it spontaneously forms two-dimensional crystals) at the cell membrane. In fact, a resolution better than 5 nm was recently obtained for the lateral surface of glutaraldehyde-fixed inner ear cells examined in buffer (Le Grimellec *et al.,* 2000). This strengthens the view that AFM can, in the near future, provide key information on the understanding of the organization of membranes at the molecular level.

For physiologists, a final important point about AFM imaging of living cells is the time required for imaging. In the best circumstances, good resolution scanning of a limited ($\sim$10-μm^2) cell region with a commercial apparatus takes about 2 min. Accordingly, this limits the studies on cell topography-function to slow processes as illustrated in Fig. 6, where the changes in the cell surface of CV-1 cells resulting from the addition, under the AFM, of dibutyryl-cAMP, are described.

VI. Imaging of Isolated Membranes

The main difficulty encountered both in AFM imaging of intact cells and in the interpretation of the images obtained lies in not only the chemical heterogeneity of cell-surface constituents but also the relative softness of the plasma membrane with regard to the other cellular elements, cytoskeleton, cytosol, and intracellular organelles, it surrounds. For instance, even by using low scanning forces, it remains difficult to assess the relative contributions of both real plasma membrane elements and subplasma membrane structures in the images of the cell surface. This also holds true when estimating the local mechanical properties of cells. The use of purified membranes adsorbed on a hard support, generally mica or glass, provides direct information about the structure of isolated biological membranes. Unroofing with filters followed by the extensive washing of support-attached cells, which results in the exposure of the cytoplamic leaflet of the basal membrane to the medium (Le Grimellec *et al.,* 1995; Ziegler *et al.,* 1998) direct examination at the tip of a pipette (Hörber *et al.,* 1995), or the examination after transfer to the support of patch-clamp membrane fragments (Lärmer *et al.,* 1997), has thus far provided AFM images resembling those obtained at the surface of intact cells. The general conditions for imaging these structures are equivalent to those followed for cells, i.e., minimizing the imaging force; except that a greater degree of freedom exists in the choice of the imaging buffer compositions, which can result in improved resolution images (Müller *et al.,* 1999).

VII. Conclusion and Perspectives

Difficulties encountered in adjusting the imaging parameters have limited the use of the AFM in cell biology and physiology. With the recent improvements in the procedure, it is now possible to image the surface of cells on a nearly routine basis. For living cells, this opens up the possibility to investigate the three-dimensional structure–function relationships of the cell surface at a mesoscopic scale, a procedure thus far inaccessible by conventional microscopic techniques. Two technical improvements would contribute to facilitate the use of AFM in cell studies and to widen the field of its applications. The first concerns the availability of short cantilevers with lower spring constants (<0.005 N/m), eventually coupled with radiation pressure compensation for thermal bending motion (Tokunaga *et al.*, 1997), which would further increase the control on the forces used for imaging. Lowering from minutes to seconds the time scale needed for one image acquisition (Walters *et al.*, 1997) is a second essential improvement which would enable the study of a much larger number of biological processes. Obtention of high-resolution images on either fixed cells or isolated membranes presents a problem regarding the identification of the structures imaged. Scanning near-field fluorescence resonance energy transfer microscopy (Vickery and Dunn, 1999) and recognition imaging by force microscopy (Hinterdorfer *et al.*, 1996) constitute the most promising approaches to solve these problems.

Acknowledgments

This work was supported by grants from La Fondation pour la Recherche Médicale, l'Association pour la Recherche sur le Cancer, la Région Languedoc-Roussillon, and l'Université Montpellier I.

References

Binnig, G., Quate, C. F., and Gerber, Ch. (1986). Atomic force microscope. *Phys. Rev. Lett.* **56**, 930–933.

Butt, H.-J. (1992). Electrostatic interaction in scanning probe microscopy when imaging in electrolyte solutions. *Nanotechnology* **3**, 60–68.

Butt, H.-J., and Jaschke, M. (1995). Calculation of thermal noise in atomic force microscopy. *Nanotechnology* **6**, 1–7.

Butt, H. J., Wolff, E. K., Gould, A. C., Dixon Northern, B., Peterson, C. M., and Hansma, P. K. (1990). Imaging cells with the atomic force microscope. *J. Struct. Biol.* **105**, 54–61.

Chen, X., Davies, M. C., Roberts, C. J., Tendler, S. J. B., Williams, P. M., Davies, J., Dawkes, A. C., and Edwards, J. C. (1998). Interpretation of tapping mode atomic force microscopy data using amplitude-phase-distance measurements. *Ultramicroscopy* **75**, 171–181.

Cleveland, J. P., Manne, S., Bocek, D., and Hansma, P. K. (1993). A nondestructive method for determining the spring constant of cantilevers for scanning force microscopy. *Rev. Sci. Instrum.* **64**, 403–405.

Han, W., Lindsay, S. M., and Jing, T. (1996). A magnetically driven oscillating probe microscope for operation in fluids. *Appl. Phys. Lett.* **69**, 4111–4114.

Haydon, P. G., Lartius, R., Parpura, V., and Marchese-Ragona. (1996). Membrane deformation of living glial cells using atomic force microscopy. *J. Microsc.* **182**, 114–120.

Hinterdorfer, P., Baumgartner, W., Gruber, H. J., Schilcher, K., and Schindler, H. (1996). Detection and localization of individual antibody-antigen recognition events by atomic force microscopy. *Proc. Natl. Acad. Sci. U.S.A.* **93**, 3477–3481.

Hoh, J. H., and Schoenenberger, C.-A. (1994). Surface morphology and mechanical properties of MDCK monolayers by atomic force microscopy. *J. Cell Sci.* **107**, 1105–1114.

Hörber, J. K. H., Häberle, W., Ohnesorge, F., Binnig, G., Liebich, H. G., Czerny, C. P., Mahnel, H., and Mayr, A. (1992). Investigation of living cells in the nanometer regime with the scanning force microscope. *Scanning Microsc.* **6**, 919–930.

Hörber, J. K. H., Mosbacher, J., Häberle, W., Ruppersberg, J. P., and Sakmann, B. (1995). A look at membrane patches with a scanning force microscope. *Biophys. J.* **68**, 1687–1693.

Jacobson, K., Sheets, E. D., and Simson, R. (1995). Revisiting the fluid mosaic model of membranes. *Science* **268**, 1441–1442.

Kasas, S., and Ikai, A. (1995). A method for anchoring round shaped cells for atomic force microscope imaging. *Biophys. J.* **68,** 1678–1680.

Lärmer, J., Schneider, S. W., Danker, T., Schwab, A., and Oberleithner, H. (1997). Imaging excised apical plasma membrane patches of MDCK cells in physiological conditions with atomic force microscopy. *Pfluegers Arch.–Eur. J. Physiol.* **434,** 254–260.

Le Grimellec, C., Giocondi, M.-C., Pujol, R., and Leniewska, E. (2000). Tapping mode atomic force microscopy allows the in situ imaging of fragile membrane structures and of intact cells surface at high resolution. *Single Molecules* **1,** 105–107.

Le Grimellec, C., Lesniewska, E., Cachia, C., Schreiber, J. P., de Fornel, F., and Goudonnet, J. P. (1994). Imaging the membrane surface of MDCK cells by atomic force microscopy. *Biophys. J.* **67,** 36–41.

Le Grimellec, C., Lesniewska, E., Giocondi, M.-C., Cachia, C., Schreiber, J. P., and Goudonnet, J. P. (1995). Imaging of the cytoplasmic leaflet of the plasma membrane by atomic force microscopy. *Scanning Microsc.* **9,** 401–411.

Le Grimellec, C., Lesniewska, E., Giocondi, M.-C., Finot, E., Vié, V., and Goudonnet, J.-P. (1998). Imaging of the surface of living cells by low-force contact-mode atomic force microscopy. *Biophys. J.* **75,** 95–703.

Müller, D. J., Fotiadis, D., Scheuring, S., Müller, S. A., and Engel, A. (1999). Electrostatically balanced subnanometer imaging of biological specimens by atomic force microscope. *Biophys. J.* **76,** 1101–1111.

Putman, C. A. J., van der Werf, K. O., de Grooth, B. G., van Hulst, N. F., Greve, J., and Hansma, P. K. (1992). A new imaging mode in atomic force microscopy based on the error signal. *Proc. Soc. Photo-Opt. Instrum. Eng.* **1639,** 198–204.

Putman, C. A. J., van der Werf, K. O., de Grooth, B. G., van Hulst, N. F., and Greve, J. (1994). Viscoelasticity of living cells allows high resolution imaging by tapping mode atomic force microscopy. *Biophys. J.* **67,** 1749–1753.

Radmacher, M., Tillmann, R. W., Fritz, M., and Gaub, H. E. (1992). From molecules to cells: Imaging soft samples with the atomic force microscope. *Science* **257,** 1900–1905.

Schauss, S. S., and Henderson, E. R. (1997). Cell viability and probe-cell membrane interactions of XR1 glial cells imaged by atomic force microscopy. *Biophys. J.* **73,** 1205–1214.

Shao, Z., Mou, J., Czajkowsky, D. M., Yang, J., and Yuan, J.-Y. (1996). Biological atomic force microscopy: what is achieved and what is needed. *Adv. Phys.* **45,** 1–86.

Simons, K., and Ikonen, E. (1997). Functional rafts in cell membranes. *Nature* **387,** 569–572.

Thimonier, J., Montixi, C., Chauvin, J.-P., Tao He, H., Rocca-Serra, J., and Barbet, J. (1997). Thy-1 immunolabeled thymocytes microdomains studied with the atomic force microscope and the electron microscope. *Biophys. J.* **73,** 1627–1632.

Tokunaga, M., Aoki, T., Hiroshima, M., Kitamura, K., and Yanagida, T. (1997). Subpiconewton intermolecular force microscopy. *Biochem. Biophys. Res. Commun.* **231,** 566–569.

Tsilimbaris, M. K., Lesniewska, E., Lydataki, S., Le Grimellec, C., Goudonnet, J. P., and Pallikaris, I. G. (2000). The use of atomic force microscopy for the observation of corneal epithelium surface. *Invest. Ophthalmol. Vis. Sci.* **41,** 680–686.

Vickery, S. A., and Dunn, R. C. (1999). Scanning near-field fluorescence resonance energy transfer microscopy. *Biophys. J.* **76,** 1812–1818.

Vié, V., Giocondi, M.-C., Lesniewska, E., Finot, E., Goudonnet, J. P., and Le Grimellec, C. (2000). Tapping mode atomic force microscopy on intact cells: Optimal adjustment of tapping conditions by using the deflection signal. *Ultramicroscopy* **82,** 279–288.

Walters, D. A., Viani, M., Paloczi, G. T., Schäffer, T. E., Cleveland, J. P., Wendman, M. A., Gurley, G., Elings, V., and Hansma, P. K. (1997). *S.P.I.E.* **3009,** 43–47.

Ziegler, U., Vinckier, A., Kernen, P., Zeisel, D., Biber, J., Semenza, G., Murer, H., and Groscurth, P. (1998). Preparation of basal cell membranes for scanning probe microscopy. *FEBS Lett.* **436,** 179–184.

CHAPTER 4

Measuring the Elastic Properties of Living Cells by the Atomic Force Microscope

Manfred Radmacher

Drittes Physics Institute
Georg-August Universität
37073 Göttingen, Germany

I. Introduction

The AFM combines high sensitivity in applying and measuring forces, high precision in positioning a tip relative to the sample in all three dimensions, and the possibility to be operated in liquids, especially physiological environments (Drake *et al.*, 1989), and therefore is capable of following biological processes *in situ* (Fritz *et al.*, 1994; Radmacher, Fritz *et al.*, 1994; Schneider *et al.*, 1997; Dvorak and Nagao, 1998; Rotsch *et al.*, 1999). One application that makes use of all three features is the investigation of cellular mechanics (Tao *et al.*, 1992; Weisenhorn *et al.*, 1993; Radmacher *et al.*, 1995, 1996; Radmacher, 1997). Technically this was first done by using the force modulation method (Radmacher, *et al.*, 1992, 1993). However, to determine elastic properties of cells quantitatively and reproducibly, force curves have to be recorded and analyzed (Radmacher, 1995) as a function of position the cell (Radmacher, Cleveland *et al.*, 1994; Radmacher *et al.*, 1996). Mechanical properties of many different cell types including glial cells (Henderson *et al.*, 1992), platelets (Radmacher *et al.*, 1996; Walch *et al.*, 2000), cardiocytes (Hofmann *et al.*, 1997), macrophages (Rotsch *et al.*, 1997), endothelial cells (Braet *et al.*, 1996, 1997, 1998; Mathur *et al.*, 2000; Miyazaki and Hayashi, 1999; Sato *et al.*, 2000), epithelial cells (Hoh and Schoenenberger, 1994; A-Hassan *et al.*, 1998), fibroblasts (Rotsch *et al.*, 1999; Rotsch and Radmacher, 2000), bladder cells (Lekka *et al.*, 1999), L929 cells (Wu *et al.*, 1998), F9 cells (Goldmann *et al.*, 1998), and osteoblasts (Domke *et al.*, 2000) have been investigated.

It has been postulated that mechanical properties play a major role in cellular processes and can thus serve as indicators for cellular processes (Elson, 1988). The mechanical properties of eucaryotic cells are determined mainly by the actin cytoskeleton (Sackmann, 1994a). The protein actin can form double-helical polymer fibers with a periodicity of 3.7 nm. A large number of actin-associated proteins control the architecture of this network (Hartwig and Kwiatkowski, 1991). There are molecules which induce bundling, cross-linking, and anchoring of actin to the cell membrane. This network is a very active cellular component which is under constant remodeling; therefore it is not only responsible for the shape of a cell but also plays a major role in dynamic processes such as cell migration (Stossel, 1993) and division (Glotzer, 1997; Robinson and Spudich, 2000).

Only a limited number of techniques are available which probe the mechanical properties of cells. Traditionally this question has been tackled with either the help of micropipettes (Evans, 1989; Discher *et al.*, 1994) or with the so-called cell poker (Felder and Elson, 1990; Petersen *et al.*, 1982; Zahalak *et al.*, 1990), which is conceptually related to the AFM. More recently, several new techniques have emerged, for example, the scanning acoustic microscope (Hildebrand and Rugar, 1984; Lüers *et al.*, 1991), optical tweezers (Ashkin and Dziedzic, 1989; Florin *et al.*, 1997), magnetic tweezers (Bausch *et al.*, 1998; Bausch, 1999), and the atomic force microscope, which will be discussed here in more detail.

In this chapter, I will discuss the principle of the measurement of elastic properties in general, the potential and possibilities when applying it to cells, and potential problems. Finally, I will present examples of measurements of the elastic properties of living cells.

A good example which proves that the investigation of living cells and hence cellular dynamics is possible by AFM can be seen in Fig. 1. Here, cardiomyocytes, which spontaneously pulse even as single cells in culture, were probed (Domke *et al.*, 1999). The AFM tip was used to monitor the mechanical contraction of a cell at different locations.

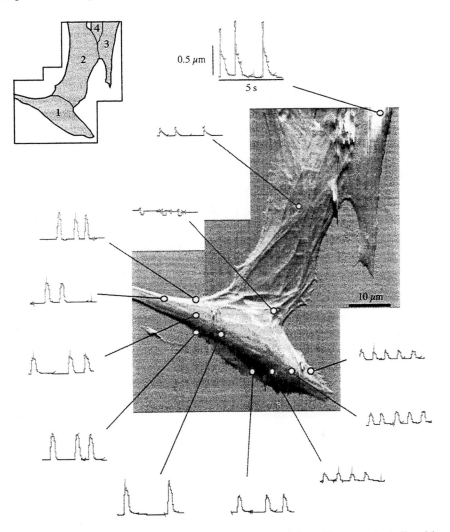

Fig. 1 Pulse mapping of a group of active cardiomyocytes. Although these cells were mechanically pulsing, it was possible to image them. In this figure the deflection images of two scans were superimposed. The cell margins are sketched in the inset on the top left. Several time series of height fluctuations at different locations on the cells were recorded. The presented sequences are scaled identically; their locations on top of the cells are marked with circular spots. Pulses on cell 1 (a myocyte) show all the same shape regardless of position. On cell 2, a fibroblast which is moved passively by the neighboring myocytes, only biphasic pulses are found. Reproduced from Domke, J., Dannohl, S., Parak, W. J., Muller, O., Aicher, W. K., and Radmacher, M. (2000). Substrate dependent differences in morphology and elasticity of living osteoblasts investigated by atomic force microscopy. *Colloids Surf.* B, **19**, 367–373, with permission from Elsevier Science.

II. Principles of Measurement

A. Force Curves on Soft Samples

In the AFM, an indentation experiment is done by employing the force curve mode (Weisenhorn *et al.*, 1989) in which the deflection of the cantilever is plotted as a function of the *z* height of the sample. On a stiff sample, the deflection is either constant as long as the tip is not in contact with the sample or proportional to the sample height while the tip is in contact. With soft samples, the tip may deform (compress) the sample when the loading force is increased (Tao *et al.*, 1992; Weisenhorn *et al.*, 1993; Radmacher *et al.*, 1995). Thus, the movement of the tip will be smaller than the movement of the sample base, the difference being the indentation of the sample. In general, this indentation may be elastic, i.e., reversible; plastic, i.e., irreversible; dynamic, i.e., viscous in nature; or a combination of the three. By tuning the experimental parameters one can determine which of the three contributions is present. Viscous effects are proportional to the velocity of the tip approaching to or retracting from the sample. By reducing the scan rate, viscous effects can be minimized and they change sign when the velocity changes its sign. So, by averaging approach and retract data viscous effects are canceled out. Since the AFM cantilever is immersed in aqueous buffer, there will always be large viscous damping by the surrounding fluid (Radmacher *et al.*, 1996). This can be seen in force curves as an increasing separation of the approach and retract curves while the tip is still off the surface. In addition, during contact, there may be viscous forces exerted by the sample itself. However, this effect seems to be smaller than the effect of the liquid; thus it will need precise data analysis procedures to separate the contributions from both sample and environment. Experimentally this problem could be solved by using small cantilevers (Schäffer *et al.*, 1996, 1997; Walters *et al.*, 1997) as soon as they are available.

The difference between elastic and plastic deformation can be seen by comparing approach and retract curves. If there are no differences between the two, and if subsequent curves are reproducible, then the deformation caused is reversible and it will be elastic in nature. Otherwise the deformation is plastic in nature or damaging. However, because of viscous effects this determination has been made at low scan velocities.

B. Range of Analysis

Typically, force curves are analyzed in a given range of loading forces. Thus, deflection values must first be converted into loading force values. Since cantilever springs are linear springs for small deflections, Hooke's law can be applied:

$$F = k_c * d. \tag{1}$$

Here k_c is the force constant of the cantilever, d is its deflection, and F is the corresponding loading force exerted by the cantilever. In experiments, the deflection is not necessarily zero when the cantilever is free, e.g., because of stresses in the cantilever, which will

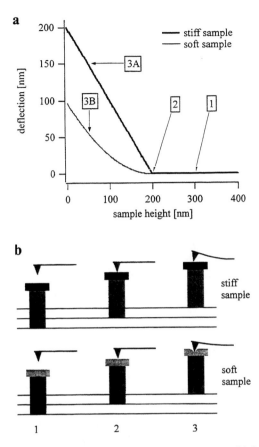

Fig. 2 Differences in force curves on stiff and soft samples. In a force curve (a) the cantilever deflection is plotted as a function of a sample base height. On a stiff sample the force curve will show two linear regimes. (i) The tip is not touching the sample yet, the deflection is constant (between points 1 and 2 in (a), position 1 in (b)) (ii) The tip is touching the sample and the deflection is proportional to the sample height (between points 2 and 3A). On a soft sample, however, due to deformation of the sample, the deflection will raise slower than the movement of the sample base height (between points 2 and 3B in (a), position 2 and 3 in (b)). In fact the elastic response of the sample will lead to a nonlinear relationship between deflection and sample base height. The difference between curve A and curve B is the indentation of the soft sample. (An animated version of this graph can be found at http://www.dpi.physik.uni-goettingen.de/~radmacher/animations.html).

deform it even without an external load. Therefore the offset d_0 must be subtracted from all deflection values. This offset can be easily obtained by calculating the mean deflection in the off-surface region (between points 1 and 2 in Fig. 2a). In principle, it would be fine to pick just one deflection value from this region, but due to noise in the data it is better to take an average. Thus, Eq. [1] transforms to

$$F = k_c * (d - d_0). \qquad [2]$$

The indentation δ is given by the difference between the sample base height z and the deflection of the cantilever d:

$$\delta = z - d. \tag{3}$$

Here again, the offsets must be considered, so I can rewrite Eq. [3] as

$$\delta = (z - z_0) - (d - d_0) \tag{4a}$$
$$\delta = z - z_0 - d + d_0, \tag{4b}$$

where d_0 is the zero deflection as above and z_0 is the z position at the point of contact. For quantifying the elastic properties of a sample, a range of data from the force curve to be analyzed and an appropriate model for analysis must be chosen. The most simple model comes from continuum mechanics and is based on the work of Heinrich Hertz (Hertz, 1882), which was extended by Sneddon (Sneddon, 1965). For an introduction in continuum mechanics one may visit the work of Fung (Fung, 1993), Johnson (Johnson, 1994), or Treloar (Treloar, 1975). The Hertzian model describes the elastic indentation of an infinitely extended sample (effectively filling out a half-space) by an indenter of simple shape. Two shapes often used in AFM are conical or parabolic indenters:

$$F_{\text{cone}} = \frac{2}{\pi} \cdot \frac{E}{(1 - v^2)} \cdot \delta^2 \cdot \tan(\alpha) \tag{5}$$

$$F_{\text{paraboloid}} = \frac{4}{3} \cdot \frac{E}{(1 - v^2)} \cdot \delta^{3/2} \cdot \sqrt{R}. \tag{6}$$

Here, F_{cone} is the force needed to indent an elastic sample with a conical indenter, whereas $F_{\text{paraboloid}}$ is the force needed to indent the sample with a parabolic indenter. For a small indentation, a spherical indenter will follow the same force deflection relation as the parabolic indenter. The indentation is denoted by δ, whereas E is the elastic or Young's modulus, v is the Poisson ratio, α is the half-opening angle of the cone, and R is either radius of curvature of the parabolic indenter or the radius of the spherical indenter, respectively. In the case of cells, the sample is virtually incompressible and therefore v can be chosen to be 0.5 (Treloar, 1975).

I can now combine Eqs. [4b] and [5] in the case of a conical indenter to obtain

$$k_{\text{c}} \cdot (d - d_0) = \frac{2}{\pi} \cdot \frac{E}{1 - v^2} * \tan(\alpha) \cdot (z - z_0 - (d - d_0))^2. \tag{7}$$

This mathematical function describes the force curve as measured on a soft sample. I can rearrange Eq. [7] to obtain

$$z - z_0 = (d - d_0) + \sqrt{\frac{k_{\text{c}} \cdot (d - d_0)(1 - v^2)}{(2/\pi) \cdot E \tan(\alpha)}}, \tag{8}$$

the mathematically more convenient form.

Most of the quantities in Eq. [8] either are known or can be measured experimentally. The force constant k_c and the half-opening angle α can be either obtained from the manufacturer's data sheet or determined before or after the experiment. A standard method for calibrating the force constant of soft cantilevers is the thermal noise method introduced by Butt and Jaschke (1995). The Poisson ratio was set to 0.5, as discussed earlier. The zero deflection d_0 can also be obtained easily from the data as described previously. The deflection d and the sample base height z are the quantities measured in the force curve, leaving only two unknown variables: the elastic modulus E and the contact point z_0. The elastic modulus E is the quantity of interest, but the procedure to obtain the point of contact z_0 needs to be discussed also.

In a stiff sample (see Fig. 2a, trace A, Fig. 3a) the contact point separates two linear regimes in the force curve with different slopes. Thus, the contact point can be obtained easily by either fitting a line to each of the two regimes or calculating the slope and looking for a discontinuity in it. The deflection rises slowly and smoothly at the contact point (Fig. 2a, trace B, Fig. 3a) for a soft sample. There is no jump in slope but only a

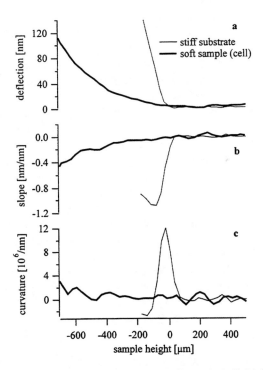

Fig. 3 Experimental force curves on a stiff substrate and a soft sample (cell) (a). In the stiff sample, the contact point can easily be obtained either from the data or by checking the slope (first derivative) for a discontinuity (b). In the soft sample, the data and the slope are continuous. The second derivative should show a jump, but this discontinuity is no longer detectable due to noise in the data (c).

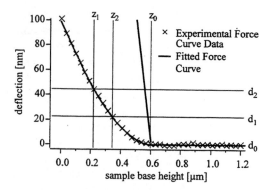

Fig. 4 Example of the fit procedure. From the experimental data (hair-crosses) the zero deflection d_0 can be obtained by averaging some part of the force curve in the flat off-surface region. The range of analysis is defined by its lower and upper limits, d_1 and d_2. The Hertzian model is fitted to the data within this range and results in values for the Young's modulus and the contact point z_0. As can be checked visually the fit follows the data very closely and yields a reasonable contact point at $z_0 = 610$ nm. (The fitted Young's modulus is $E = 5100$ Pa.)

continuous change in slope (Fig. 3b). Although the second derivative should then show a discontinuity at the contact point for soft samples, this is often buried in the noise (Fig. 3c). Therefore, it is very difficult or even impossible to determine the contact point in force curves of soft samples in this simple way from experimental data.

Another simple, straightforward, reliable, and robust method to obtain the contact point was established in previous work (Rotsch *et al.*, 1999). Briefly, the elastic modulus is obtained by fitting Eq. [8] within a given range to the experimental data. In addition to using E as a fit parameter the contact point z_0 can also be included as a fit parameter. In fact, this yields a very good value for the contact point, and in our experience this procedure is more stable, reliable, and robust than any other method in use. Figure 4 shows the procedure in more detail. A range of analysis is chosen by defining an upper and a lower limit of deflection values (d_1 and d_2), which correspond to a range of loading forces F_1 and F_2, given by $F_1 = k_c * d_1$ and $F_2 = k_c * d_2$. This also defines a range of analysis in terms of z height, given by z_1 and z_2. By employing a Monte Carlo fit, optimized values, which fit the data best, for E and z_0 are obtained. Although z_0 will always be outside of the range of analysis, it turned out to be a very reliable quantity. There are two major advantages in this procedure. (i) Because the contact point is obtained by fitting a range of data, noise will average out. (ii) Because in the range of analysis the tip is in contact, noise will be smaller compared to data recorded off the surface.

III. Application to Cells

The elastic response of cells in indentation experiments could stem from several cellular compartments. Since the tip approaches from the extracellular medium, it will first encounter the glycocalix, then the membrane, and then either the intracellular organelles or the cytoskeleton. The glycocalix and the membrane, in the case of eucaryotic cells, turn out to be very soft and can be neglected in AFM experiments. First, I want to estimate the elastic response of these components.

The glycocalix is a soft polymer brush, whose elastic properties can be estimated in the framework of a worm-like chain model. In this model the elastic response of a polymer molecule with contour length L is determined by its persistence length l_p. The persistence length is the length at which the orientation of the molecules becomes uncorrelated due to thermal bending. Recently the relation between force F_{wlc} and extension x for a single polymer chain was derived (Marko and Siggia, 1995);

$$F_{wlc} = \frac{kT}{l_p} * \left(\frac{x}{L} + \frac{1}{4(1 - \frac{x}{L})^2} - \frac{1}{4} \right). \qquad [9]$$

For an extension of 50% of the contour length and a persistence length of 3 Å, which is a reasonable value for a polysaccharide chain (Rief et al., 1997), a force of about 10 pN is obtained. This is about the sensitivity of a state-of-the-art AFM using the softest cantilevers available. Since the AFM tip may be in contact with several polymer chains at the same time, it is conceivable that the elasticity of the glycocalix could be detectable. However, since the glycocalix is supported by the cell membrane, the membrane's elastic resilience must be evaluated first.

The lipid membrane is much softer than AFM cantilevers as can be seen by the following argument. The AFM cantilever is a thermodynamic system with one degree of freedom, which will fluctuate in position in thermodynamic equilibrium. The average energy of this fluctuation will be given by

$$1/2 \, k_b \, T = E_{avg} = 1/2 \, k_c * <x^2>, \qquad [10]$$

the equipartition theorem, where k_b is the Boltzmann constant, T is the absolute temperature, k_c is the force constant of the cantilever, and $<x^2>$ is the time average of the mean-square displacement. Typical values for the displacement will be several Ångstrøms in the case of very soft cantilevers (10 mN/m). In fact Eq. [10] is the basis for the method mentioned previously for calibrating force constants of AFM cantilevers. For softer cantilevers the fluctuations will be larger. Therefore, this method works best with ultrasoft cantilevers. Lipid membranes will also show thermal fluctuations in which the restoring force stems from the bending modulus of the lipid bilayer membrane (Helfrich, 1973). In cellular membranes (like the membrane of erythrocytes), the fluctuations can be on the order of 100 nm, therefore they are detectable in the optical microscope (Sackmann, 1994b; Svoboda et al., 1992; Zeman et al., 1990). Equation [10] demonstrates that lipid bilayers are several orders of magnitude softer than AFM cantilevers. A similar result was obtained from a more thorough theoretical estimation of the response of cellular membranes to the indentation by AFM cantilevers (Boulbitch, 1998). Thus, the elastic response of the membrane is not detectable. Consequently the elastic response of the glycocalix was not observed, since its supporting structure, the cell membrane, was too soft. This is only true in the case of eucaryotic cells with soft cell membranes. For other types of cells, such as plant cells or bacteria with different cell wall compositions, the cell wall may become very stiff and will become the determining factor in cellular deformability (Arnoldi *et al.*, 1997, 2000; Arnoldi, Boulbitch, 1998; Xu *et al.*, 1996).

I will now discuss the actin cytoskeleton as a cellular compartment. The mechanical properties of single actin filaments were determined experimentally (Käs *et al.*, 1996). Rheological data from actin gels *in vitro* give values of 200 Pa for the elastic modulus for concentrations of 2 mg/ml (Janmey *et al.*, 1994). Since the concentration of actin can be higher in cells and since there are several possibilities of enhancing stiffness e.g., by bundling of filaments, elastic modules of several kilopascals are conceivable in cells. Let us assume a typical loading force of 50 pN, which can easily be set and detected by AFM, and a Young's modulus of 1 kPa. Using a conical tip with an opening angle of 35°, an indentation of 200 nm is obtained (cf. Eq. [5]). For a soft 10 mN/m cantilever a 500-nm deflection is needed to achieve a 50-pN force. Thus the sample movement will be 700 nm. Deflection, indentation, and sample movement are all of the same order of magnitude; thus all quantities are easily and accurately detectable. Therefore, I can conclude that the softest AFM cantilevers available can easily detect and measure the elastic properties of the cytoskeleton of eucaryotic cells.

A. Elasticity of the Cytoskeleton

Figure 5a shows a contact mode image of an osteoblast cell adhered to a plastic Petri dish (Domke *et al.*, 2000). This well-spread cell exhibits many cable-like structures

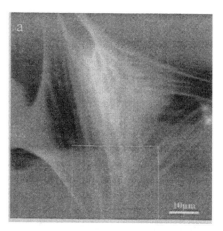
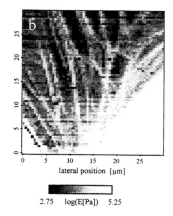

lateral position [μm]

2.75   log(E[Pa])   5.25

Fig. 5 (a) Contact mode AFM image of a living osteoblast. The cable-like cytoskeletal structures are very prominent in this image. (b) The elasticity map of a part of this osteoblast denoted by the square in (a). The local elastic modulus was obtained from force curves by fitting the Hertz model to the data as described in the text. Reproduced from Domke, J., Dannohl, S., Parak, W. J., Muller, O., Aicher, W. K., and Radmacher, M. (2000). Substrate dependent differences in morphology and elasticity of living osteoblasts investigated by atomic force microscopy. *Colloids Surf.* B, **19**, 367–373, with permission from Elsevier Science.

running along the entire cell. The cable-like structures are actin bundles which will be proven in the following. Actin is a very abundant protein, whose concentration can be up to 20 mg/ml. *In vitro* actin gels can self-assemble in the presence of calcium and ATP at elevated concentrations. In cells, there is a multitude of actin-binding proteins which can trigger or inhibit the formation of filaments, disassemble filaments, or determine the architecture of the network. Prominent examples of actin-binding proteins are α-actinin (bundler), gelsolin (severs filaments, binds to monomers), filamin (cross-links filaments), myosin (slides along filaments), myosin II (cross-links filaments), and spectrin (attaches filaments to the plasma membrane) (Hartwig and Kwiatkowski, 1991). In Fig. 5b the local elastic modulus, as calculated from a two-dimensional array of force curves, is plotted. Each force curve was analyzed off-line as described earlier and the local elastic modulus was obtained at this location. Since variations in the elastic modulus are substantial and range from 500 to over 10,000 Pa, the logarithm of the elastic modulus was encoded in gray shades. Typical parameter settings have an array size of 64*64 force curves, each consisting of 100 data points on approach and 100 data points on retract. Due to hydrodynamic interactions and travel ranges in the 1- or 1.5-μm force curve, the scan rate had to be limited to 10–20 force curves per second, which gives a total acquisition time of 64*64*2/10 = 820 s = 14 min, which makes it difficult to follow fast cellular processes in this standard mode of operation.

The local mechanical properties and the contact point in the force curve can be obtained from the force curves. This corresponds to the height of the cell at zero loading force. In addition topographic images for any other force value can be reconstructed from the force curve. Such a stack of the topographic images of the cell for a whole range of loading forces is presented in Fig. 6. The smooth appearance of the cell at zero loading force ("true" topography) becomes increasingly structured when raising the loading force until images, at forces of around 500 pN, that are reminiscent of standard contact mode AFM images are obtained. In these images at elevated forces the cytoskeletal structures are nicely visible, which was not the case for smaller or vanishing loading forces. Figure 7 illustrates the explanation for this behavior. The cytoskeletal structures are stiffer than the surrounding cytosol. By applying a force the softer parts of the cell will be compressed until the tip senses the underlying stiff, cytoskeletal structures (Henderson *et al.*, 1992; Rotsch *et al.*, 2000b), which will then be the most prominent features in standard topographic AFM images. In a strict sense these images at elevated forces do not reflect the true topography of the cell. However, these artificial images show much more information regarding the intracellular structures of the cell such as the cytoskeleton. Thus, from a scientific point of view these artificial images are much more interesting than the true topographies. However, there are applications where the true topography of the cell is an issue, for example, when determining the cell volume (Korchev *et al.*, 2000; Schneider *et al.*, 2000). The cell controls its volume and there are processes in which cell volume changes occur that are of major importance, such as the adaptation to osmotic pressure.

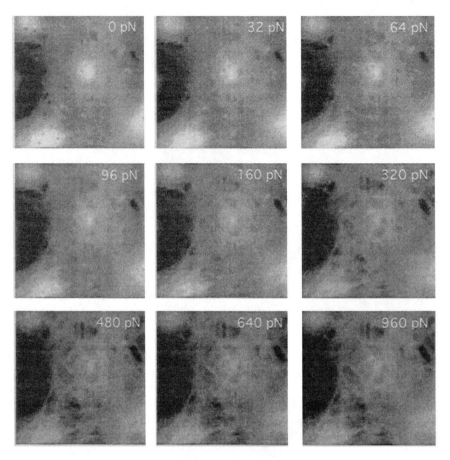

Fig. 6 Stack of constant force images calculated from force curve data. As can be seen in this series the cell exhibits a rather smooth surface at zero or low loading forces, whereas the cytoskeletal structures become visible at elevated forces. (An animated version of this graph can be found at: http://www.dpi.physik.uni-goettingen.de/~radmacher/animations.html)

B. Are Actin Filaments the Major Contributors to Cellular Mechanics?

In the data presented up to this point, I have not yet proven that the mechanical properties apparent in AFM measurements are in fact caused by the actin cytoskeleton. There are several possibilities of achieving the correlation between mechanical properties and biochemical composition. One is to use labeling techniques (Henderson *et al.,* 1992; Braet *et al.,* 1998; Rotsch and Radmacher, 2000). In Fig. 8, I present data of such a labeling experiment. Figures 8a and 8b are topographic images of a fibroblast cell. Figure 8c is the true topography at zero loading force as obtained from the force data. Figure 8f is the elasticity map obtained from the same data set. After the AFM investigations, the cell was fixed with glutaraldehyde and the actin filaments were

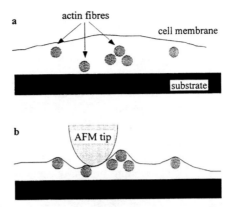

Fig. 7 In AFM contact mode imaging, cytoskeletal structures are emphasized due to the effect of high loading forces. The softer regions between supporting fibers are compressed to a larger degree, thus showing the fibers more pronounced.

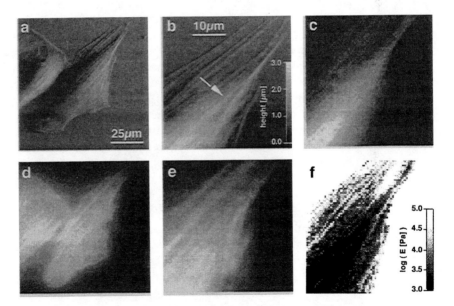

Fig. 8 Correlation of AFM images, elasticity maps, and fluorescence images to identify fibrous structures like stress fibers: (a) AFM deflection image of a living NRK fibroblast; (b) height image at higher magnification, with the arrow pointing to fibrous structures, presumably stress fibers; (c) height image at zero loading force calculated from a force map of the same region, no fibrous structures present proving these structures to be underneath the membrane; (d, e) fluorescence images of the regions shown in (a) and (b), actin network labeled with rhodamin-phalloidin after the AFM experiment; (f) elasticity map corresponding to (c). Fibrous structures possess increased Young's modulus. Reproduced from Rotsch, C., and Radmacher, M. (2000). Drug-induced changes of cytoskeletal structure and mechanics in fibroblasts—An atomic force microscopy study. *Biophys. J.* **78**, 520–535, with permission from the Biophysical Society.

labeled with phalloidin-rhodamine. The fluorescence images can be seen in Figs. 8d and 8e. There is a strong correlation between fiber-like structures in Fig. 8e, the stiff structures in the elasticity map in Fig. 8f, and the cable-like structures in the contact mode topographic images in Figs. 8a and 8b which are now definitely identified as actin filaments.

Another experimental way of determining the relevance of filamentous structures in the mechanical response of cells is the application of drugs disassembling these filaments. A prominent example is cytochalasin (cytochalasin B and D); other examples are jasplakinolide and latrunculin A. All these drugs, which are known to disrupt actin filaments, had a dramatic effect on the stiffness of cells, which was decreased by a factor of 2–3, depending on the concentration and location on the cell (Rotsch and Radmacher, 2000). Other drugs, such as colchicine, colcemide, and nocodazole, which are known to disassemble microtubules showed no effect on the mechanical properties of cells as measured by AFM. This finding may be very surprising, since it is known that microtubules are rather stiff structures (persistence length on the order of millimeters) (Felgner et al., 1996) in comparison to the semiflexible actin filaments (persistence length in micrometers) (Käs et al., 1994, 1996). Therefore, one would expect that only a small number of microtubules may render the cell much stiffer than the abundance of actin filaments. The difference between the mechanics of single filaments (persistence length) and the mechanics of the cell may come from the difference in function of the two. Actin filaments build up the supporting structures of cells and are responsible for cellular processes like migration, adhesion to substrates, and generation of forces and stresses. Microtubules are transport structures, e.g., for either vesicles in axons of neuron cells or chromosomes during cell division; thus, they may be buried deeper in the cell and surrounded by an actin gel. Therefore, they may just not show up in mechanical experiments, even if they may also play a central role in cellular mechanics (Ingber, 1993). Another possible explanation is that the cell can use the cross-linker myosin II to build up tension in the cytoskeleton, which will consequently appear much stiffer. This possibility has not been tested in the single-molecule experiments on actin filaments.

C. Influence of Mechanical Properties on Resolution in Imaging

As mentioned previously, the mechanical properties of cells determine the appearance of cells in the topographic AFM images of cells. At elevated loading forces, soft structures will be compressed and stiffer structures will therefore be both enhanced and clearly visible. There is also a more general influence on imaging, since the resolution of images will also be a function of loading force, tip shape, and elastic properties of the sample. Since a soft sample can be indented by several hundred nanometers by the AFM tip, the contact area between tip and sample will be large. The resolution to be obtained will depend on the contact area, and its value can serve as a measure for the best resolution to be achievable under certain experimental conditions. For a conical indenter the contact radius r can be estimated by geometric reasoning from the indentation δ and the tip opening angle α:

$$r_{cone} = \delta * \tan(\alpha). \tag{11}$$

Replacing δ in Eq. [5] by this expression and solving for the contact radius yields

$$r_{cone} = \sqrt{F * \frac{\pi}{2} * \frac{1 - v^2}{E} * \tan \alpha}.$$ [12]

Although this expression is not exact, it will be acceptable as an order of magnitude estimation. For a parabolic tip, Sneddon found the following exact formula (Sneddon, 1965):

$$r_{paraboloid} = \sqrt[3]{\frac{\frac{3}{4} * (1 - v^2)}{E * R * F}}.$$ [13]

In Fig. 9, the diameter of the contact area is plotted as a function of the elastic modulus for several loading forces. With the softest AFM cantilevers available (10 mN/m force constant) state-of-the-art AFMs can achieve a 10-pN force resolution. However, in the future due to both instrumental improvements and to the availability of softer cantilevers, operation at even smaller forces may be feasible. Typical values of loading forces in the imaging of cells may be between 100 pN and 1 nN. These forces result in substantial sample deformations, hence contact areas of several tens or even several hundreds of nanometers. High-resolution images of cells are possible only in either very stiff areas of cells or when minimizing the force to below 100 pN. Obviously a possible route would be to stiffen cells by chemical fixation (Braet *et al.*, 1997). However, this will destroy one of the major advantages of AFMs, since after fixation the observation of dynamical processes is no longer possible. The cells will be dead after fixation. The

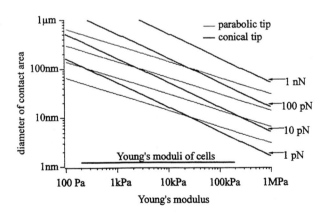

Fig. 9 Predicted diameter of the contact area between the AFM tip and a soft substrate, as predicted by the Hertz model for two different tip shapes: a cone with an opening angle of 30° and a parabolic tip with a radius of curvature *R* 50 nm press against a flat, soft surface. In both cases, the tip is assumed to be infinitely stiff and the Poisson ratio of the sample is taken to be 0.5. The calculated diameter of the contact area serves as a rough estimate of the resolution achievable in AFM of living cells. Typical values for the Young's modulus of cells can range from only a few 100 Pa up to more than 10 kPa. Only in the latter case does a resolution on the order of 10 nm seem possible.

minimization of the loading force was reported when imaging living cells, but it will only work sufficiently in very flat regions of the cell (Grimellec *et al.,* 1994, 1998).

D. Limitations and Problems

Measuring mechanical properties of cells makes the most sense if it is quantitative. This includes that the measured quantities are independent of the experimental details, for example, tip shapes and cantilever force constants. Therefore, the force indentation values must be analyzed to obtain values for the elastic modulus of the sample. This is typically done by using the Hertz model as described previously, although other procedures have been suggested (A-Hassan *et al.,* 1998). The Hertz model describes the elastic deformation of a perfectly linear elastic material, which is homogeneous and infinitely large, by a well-defined (for example, conical) indenter in the limit of small indentations. All prerequisites are not given in a strict sense with cells and therefore need either some discussion or even modification of the model.

E. Tip Shape

The Hertz model describes the indentation of a soft sample for simple indenters, such as a paraboloid, a sphere, or a cone. AFM tips are four-sided pyramids which are rounded to some degree at their apex. It can be shown that the indentation behavior of a four-sided pyramid is identical to the behavior of a cone with a slightly different opening angle (Domke and Radmacher, 2000). In the case of AFM tips with an opening angle of $35°$, the equivalent cone angle is $38.3°$. The change in geometry from a spherical to a conical indenter, however, cannot be included within the framework of the Hertz model; however, it is possible to discuss two limiting cases within the Hertz model. If the indentation is smaller than the radius of curvature at the apex of the tip, the Hertz model, for a spherical indenter, will nicely match the data. If the indentation is very large, it was experimentally found that the model of a cone matches the data very well (Radmacher *et al.,* 1995). There will be a transition regime, where neither model will match. Obviously, there is a need for improving the model (Domke and Radmacher, 2002).

F. Sample Thickness

Many interesting regions of adherent cells tend to be very thin. For example, lamellipodia show a thickness of less than 500 nm. Here, at typical indentations of 100 or even 200 nm, it is conceivable that the AFM tip senses the stiffness of the underlying substrate. Applying a Hertzian model to fit the data will then produce an artificially high value for the elastic modulus, or, in other words, a strong correlation between sample thickness and apparent Young's modulus will be found (Rotsch *et al.,* 2000a). This effect was characterized experimentally by measuring the response of a model sample with tunable thickness and elastic modulus: gelatin (Domke and Radmacher, 1998). Again, there is a clear need for the model extension to include the effect of sample thickness. Both points could be considered by simulation, e.g., employing finite element methods.

This model extension is in fact underway in our group but has not yet been completed (Domke and Radmacher, 2002).

G. Homogeneity of the Cell

Cells are structured entities with large local differences in their composition and hence in their mechanical behavior. This is in fact motivation for using the AFM to measure the mechanics of cells. The AFM will always measure an average elastic modulus over a region roughly given by the contact area. The measurement area may even be larger since the deformation field will extend over a much larger area (Boulbitch, 1998). In experiments, it will always be necessary to find a compromise between resolution and possibility to interpret the data. One extreme case is using a very sharp, needle-like tip which will penetrate through the membrane and will even be smaller than the mesh size of the actin cytoskeleton. Here, information on the elasticity of single filaments, but not on the elasticity of cells, is obtained. The other extreme case is to use a very blunt indenter, e.g., a 100-μm sphere or a flat punch. This is the experimental setup introduced before the advent of the AFM and dubbed the cell poker (Zahalak *et al.*, 1990). In this case, the measurement averages over the entire cell and basically one number for the cell is obtained. Here, no information is obtained on subcellular variations in cell elasticity. The meaningful and interesting range of experiments lies directly inbetween, using standard

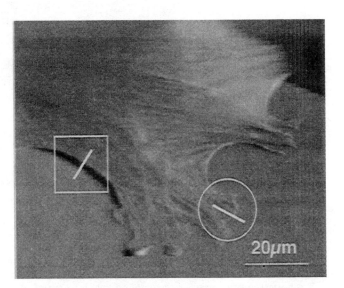

Fig. 10 AFM image of a migrating fibroblast. The extending flat lamellipodium (circle) can be distinguished from the stable edge (square) in the rear of the cell. Reprinted with permission from Rotsch, C., Jacobson, K., and Radmacher, M. (1999). The dynamics of active and stable edges in motile fibroblasts investigated by atomic force microscopy. *Proc. Nat. Acad. Sci.* **96,** 921–926. Copyright (1999) *National Academy of Sciences, U.S.A.*

AFM tips, where the measurement averages over several mesh sizes of the actin network, yet local differences between different areas of the cell are still visible.

IV. Mechanics of Cellular Dynamics

I have shown that the AFM can be used to determine the elastic properties of soft samples, including cells. In addition it can be operated under physiological conditions

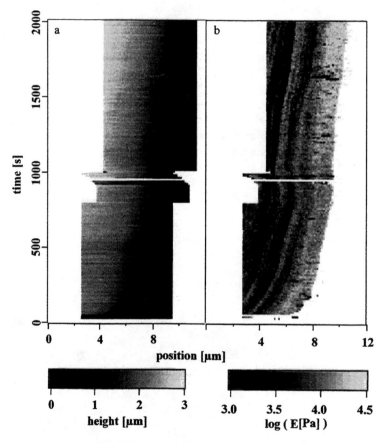

Fig. 11 (a, b) Successive single line scans of height (a) and Young's modulus (b) across the protruding edge of a motile 3T3 cell in which values are encoded in a gray scale. (c, d) Successive single-line scans of height (c) and Young's modulus (d) of the stable edge of a 3T3 cell in which values are encoded in a gray scale. Inset in panel (c) shows position versus time for a point on the margin. The band structure in the Young's modulus is presumably due to stress fibers crossing the scan line. Note the gradual extension. Reproduced with permission from Rotsch, C., Jacobson, K., and Radmacher, M. (1999). The dynamic of active and stable edges in motile fibroblasts investigated by atomic force spectroscopy. *Proc. Nat. Acad. Sci.* **96**, 921–926. Copyright (1999) *National Academy of Sciences, U.S.A.*

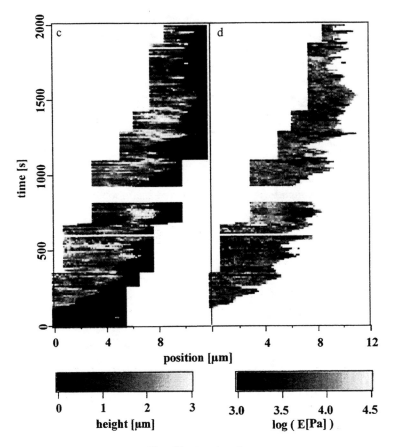

Fig. 11   (*continued*).

and therefore it is capable of following biological processes *in situ*. Examples for cellular processes in which the mechanical properties play a major role are cell migration (Stossel, 1993) and cell division (Glotzer, 1997). Figure 10 shows an AFM image of a locomoting fibroblast (Rotsch *et al.,* 1999). In fibroblasts, one can distinguish active, motile (the extending lamellipodium), and stable regions, which will stay in place for an extended period of time. Only after extensive protrusion of the lamellipodium, will the rear, stable parts of the cell follow the front part with a jump to find a new resting position (Lee *et al.,* 1999). To increase temporal resolution we measured the mechanical response of the cell along one scan line which crosses the edge of the cell, as indicated in Fig. 10 by the lines inside the circular and square regions, respectively. Figure 11 shows a time series of the topography and an elastic modulus along these scan lines, which either crosses the edge of the cell either at the protruding lamellipodium (Figs. 11a and 11b) or at the stable edge (Figs. 11c and 11d). The extension of the cell can be followed in these traces, and the two regions of the cell can be distinguished by their mechanical properties. The extending lamellipodium is somewhat softer and less variable in time compared to the stable edge, which shows a well-pronounced and stable pattern of stress fibers running parallel to the edge.

Another very exciting process is cell division, especially the formation of the cleavage furrow, which will eventually separate the two daughter cells and may involve mechanical changes in the mother cell. This process was first investigated by AFM by Dvorak and Nagao (Dvorak and Nagao, 1998), who took a single force curve at one position, which was very undefined, since dividing cells change shape and tend to move and rotate on the substrate. Therefore, it is difficult to stay at the same location of a cell. We used the AFM in line scan mode as described previously to measure the mechanics along a single scan line, which is positioned such that it will be crossing the newly formed cleavage furrow. Here we could unambiguously show that the furrow region stiffens before the furrow is visible in the topography (Matzke et al., 2000).

V. Summary

In this chapter I discussed the possibility of measuring elastic properties of living cells by AFM. One reason for using the AFM for this purpose is its ability to both measure locally the mechanics of a cell and to distinguish different regions of the cell. Since the AFM can be operated under physiological conditions cellular processes can be followed, for example, cytokinesis and the investigation of the migration of cells.

List of Symbols

α	half-opening angle of conical indenter
δ	indentation of (soft) sample
d	deflection of AFM cantilever
d_0	zero deflection of the free AFM cantilever (off the surface)
d_1	lower limit of deflection in range of analysis
d_2	upper limit of deflection in range of analysis
E	elastic or Young's modulus of sample
F	loading force of AFM cantilever tip
F_1	lower limit of loading force in range of analysis
F_2	upper limit of loading force in range of analysis
F_{cone}	loading force predicted from the Hertz model for a conical indenter
$F_{paraboloid}$	loading force predicted from the Hertz model for a parabolic indenter
F_{WLC}	force needed to extend a polymer molecule within the framework of the worm-like chain model
k_b	Boltzmann's constant
k_c	force constant of AFM cantilever
L	contour length of polymer molecule

l_p	persistence length
ν	Poisson ratio
r_{cone}	radius of contact area between conical tip and sample
$r_{paraboloid}$	radius of contact area between parabolic tip and sample
R	radius of curvature for parabolic or spherical indenter
T	absolute temperature
x	extension of polymer molecule
z	sample base height
z_1	lower limit of sample base height in range of analysis
z_2	upper limit of sample base height in range of analysis
z_0	sample base height at point of contact between tip and sample

References

A-Hassan, E., Heinz, W. F., Antonik, M. D., D'Costa, N. P., Nagaswaran, S., Schoenenberger, C-A., and Hoh, J. H. (1998). Relative micro-elastic mapping of living cells by atomic force microscopy. *Biophys. J.* **74**(3), 1564–1578.

Arnoldi, M., Fritz, M., Bäuerlein, E., Radmacher, M., Sackmann, E., and Boulbitch, A. (2000). Bacterial turgor pressure can be measured by atomic force microscopy. *Phys. Rev. E,* vol. 62(1), 1034–1044.

Arnoldi, M., Kacher, C., Bäuerlein, E., Radmacher, M., and Fritz, M. (1997). Elastic properties of the cell wall of Magnetospirillum Gryphiswaldense investigated by atomic force microscopy. *Appl. Phys. A* **66**, S613–S617.

Ashkin, A., and Dziedzic, J. M. (1989). Internal cell manipulation using infrared laser traps. *Proc. Natl. Acad. Sci. U.S.A.* **86**, 7914–7918.

Bausch, A., Möller, W., and Sackmann, E. (1999). Measurement of local viscoelasticity and forces in living cells by magnetic tweezers. *Biophys. J.* **76**, 573–579.

Bausch, A. R., Ziemann, F., Boulbitch, A. A., Jacobson, K., and Sackmann, E. (1998). Local measurements of viscoelastic parameters of adherent cell surfaces by magnetic bead microrheometry. *Biophys. J.* **75**, 2038–2049.

Boulbitch, A. A. (1998). Deflection of a cell membrane under application of a local force. *Phys. Rev. Lett.* **57**(2), 2123–2128.

Braet, F., Kalle, W. H. J., Zanger, R. B. D., Grooth, B. G. D., Raap, A. K., Tanke, H. J., and Wisse, E. (1996). Comparative atomic force and scanning electron microscopy: An investigation on fenestrated endothelial cells in vitro. *J. Microsc.* **181**(1), 10–17.

Braet, F., Rotsch, C., Wisse, E., and Radmacher, M. (1997). Comparison of fixed and living endothelial cells by atomic force microscopy. *Appl. Phys. A* **66**, S575–S578.

Braet, F., Seynaeve, C., De Zanger, R., and Wisse, E. (1998). Imaging surface and submembraneous structures with the atomic force microscope: A study on living cancer cells, fibroblasts and macrophages. *J. Microsc.* **190**(3), 328–338.

Butt, H.-J., and Jaschke, M. (1995). Thermal noise in atomic force microscopy. *Nanotechnology* **6**(1), 1–7.

Discher, D. E., Mohandas, N., and Evans, E. A. (1994). Molecular maps of red cell deformation: Hidden elasticity and in situ connection. *Science* **266**, 1032–1035.

Domke, J., Dannöhl, S., Parak, W. J., Müller, O., Aicher, W. K., and Radmacher, M. (2000). Substrate dependent differences in morphology and elasticity of living osteoblasts investigated by atomic force microscopy. *Colloids Surf.* B, vol. 19, pp. 367–379.

Domke, J., Parak, W. J., George, M., Gaub, H. E., and Radmacher, M. (1999). Mapping the mechanical pulse of single cardiomyocytes with the atomic force microscope. *Eur. Biophys. J.* **28**, 179–186.

Domke, J., and Radmacher, M. (1998). Measuring the elastic properties of thin polymer films with the AFM. *Langmuir* **14**(12), 3320–3325.

Domke, J., and Radmacher, M. (2002). The elastic indentation of thin films—A parametric model for the application in atomic force microscopy, part I: Theory. In preparation.

Drake, B., Prater, C. B., Weisenhorn, A. L., Gould, S. A. C., Albrecht, T. R., Quate, C. F., Cannell, D. S., Hansma, H. G., and Hansma, P. K. (1989). Imaging crystals, polymers and biological processes in water with AFM. *Science* **243,** 1586–1589.

Dvorak, J. A., and Nagao, E. (1998). Kinetic analysis of the mitotic cycle of living vertebrate cells by atomic force microscopy. *Exp. Cell Res.* **242,** 69–74.

Elson, E. L. (1988). Cellular mechanics as an indicator of cytoskeletal structure and function. *Ann. Rev. Biophys. Biophys. Chem.* **17,** 397–430.

Evans, E. (1989). Structure and deformation properties of red blood cells: Concepts and quantitative methods. *Methods Enzymol.* **173,** 3–35.

Felder, S., and Elson, E. L. (1990). Mechanics of fibroblast locomotion: Quantitative analysis of forces and motions at the leading lamellas of fibroblasts. *J. Cell Biol.* **111** (6(1)), 2513–2526.

Felgner, H., Frank, R., and Schliwa, M. (1996). Flexural rigidity of microtubules measured with the use of optical tweezers. *J. Cell Sci.* **109,** 509–516.

Florin, E. L., Pralle, A., Hörber, J. K. H., and Stelzer, E. H. K. (1997). Photonic Force Microscope Based on Optical Tweezers and Two-Photon Excitation for Biological Applications. *Journal of Structural Biology* **119,** 202–211.

Fritz, M., Radmacher, M., and Gaub, H. E. (1994). Granula motion and membrane spreading during activation of human platelets imaged by atomic force microscopy. *Biophys. J.* **66**(5), 1328–1334.

Fung, Y. C. (1993). "Biomechanics—Mechanical Properties of Living Tissues." Springer, New York.

Glotzer, M. (1997). The mechanism and control of cytokinesis. *Curr. Opin. Cell Biol.* **9,** 815–823.

Goldmann, W., Galneder, R., Ludwig, M., Xu, W., Adamson, E. D., Wang, N., and Ezzell, R. M. (1998). Difference in elasticity of vinculin-deficient F9 cells measured by magnetometry and atomic force microscopy. *Exp. Cell Res.* **239**(2), 235–242.

Grimellec, C. L., Lesniewska, E., Cacchia, C., Schreiber, J. P., Fornel, F. D., and Goudonnet, J. P. (1994). Imaging of the membrane surface of MDCK cells by atomic force microscopy. *Biophys. J.* **67,** 36–41.

Grimellec, C. L., Lesniewska, E., Giocondi, M.-C., Finot, E., Vié, V., and Goudonnet, J.-P. (1998). Imaging of the surface of living cells by low-force contact-mode atomic force microscopy. *Biophys. J.* **75**(2), 695–703.

Hartwig, J. H., and Kwiatkowski, D. J. (1991). Actin-binding proteins. *Curr. Op. Cell Biol.* **3,** 87–97.

Helfrich, W. (1973). Elastic properties of lipid bilayers: theory and possible experiments. *Z. Naturforsch.* **28c,** 693.

Henderson, E., Haydon, P. G., and Sakaguchi, D. S. (1992). Actin filament dynamics in living glial cells imaged by atomic force microscopy. *Science* **257,** 1944–1946.

Hertz, H. (1882). Über die Berührung fester elastischer Körper. *J. Reine Angew. Mathematik* **92,** 156–171.

Hildebrand, J. A., and Rugar, D. (1984). Measurement of cellular elastic properties by acoustic microscopy. *J. Microsc.* **134**(3), 245–260.

Hofmann, U. G., Rotsch, C., Parak, W. J., and Radmacher, M. (1997). Investigating the cytoskeleton of chicken cardiocytes with the atomic force microscope. *J. Struct. Biol.* **119,** 84–91.

Hoh, J. H., and Schoenenberger, C.-A. (1994). Surface morphology and mechanical properties of MDCK monolayers by atomic force microscopy. *J. Cell Sci.* **107,** 1105–1114.

Ingber, D. E. (1993). Cellular tensegrity: defining new rules of biological design that govern the cytoskeleton. *J. Cell Sci.* **104,** 613–627.

Janmey, P. A., Hvidt, S., Käs, J., Lerche, D., Maggs, A., Sackmann, E., Schliwa, M., and Stossel, T. P. (1994). The mechanical properties of actin gels-elastic modulus and filament motion. *J. Biol. Chem.* **269**(51), 32,503–32,513.

Johnson, K. L. (1994). "Contact Mechanics." Cambridge University Press, Cambridge.

Käs, J., Strey, H., and Sackmann, E. (1994). Direct imaging of reptation for semiflexible actin filaments. *Nature* **368,** 226–229.

Käs, J., Strey, H., Tang, J. X., Finger, D., Ezzell, R., Sackmann, E., and Janmey, P. A. (1996). F-actin, a model polymer for semiflexible chains in dilute, semidilute, and liquid crystalline solutions. *Biophys. J.* **70,** 609–625.

Korchev, Y. E., Gorelik, J., Lab, M. J., Sviderskaya, E. V., Johnston, C. L., Coombes, C. R., Vodyanoy, I., and Edwards, R. W. (2000). Cell volume measurement using scanning ion conductance microscope. *Biophys. J.* **78,** 451–457.

Lee, J., Ishihara, A., Oxford, G., Johnson, B., and Jacobson, K. (1999). Regulation of cell movement is mediated by stretch-activated calcium channels. *Nature* **400**(6742), 382–386.

Lekka, M., Laidler, P., Gil, D., Lekki, J., Stachura, Z., and Hrynmiewicz, A. Z. (1999). Elasticity of normal and cancerous human bladder cells studied by scanning force microscopy. *Eur. Biophys. J.* **28**(4), 312–316.

Lüers, H., Hillman, K., Litniewski, J., and Bereiter-Hahn, J. (1991). Acoustic microscopy of cultured cells. Distribution of forces and cytoskeletal elements. *Cell Biophys.* **18,** 279–293.

Marko, J. F., and Siggia, E. D. (1995). Stretching DNA. *Macromolecules* **28,** 8759–8770.

Mathur, A. B., Truskey, G. A., and Reichert, W. M. (2000). Atomic force and total internal reflection fluorescence microscopy for the study of force transmission in endothelial cells. *Biophys. J.* **87**(4), 1725–1735.

Matzke, R., Jacobson, K., and Radmacher, M. (2000). Direct, high resolution measurement of furrow stiffening during the division of adherent cells. In preparation.

Miyazaki, H., and Hayashi, K. (1999). Atomic force microscopic measurement of the mechanical properties of intact endothelial cells in fresh arteries. *Med. Biol. Eng. Comp.* **37,** 530–536.

Petersen, N. O., McConnaughey, W. B., and Elson, E. L. (1982). Dependence of locally measured cellular deformability on position on the cell, temperature and cytochalasin B. *Proc. Natl. Acad. Sci. U.S.A.* **79,** 5327–5331.

Radmacher, M. (1997). Measuring the elastic properties of biological samples with the atomic force microscope. *IEEE Eng. Med. Biol.* **16**(2), 47–57.

Radmacher, M., Cleveland, J. P., Fritz, M., Hansma, H. G., and Hansma, P. K. (1994). Mapping interaction forces with the atomic force microscope. *Biophys. J.* **66**(6), 2159–2165.

Radmacher, M., Fritz, M., Hansma, H. G., and Hansma, P. K. (1994). Direct observation of enzyme activity with the atomic force microscope. *Science* **265,** 1577–1579.

Radmacher, M., Fritz, M., and Hansma, P. K. (1995). Imaging soft samples with the atomic force microscope: Gelatin in water and propanol. *Biophys. J.* **69**(7), 264–270.

Radmacher, M., Fritz, M., Kacher, C. M., Cleveland, J. P., and Hansma, P. K. (1996). Measuring the elastic properties of human platelets with the atomic force microscope. *Biophys. J.* **70**(1), 556–567.

Radmacher, M., Tillmann, R. W., Fritz, M., and Gaub, H. E. (1992). From molecules to cells—Imaging soft samples with the AFM. *Science* **257,** 1900–1905.

Radmacher, M., Tillman, R. W., and Gaub, H. E. (1993). Imaging viscoelasticity by force modulation with the atomic force microscope. *Biophys. J.* **64,** 735–742.

Rief, M., Oesterheld, F., Berthold, M., and Gaub, H. E. (1997). Single molecule force spectroscopy on polysaccharides by atomic force microscopy. *Science* **275,** 1295–1297.

Robinson, D. N., and Spudich, J. A. (2000). Towards a molecular understanding of cytokinesis. *Trends Cell Biol.* **10**(6), 228–237.

Rotsch, C., Braet, F., Wisse, E., and Radmacher, M. (1997). AFM imaging and elasticity measurements of living rat liver macrophages. *Cell Biol. Int.* **21**(11), 685–696.

Rotsch, C., Jacobson, K., and Radmacher, M. (1999). The dynamics of active and stable edges in motile fibroblasts investigated by atomic force microscopy. *Proc. Natl. Acad. Sci. U.S.A.* **96,** 921–926.

Rotsch, C., Jacobson, K., and Radmacher, M. (2000a). EGF-stimulated lamellipod extension in mammary Adenocarcinoma cells. *Ultramicroscopy.* Submitted for publication.

Rotsch, C., Jacobson, K., and Radmacher, M. (2000b). Investigating living cells with the atomic force microscope. *Scanning Microsc.* in press.

Rotsch, C., and Radmacher, M. (2000). Drug-induced changes of cytoskeletal structure and mechanics in fibroblasts—An atomic force microscopy study. *Biophys. J.* **78,** 520–535.

Sackmann, E. (1994a). Intra- and extracellular macromolecular networks: Physics and biological function. *Macromol. Chem. Phys.* **195,** 7–28.

Sackmann, E. (1994b). Membrane bending energy concept of vesicle- and cell-shapes and shape-transitions. *FEBS Lett.* **346,** 3–16.

Sato, N. K., Kataoka, N., Sasaki, M., and Hane, K. (2000). Local mechanical properties measured by atomic force microscopy for cultured bovine endothelial cells exposed to shear stress. *J. Biomech.* **33**, 127–135.

Schäffer, T. E., Cleveland, J. P., Ohnesorge, F., Walters, D. A., and Hansma, P. K. (1996). Studies of vibrating atomic force microscope cantilevers in liquid. *J. Appl. Phys.* **80**(7), 3622–3627.

Schäffer, T. E., Viani, M., Walters, D. E., Drake, B., Runge, E. K., Cleveland, J. P., Wendman, M. A., and Hansma, P. K. (1997). An atomic force microscope for small cantilevers, "SPIE Proceedings," pp. 49–52. San Jose, CA.

Schneider, S. W., Pagel, P., Rotsch, C., Danker, T., Oberleithner, H., Radmacher, M., and Schwab, A. (2000). Volume dynamics in migrating cells measured with atomic force microscopy. *Pfluegers Arch.* **439**(2), 297–303.

Schneider, S. W., Sritharan, S. W., Geibel, J. P., Oberleithner, H., and Jena, B. (1997). Surface dynamics in living acinar cells imaged by atomic force microscopy: Identification of plasma membrane structures involved in exocytosis. *Proc. Natl. Acad. Sci. U.S.A.* **94**(1), 316–321.

Sneddon, I. N. (1965). The relation between load and penetration in the axisymmetric Boussinesq problem for a punch of arbitrary profile. *Int. J. Eng. Sci.* **3**, 47–57.

Stossel, T. P. (1993). On the crawling of animal cells. *Science* **260**, 1086–1094.

Svoboda, K., Schmidt, C. F., Branton, D., and Block, S. M. (1992). Conformation and elasticity of the isolated red blood cell membrane skeleton. *Biophys. J.* **63**, 784–793.

Tao, N. J., Lindsay, N. M., and Lees, S. (1992). Measuring the microelastic properties of biological material. *Biophys. J.* **63**, 1165–1169.

Treloar, L. R. (1975). "The Physics of Rubber Elasticity." Clarendon Press, Oxford.

Walch, M., Ziegler, U., and Groscurth, P. (2000). Effects of streptolysinO on the microelasticity of human platelets analyzed by atomic force microscopy. *Ultramicroscopy* **82**(1–4), 259–267.

Walters, D. E., Viani, M., Paloczi, G. T., Schäffer, T. E., Cleveland, J. P., Wendman, M. A., Gurley, G., Elings, V., and Hansma, P. K. (1997). An atomic force microscope using small cantilevers. "SPIE Proceedings," pp. 43–47. San Jose, CA.

Weisenhorn, A. L., Hansma, P. K., Albrecht, T. R., and Quate, C. F. (1989). Forces in atomic force microscopy in air and water. *Appl. Phys. Lett.* **54**, 2651–2653.

Weisenhorn, A. L., Khorsandi, M., Kasas, S., Gotozos, V., Celio, M. R., and Butt, H. J. (1993). Deformation and height anomaly of soft surfaces studied with the AFM. *Nanotechnology* **4**, 106–113.

Wu, H. W., Kuhn, T., and Moy, V. T. (1998). Mechanical properties of L929 cells measured by atomic force microscopy: Effects of anticytoskeletal drugs and membrane crosslinking. *Scanning* **20**(5), 389–397.

Xu, W., Mulhern, P. J., Blackford, B. L., Jericho, M. H., Firtel, M., and Beveridge, T. J. (1996). Modeling and measuring the elastic properties of an archael surface, the sheath of methanospirillum hungatei, and the implication for methane production. *J. Bacteriol.* **178**(11), 3106–3112.

Zahalak, G. I., McConnaughey, W. B., and Elson, E. L. (1990). Determination of cellular mechanical properties by cell poking, with an application to leukocytes. *J. Biomech. Eng.* **112**, 283–294.

Zeman, K., Engelhard, H., and Sackmann, E. (1990). Bending undulations and elasticity of the erythrocyte membrane: Effects of cell shape and membrane organization. *Eur. Biophys. J.* **18**, 203–219.

CHAPTER 5

Cell Adhesion Measured by Force Spectroscopy on Living Cells

Martin Benoit

Center for Nanoscience
Ludwig-Maximilians-Universität München
D-80799 München, Germany

I. Introduction

Cell-to-cell adhesion is essential for multicellular development and arrangement. Cells may carry several different adhesion molecules (Kreis and Vale, 1999), resulting in a high variability of the molecular repertoire of the cell surfaces. This variability is reflected in the broad pattern of adhesion-controlled cellular functions during development and adult life (Fritz *et al.,* 1993; Springer, 1990; Vestweber and Blanks, 1999).

To determine cell adhesion many techniques have been evolved, such as functionalized latex beads moved with optical tweezers (Choquet *et al.,* 1997), microfluorescence assays or interferrometric techniques (Bruinsma *et al.,* 2000), and centrifugation experiments, e.g., with cell spheroids (John *et al.,* 1993; Suter *et al.,* 1998). Viscoelastic properties of cells were measured by cell poking and even with spatial resolution by an atomic force microscope (AFM) in either force modulation mode or more recently by force volume techniques (Domke *et al.*, 2000; Goldmann *et al.*, 1998; Hoh and Schoenenberger, 1994; Radmacher *et al.*, 1996; Zahalak *et al.*, 1990). Adhesion between single cells, e.g., granulocytes and target cells, was measured in the past using mechanical methods, such as micropipette manipulations (Evans, 1985, 1995) or hydrodynamic stress (Chen and Springer, 1999; Curtis, 1970). With the development of piconewton instrumentation based on AFM technology (Binnig *et al.*, 1986), the force resolution and the precision of positioning have allowed measurements at the single-molecule level (Gimzewski and Joachim, 1999; Müller *et al.*, 1999; Oesterhelt *et al.*, 2000). Forces for conformational transitions in polysaccharides (Marszalek *et al.*, 1999; Rief, Oesterhelt *et al.*, 1997) for the unfolding of proteins (Oberhauser *et al.*, 1998; Rief, Gautel *et al.*, 1997; Smith *et al.*, 1999) and for stretching and unzipping of DNA (Rief *et al.*, 1999; Strunz *et al.*, 1999) were measured. Unbinding forces of individual ligand–receptor pairs were determined (Baumgartner *et al.*, 2000; Florin *et al.*, 1994; Hinterdorfer *et al.*, 1996; Müller *et al.*, 1998) and the basic features of the binding potentials were reconstructed (Grubmüller *et al.*, 1995; Merkel *et al.*, 1999). Recently, the first steps toward cell adhesion measurements with AFM technology were made (Domke *et al.*, 2000; Razatos *et al.*, 1998; Sagvolden *et al.*, 1999).

Several theories have been developed to describe the processes which are involved while separating cells by either modeling single independent contacts or picturing more elaborate mechanisms such as molecular clustering (Evans and Ritchie, 1997; Kuo *et al.*, 1997; Ward *et al.*, 1994; Ward and Hammer, 1993).

In this section a new AFM-based experimental platform to investigate cell-to-cell interactions *in vivo* down to the molecular level will be described, immobilizing living cells to a force sensor. Epithelial cells (RL95-2 and HEC-2-A) from human endometrium as a substrate for an artificially rebuilt human trophoblast (JAR) are used to distinguish molecular adhesion processes involved not only in embryo–uterus interactions but also between individual cells of *Dictyostelium discoideum* to measure the adhesion force of single-contact site A proteins.

To obtain reproducible results, the complexity of living cells demands recording, estimating, and pinpointing a large variety of parameters. Therefore the contact-force is controlled down to 30 pN during the contact between well-studied cell types in a defined cell culture environment.

II. Instrumentation

The cell adhesion force spectrometer with an integrated optical microscope is specialized for force measurements on living cells. As a force sensor, a standard AFM cantilever is placed underneath a Perspex holder. The force signal is obtained from the deflection of the laser beam (Fig. 1) and plotted as force versus piezo position (e.g., Fig. 5). The spring constant of the cantilever in each experiment is determined using the thermal noise technique reported earlier (Florin *et al.*, 1995). By using sensors with a low spring constant, less force is applied to a cell when touched. The force resolution lies between 20

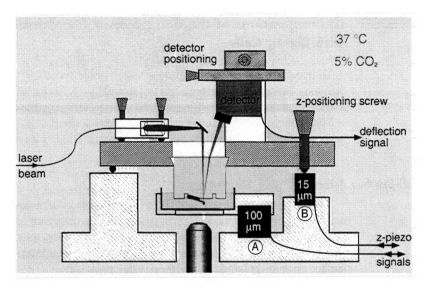

Fig. 1 Schematic of the adhesion force spectrometer with a light microscope below the Petri dish. The sensor mounted on a Perspex holder is placed from above in the Petri dish with the detecting laser unit. Two versions of cell adhesion force spectrometers: (A) long-range (100 μm) piezo moving the Petri dish and (B) short-range (15 μm) piezo moving the force sensor.

and 3 pN and is recorded together with the piezo position at a precision of 1 Å in either 256 pts (12 bit) or 32,768 pts (16 bit) per trace. The frequency of data collection is 60 kHz and the noise can be reduced by either filtering or averaging. For positioning, the sample is manually driven by an x-y stage mounted on a high-precision z piezo-actuator (100 μm)[1] with a strain gauge for long-range cell interactions (Thie *et al.*, 1998) (Fig. 1A). To detect shorter range interactions the Perspex holder is moved by a high-precision z piezo actuator (15 μm) which is equipped with a strain gauge (Dettmann *et al.*, 2000) (Fig. 1B). The z piezo velocity was typically set between 1 and 7 μm/s.

Slower velocities often interfere with drift effects basically caused by cell movement, while at higher velocities hydrodynamics influence the measurement. The lateral sample displacement is disabled during most of the experiments reported here. The approach of the sensor to the surface stops automatically if a certain threshold force is reached. This force can be kept constant within a certain range by a feedback loop compensating movements of the cells or piezo drift, especially if contacts last several minutes. Measurements are performed in an appropriate medium for living cells in a cell culture dish. To achieve long-time measurements standard cell culture conditions at 37°C in CO_2 (5% v/v) can be applied. The cells are monitored using the light microscope during the entire experiment.

[1]Especially nerve cells tend to form strongly adhering membrane tethers (Dai and Sheetz, 1998) over distances of millimeters. Even 100 μm is not enough to separate these cells from each other.

III. Preparations of the Force Sensor for Measurements with Living Cells

To immobilize cells on the force sensor without harming them is most crucial for operating the cell adhesion force spectrometer (Fig. 1). Here a single cell or, alternatively, a whole monolayer of epithelial cells will be immobilized to the sensor (Figs. 2A and 2B).

Since most cells express adhesion molecules on their surface, a very gentle method of immobilization is establishing a matching connection to these molecules.

A. Cell–Surface Adhesion Force Measurements

To determine which proteins to use for immobilizing the respective cells, as a first step, we characterized the adhesion forces by probing the cells with differently functionalized sensors. To distinguish the adhesion of the coating to be tested from the nonspecific interaction between surface and cell, a treatment had to be found to inactivate the sensor surface prior to applying the functionalizing molecules.

1. Immobilization of a Sphere to the Sensor

To better define the contact area between sensor and cells, a sphere of 60 μm in diameter from either sephacryl or glass is fixed at the end of a cantilever. The spheres are mounted to the cantilevers (DNP-S Digital Instruments, Santa Barbara, CA; or Microlever, Park Scientific Instruments, Sunnyvale, CA) in the following manner.

A tiny spot of epoxy glue (UHU plus endfest 300, Bühl, Germany) is applied to the tip of a cantilever using a patch-clamp glass electrode. Then a single Sephacryl S-1000 sphere (Pharmacia, Freiburg, Germany) or a glass sphere (G 4649; Sigma, Deisenhofen/Germany), about 60 μm in diameter, which sticks electrostatically to a cannula (Terumo No. 20, Leuven/Belgium) is placed on the epoxy. To cure the epoxy, the microsphere-mounted cantilever is then heated at 90°C for 45 min. Another method is described in Holmberg *et al.* (1997). Before use, the cantilevers were sterilized in 70% ethanol for 2 h and washed thoroughly in distilled water. Sensor tips and spheres fixed to the force sensor (Fig. 3) with various coatings were tested on the cells of interest.

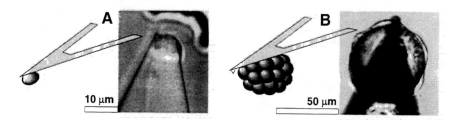

Fig. 2 Schematics and light microscopic image of (A) a single-cell (*Dictyostelium discoideum*; the cells on the cover slide are out of focus) and (B) a layer of cells (osteoblasts) on a glass sphere immobilized on a force sensor.

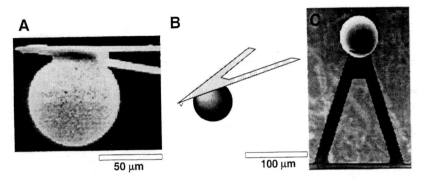

Fig. 3   Schematics (B) and images of a sephacryl (A) and a glass (C) sphere (diameter 60 μm) glued to a force sensor.

2. Passivated Force Sensors

The following protocol, derived from Johnsson *et al.* (1991), proved useful for preparing sensors with a sufficiently low nonspecific interaction with cells.

First the Si–OH layer of either a SiO_2 or a Si_3N_4 surface is ammino-silanized with N'-(3-(trimethoxysilyl)-propyl)-diethylentriamin (Aldrich) at 80°C for 10 min to obtain an amino-functionalized surface. It is then washed in ethanol and completely crosslinked for 10 min in water at 80°C. A phosphate-buffered saline (pH 7.4) (PBS Sigma) solution of 10 mg/ml of carboxymethylamylose (Sigma) is activated with 20 mg/ml N-hydroxy-succinimide (NHS, Aldrich) and 20 mg/ml 1-ethyl-3-(3-dimethylaminopropyl)carbodi-imide (EDC, Sigma) for 2 min. The tip is then incubated with the NHS-activated amylose for 15 min, rinsed three times in PBS, incubated with 0.5 mg/ml ethanolamine (Sigma) in PBS for 1–2 h, and intensively rinsed in PBS. Other preparation techniques with PEG have also proved to sufficiently passivate surfaces for proteins and cells (Bruinsma *et al.*, 2000; Willemsen *et al.*, 1998).

3. Results

From the "deadhesion force versus piezo position traces" (e.g., Fig. 4 or Fig. 5) adhesion can be characterized in an initial approach by measuring the maximum adhesion force. As shown later, other adhesion parameters will be derived from these traces. If a sphere is lowered onto a soft cell surface, the area of interaction increases with the indentation which leads to an enhancement of the adhesion signal. The adhesion strength is not only dependent on the indentation force (here 3 ± 1 nN) while the cells are brought and held in contact as mentioned earlier but also, as shown in Fig. 5, on the duration of the contact. This is probably due to the fact that the cell shape adapts to the sphere's surface and more and more molecules can interact with this surface with time. The adhesion to the sepacryl spheres is enhanced by at least 50% compared to that to a glass sphere, in agreement with their structured and therefore larger surface. Changing the velocity of retraction leads to a fairly linear relation between separation speed and adhesion in the range between 2 and 27 μm/s. However, for low velocities the influence

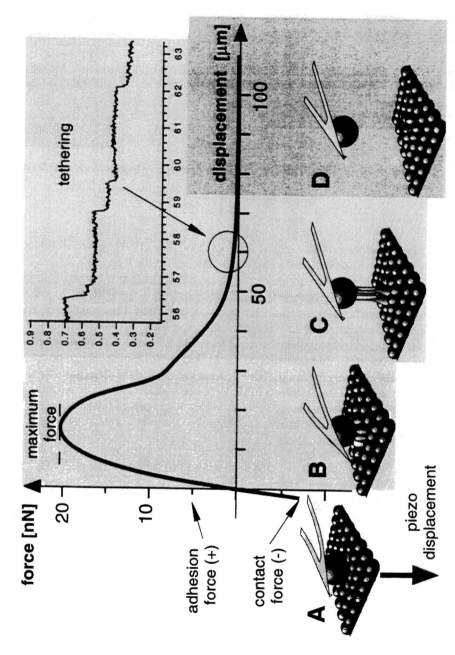

Fig. 4 A typical force versus piezo displacement trace on an epithelial cell layer with schematics of the experimental situation of the cell contact (A), the maximum force regime (B), the tethering regime (C), and the complete separation (D). Contact force indicated with negative values and adhesion force (sensor is elongated downwards) indicated with positive values.

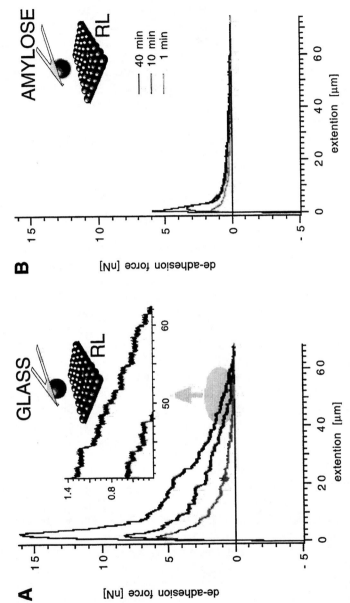

Fig. 5 Two sets of deadhesion force traces recorded with a plain glass surface on the sensor (A) and an amylose-passivated surface (B) after contacts of 1, 10, and 40 min at 5 nN on a confluent monolayer of epithelial cells (RL95-2). The contact area was about 500 μm^2. Inset (A) zooms into the traces where indicated by the circle revealing single rupture steps (Thie et al., 1998). (See Color Plate.)

of the cells on both laser reflex and drift becomes prominent, whereas for high velocities hydrodynamics come into play. To compare the results from different experiments in all the measurements with spheres presented here, the velocity was held constant at 7 μm/s.

Scrutinizing the deadhesion force curves more closely one will find more or less pronounced single steps in the order of 100 pN in the regions of descending adhesion (inset of Fig. 5A) indicating ruptures on the molecular level. Counting them could give a very rough estimate on the number of molecules involved. But unfortunately these steps are only resolved in the descending shoulder of the traces far away from the maximum force.

Nevertheless the results shown in Fig. 5 clearly point out the passivation of the originally adhesive glass surface with amylose even for long contact times.

Carboxymethylamylose also can be used for crosslinking certain molecules to specifically activate the passivated surface. Before saturating the amylose with ethanolamine, a protein with a NH_2 group can be bound covalently to the carboxy groups of the amylose activated with EDC and NHS.

4. Specifically Functionalized Force Sensors

Protein is immobilized on the sensor surface by carboxy-methylated amylose (Grandbois *et al.*, 1999; Johnsson *et al.*, 1991). Alternatively the surface treatment with PEG by Hinterdorfer *et al.* (1996) is recommended for such activations (Willemsen *et al.*, 1998). Initially the sensor is functionalized in the same way as described earlier: First, the standard commercially available Si_3N_4 cantilevers (with glass beads) are silanized. Second, 10 mg/ml of carboxymethylamylose is activated with 2–10 mg/ml NHS and 5–10 mg/ml EDC[2] for 1–5 min in PBS solution. As described earlier, the sensor is incubated with the activated amylose for 10–15 min. After rinsing three times in PBS it is immediately incubated with 0.05–0.5 mg/ml of the molecule of interest in PBS for 2 h—optionally fresh EDC and NHS have to be added with the molecule if there is a time delay of more than 15 min. The only restriction to this molecule is that it must favorably exhibit an exposed NH_2 group, preferably far away from its binding pocket.[3] Intensive rinsing in PBS removes the unbound molecules.

5. Results and Discussion

It is now, for instance, possible to probe a cell surface by an AFM tip specifically functionalized with a lectin (Grandbois *et al.*, 2000). This technique might be used to monitor locally the time course of extracellular membrane molecule expression because it is harmless to the cells and does not block the receptors by labeling. With this setup various molecules were tested for their ability to immobilize cells on a sensor. Even though there are differences in adhesion between the various cell types, in general, the spontaneous adhesion, after short contact periods of less than a minute, either to NH_2 groups of just amino-silanized spheres or to aldehyde groups was found to be extremely

[2]For lower concentration the free segments of the amylose chain become longer.

[3]If the binding pocket is potentially inactivated by binding the amylose with NH_2 groups too close to the pocket, a soluble binding partner lacking any NH_2 groups could be added during activation to protect the binding site. The ligand must then be washed out carefully before the experiment.

strong for all cells at the initial contacts. But the adhesion decreased very rapidly with each additional experiment, especially when measuring in nutrient medium with protein-rich additives, indicating that the reactive groups were saturated irreversibly with bound molecules. Lower adhesion strengths were measured for certain lectins, such as either wheat germ agglutinin (WGA, Sigma) or concavalin A (con A, Sigma), which bind to the glycocalyx of assigned cells. Due to the lectins' specificity to certain glycoproteins, they were not significantly affected by nutrient medium and did not lose their ability to bind. In addition they self-cleaned due to their offrate (10^{-3-5}/s) if free molecules were caught in the binding pocket.

In measuring adhesion on epithelial cells, one must consider their habit to polarize. Thus, in a monolayer, epithelial cells differ among other properties in the proteins expressed on both the apical (free) membrane and the membranes presented to the substrate or to their neighbors. Interestingly, even the spontaneous adhesion of epithelial cells—whose apical surface is not supposed to adhere—to plain cantilevers or glass surfaces is still prominent: however, it does not differ much from measurements on surfaces functionalized by BSA, or just incubated in polylysins, fibronectin, or laminin (data not shown). Since this adhesion is independent of the coating, it is assumed to be a non-specific surface interaction, whereas this "nonspecific" adhesion of single *Dictyostelium discoideum* cells e.g. to the plain force sensor (data not shown) is useful for locomotion or ingestion.

B. Adhesion Force Measurements between Cell Layers

To apply this technique to force measurements between epithelial cell layers (epithelium) the already well-investigated human embryo–uterus interaction was chosen. JAR cells represent the invasive trophoblast, RL cells, and the receptive uterine epithelium. From centrifugation experiments (John *et al.*, 1993) HEC cells are supposed to represent the nonreceptive uterine epithelium.

1. Immobilizing a Monolayer of Cells to a Force Sensor

Before seeding cells, the sensor is carefully rinsed in alcohol and water, precoated with polylysins, laminins, fibronectins, or other adhesive coatings,[4] and placed upside-down in nutrient medium.

Of course surface passivation is neither helpful for seeding cells nor necessary, since the whole surface will be covered with cells. To grow a monolayer of cells on a force sensor and to predefine the later contact area, it is also useful to glue either a glass or, better, a sephacryl sphere underneath the cantilever as described earlier. Then a drop of suspended cells is released onto the sphere (Fig. 6). From time to time (3–14 h) this procedure is repeated until one or more cells are attached to the sphere. Cells growing on the sensor arms are not found to disturb the experiments as long as they neither move nor cover the part of the sensor where the detecting laser beam is reflected. Even then, measurements are possible although occasionally suspicious malformed traces should

---

[4]We recommend using the customary surface treatment techniques for the force sensor's surface because the cells must grow on the surfaces for several days until a proper monolayer is established.

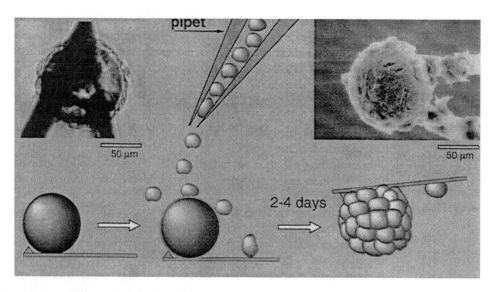

Fig. 6 Schematic of growing cells on a sensor sphere [developed by R. Röspel (Thie *et al.*, 1998)]. The result is shown as an LM image and a SEM image.

be neglected.[5] Vesicles or other cellular compartments, which tend to disturb the laser beam, easily contaminate the fluid when working with living cells. Therefore the medium should be exchanged freshly several times before starting the measurement.

When the cells finally cover the sphere confluently, the lever is placed in the force spectrometer and moved onto a surface of interest.

With such prepared sensors (Fig. 6) several surfaces (e.g., potential implants) can either be probed or, as shown here, be used to investigate certain aspects of embryo–uterus interactions.

With this preparation, the mechanisms of the interaction between human trophoblast-type JAR cells (JAR) with the two human uterine epithelial cell lines RL95-2 (RL) and HEC-1-A (HEC) were investigated. RL cells, in contrast to HEC cells, are supposed to respond to the contact with JAR cells in a specific way (John *et al.*, 1993; Thie *et al.*, 1997, 1998).

2. Results

JAR-coated force sensors were brought into contact with confluent monolayers of HEC and RL cells. Figure 7 shows de-adhesion force curves for the different cell types after various contact duration. Despite the variation of the de-adhesion curves due to the various radii of the spheres ($65 \pm 10 \ \mu$m) and some cells occasionally disturbing the

[5]To be on the safe side, one could think of passivating the sensor prior to gluing the sphere, but this might affect the strength of the epoxy glue.

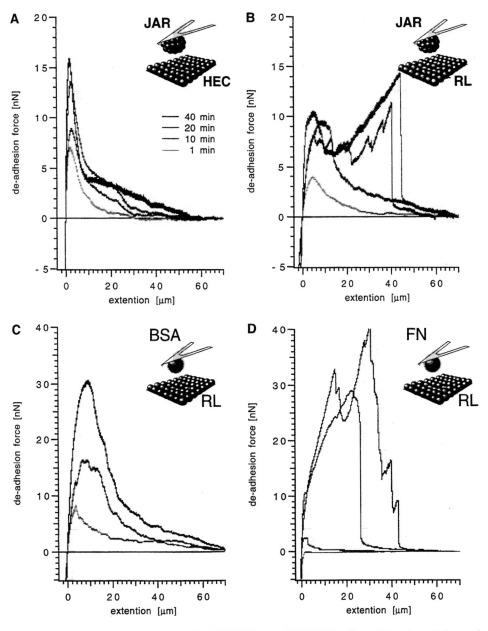

Fig. 7 Typical adhesive force curves for (A) HEC-1-A and (B) RL95-2 cells resulting when a JAR-coated sensor, (C) a bovine serum albumin (BSA)-coated sensor, or a (D) fibronectin (FN)-coated sensor is retracted after periods of 1–40 min time of contact at approximately 5 nN (Thie *et al.*, 1998).

Table I
Listing of JAR Experiments[a]

	HEC/JAR		RL/JAR	
Time (min)	Continuous	Discontinuous	Continuous	Discontinuous
1	87	0	154	0
10	55	0	67	0
15	24	0	38	0
20	24	0	19	5[b]
30	11	0	6	4
40	13	0	5	14[b]

[a] HEC never showed discontinuous deadhesion, while RL does.
[b] After one of these experiments, a cell was found loosened on the monolayer.

laser reflex, typical features can be recognized. As in Fig. 5, the adhesion increases with increasing contact time in Fig. 7. For contact times of less than 20 min, the response of both cell types is virtually indistinguishable within the experimental error of the variations between different preparations. However, for prolonged contact duration, the two cell types (Fig. 5A HEC, Fig. 5B RL) showed a marked difference in the deadhesion profile. This difference becomes particularly prominent after the maximum adhesion force. Here, the contact between JAR and RL cells was found to be significantly different from JAR and HEC cells, since the de-adhesion often propagates discontinuously after cell contacts of at least 20 min. The force builds up upon separation until the cells detach with remarkably large force steps from 0.5 to far more than 5 nN (Table I), while the force between JAR and HEC decreases continuously in small steps (inset Fig. 5A) of less than 200 pN upon separation in all cases.

It has been previously postulated that there is a local cross-talk between trophoblast and uterine epithelium leading to specific cell–cell binding, i.e., a redistribution/upregulation/activation of adhesion systems at the free cell pole (Albers *et al.,* 1995; Denker, 1994; Thie *et al.,* 1995, 1996). In this context, it is of interest that a discontinuous JAR–RL interaction is observed only after prolonged contact of both partners. This could be due to the time needed to build up cooperative islands of interacting adhesion molecules (molecular clusters). In contrast, the initial interaction observed after short contacts (<20 min) is caused by adhesion of independently adhering molecules.

To simulate this independent (noncooperative) adhesion mechanism, we probed both RL and HEC cells with pure spheres and spheres that were coated with bovine serum albumin (BSA) at different contact times. Since none of these experiments ($N = 250$) show the significantly discontinuous de-adhesion, they should reproduce, solely, the noncooperative molecular interactions. Figure 7C shows a typical set of the resulting de-adhesion curves for various contact times. As can be seen, there is very good agreement with the detachment traces recorded for HEC cells. But discontinuous de-adhesion at prolonged contact times is no longer detectable.[6] An experiment equivalent to JAR spheres

[6] Notable is the increased maximum adhesion compared to the measurement in Fig. 5A due to the fact that here a sephacryl sphere was coated with BSA and not a glass sphere.

on RL monolayers, carried out with a fibronectin-coated sphere (with reduced adhesion background by the amylose), also results in discontinuous de-adhesion in 14 of 20 sets (as can be seen in Fig. 7D). As in the case of the JAR–RL interaction after long cell contacts, the discontinuous de-adhesion signature appears only after contacts for more than 20 min. On the contrary, the noncooperative adhesion of the FN spheres is drastically reduced for short contact duration, but the FN reaction is enhanced[7] compared to the measurements on JAR/RL after long contacts. Since RL cells are known to interact with JAR cells via fibronectin-binding proteins (certain integrins; Thie and Ramunddal, unpublished data), these experiments corroborate our assumption that this discontinuous adhesion is due to a specific interaction. This conclusion is supported by previous work of other groups which had investigated integrin–cytoskeleton interactions in other cell types where integrin–ligand binding promotes redistribution of the integrins to molecular clustering (Felsenfeld *et al.*, 1996; Yauch *et al.*, 1997).

3. Discussion

The first set of findings here demonstrates that force spectroscopy is a technique, which allows the uncovering of contact time-dependent adhesion processes and the scrutinization of specific receptor-mediated interactions as well as nonspecific cell–cell or cell–substrate adhesion.

The adhesion to sephacryl spheres after some minutes of contact with cells (which is identical to BSA-coated sephacryl spheres Fig. 7D) is rather high compared to that to glass spheres. The cells may gain adhesion strength when creeping into the structured surface (Fig. 3A) of the sephacryl spheres enlarging the interacting area and possibly snatching the hooks and grips. These surfaces are then good for culturing cells on them, but for passivation smooth surfaces like glass are recommended. The adhesion between JAR and HEC is in most cases slightly higher than that between JAR and RL which might be due to the larger surfaces of the microvilli-rich HEC cells. Assumably the cell body becomes elongated in the ascending part of the deadhesion trace until the cytoskeleton is stretched out after 10–20 μm depending on the thickness of the layers and the type of cells. In the region of maximum force the likelihood of braking adhesion links due to the large load is very high. The interconnection does not break up to 80 μm despite overstretching the cells further than 40 μm, indicating an adhesion concept which does not include the cytoskeleton. The membrane tethers must mediate this adhesion, as visible by the characteristic step pattern indicating single de-adhesion events on the molecular level. The small slope prior to the stepwise ruptures also confirms the presence of tethers in this region. Here the HEC cells show a larger tendency to form tethers due to the microvilli-rich surface compared to the smoother surface of RL cells. Tether formation in the case of RL/JAR cells after long contacts is outranged by the strong adhesion concept[8] of cooperating molecules clustering in adhesion islands. Furthermore, from

---

[7]Even though using passivated glass spheres, the adhesion is significantly higher than all other measurements, probably due to high density of fibronectin on the surface.

[8]Despite the fact that in the LM no damages of the cells could be resolved, the fact that in three cases single cells were found pulled out from the monolayer suggests an injury of the membranes after such strong de-adhesion events.

the fact that the adhesion forces increase after the first force maximum upon further stretching prior to the ruptures, an involvement of the cytoskeleton must be considered.

The embryo passing to the site of implantation always stays in contact with cells of the guiding tissue, i.e., oviductal and uterine epithelium. This can be guaranteed by the adhesion concept of the tethers (Chen and Springer, 1999) as shown by the JAR/RL adhesion measurements. Slow movement of the sphere is also provided, since the adhesion forces of former contacts disappear while moving away while new connections are formed in the direction of motion. During this "damped walk" the cells have time to "communicate" and to form strong adhesion clusters when the target area is reached. Then the force increases when trying to move farther—the motion will be stopped by this adhesion concept as shown by the JAR/RL adhesion measurements.

C. Cell–Cell Adhesion Force Measurements

We will now consider the interaction between single cells by combining single-molecule force spectroscopy with genetic manipulation for the measurement of de-adhesion forces at the resolution of individual cell-adhesion molecules. In general single cells behave different from cells in tissue. But, to resolve single-molecule events instead of cooperative molecular effects, it is necessary to minimize contact area, contact time, and contact force. Two steps in the refinement of the technique are crucial for measuring discrete de-adhesion forces at molecular resolution: (i) reduction of the contact force which results in a further decrease of the contact area and (ii) shortening of the contact time which reduces the number of contacts established. The cell adhesion force spectrometer is therefore able to control the contact force down to 30 pN when using soft cantilevers (5 mN/m). A single cell on the cantilever reduces the contact area to a minimum. The measurements focus on contact site A (csA) as a prototype of cell adhesion proteins. The csA glycoprotein is specifically expressed in aggregating cells of *Dictyostelium discoideum* which are engaged in the process of building up a multicellular organism.

The eukaryote *D. discoideum* offers the advantage that one particular type of adhesion molecule, the developmentally regulated (csA) glycoprotein, can be singled out by genetic manipulations (Faix, 1999). In *D. discoideum*, csA participates in cell aggregation, the transition from the single-cell to the multicellular stage (Ponte *et al.*, 1998). Thus, csA is undetectable in growth-phase cells but is expressed upon starvation (Murray *et al.*, 1981). In developing cells of the aggregation stage, csA covers roughly 2% of the total cell surface area (Beug, Katz, and Gerish, 1973). CsA molecules react with each other (homophilic interaction), forming noncovalent bonds linking the surfaces of adjacent cells (Kamboj *et al.*, 1988), which are anchored in the plasma membrane by a ceramide-based phospholipid (Stadler *et al.*, 1989).

1. Immobilizing a Single Living Cell to a Force Sensor

As shown by Razatos *et al.* (1998) even bacteria can be fixed to an AFM tip. With an appropriately functionalized force sensor, a living cell loosely sitting on a cell culture dish is tethered to the sensor (see following). Therefore the very end of the lever is lowered onto the cell at a force of a few nanonewtons and held in contact for approximately 30 s to allow the specific molecules on the lever to bind before lifting the cell off the bottom of the dish. When the cell sticks to the sensor, it can be moved to a cell or surface of interest. Typically, the interaction strength between cell and cantilever increases with time, due to the assimilating cell surface. If the cell on the dish already sticks to the substrate too strongly, it can be pushed gently from the side with the edge of the lever before lifting it up. It would be helpful if at least the z piezo could be moved manually for this purpose. The best results were obtained with tipless cantilevers, as the tip either is likely to interfere with the adhesion measurements if it surmounts the cell or hinders the cell adhesion. Unfortunately, the spring constants of commercial tipless cantilevers are very stiff (Digital Instruments, 60 mN/m) compared to the soft cells (Radmacher, 1997). To obtain compliant and tipless force sensors, the cantilevers had to be modified destructively with thin tweezers, as follows, prior to functionalization. A curvature on one of the tweezers' fingers (Fig. 8) effects a grinding while gently squeezing the cantilever between the tweezers.

2. Results

As described previously, a *Dictyostelium* cell is picked up with a tipless AFM cantilever whose end had been covalently functionalized with wheat germ agglutinin (WGA lectin, Sigma). A target cell resting at the bottom of a Petri dish is positioned under the cell on the cantilever and approached until a predefined repulsive contact force is established (Fig. 9). This contact force is held constant for a defined time interval to allow cell adhesion to become established.

Force traces of the de-adhesion process between the two cells are shown in Fig. 10. A trace, typical of growth-phase cells that had been allowed to interact for 20 s at a contact force of 150 pN, is shown at the bottom of Fig. 10A. The adhesion between these cells gave rise to unbinding forces on the order of 1 nN, caused by several molecular

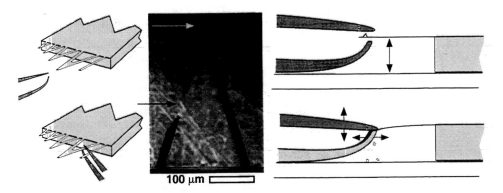

100 μm

Fig. 8 Scheme of the "surgery" on a cantilever with tweezers and a SEM image of such a sensor. (See Color Plate.)

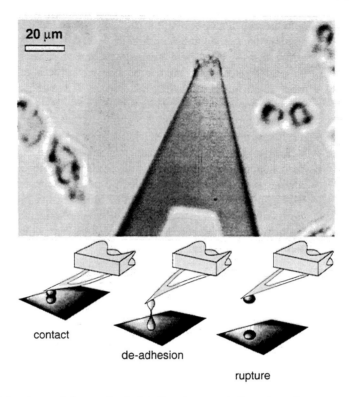

contact

de-adhesion

rupture

Fig. 9 Light microscopic image of a single cell on the sensor (cells on the surface are out of focus) and schematics of a force experiment (Benoit *et al.*, 2000).

interactions. Particularly in the last part of this deadhesion force trace the typical pattern for tether formation appears (Hochmuth *et al.*, 1996). Adhesion of the nondeveloped cells used in this experiment is known to be Ca^{2+} dependent (Beug, Katz, Stein, *et al.*, 1973). To test this Ca^{2+} sensitivity, 5 mM EDTA, a chelating agent, was added to the buffer. As illustrated (at the bottom of Fig. 10B) the adhesion is drastically reduced. Within the duration of the experiments this low amount of EDTA did not affect the cells' integrity. Since the cells tend to move on the surface of the dish it is necessary to check the cell contact by the built-in light microscope and readjust the positioning of the cells.

After growth-phase cells were brought together by contact forces of 30–40 pN applied for only 0.2 s, less than 20% of the de-adhesion traces showed binding between the cells (Fig. 10A). The histogram of the deadhesion forces showed a broad distribution with a maximum at about 50 pN. The low frequency of these de-adhesion events implies that, based on Poisson statistics, more than 90% of the contacts should reflect single binding events. Thus, the width of the force distribution most likely reflects a multitude of molecular species involved in the Ca^{2+}-dependent adhesion. In the presence of 5 mM EDTA, 96% of the cells did not establish detectable adhesion within 0.2 s, even when they were brought into contact with an increased force of 90 pN (Fig. 10B). On the basis

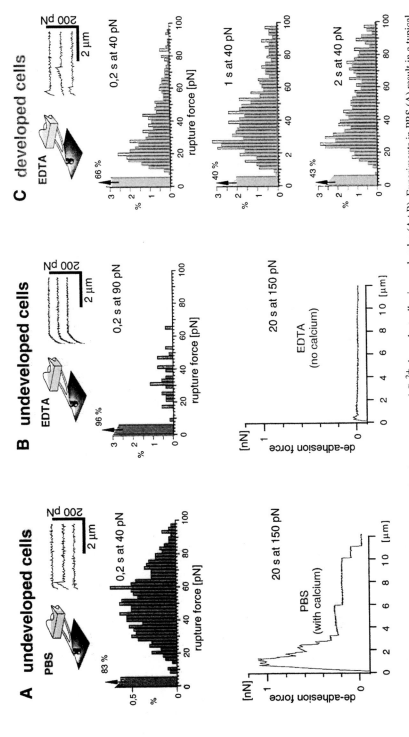

Fig. 10 Undeveloped cells lacking the CSA molecule express several Ca^{2+}-dependent adhesion molecules (A+B). Experiments in PBS (A) result in a typical rupture force spectrum derived from 5760 traces (inset) after contact for 0.2 s at 35 pN. Below: a representative trace from a prolonged contact for 20 s at 150 pN. Experiments in 5 mM EDTA (B) result in a force spectrum with reduced adhesion (only 4%) from 960 traces (inset) even though there was an increased contact force of 90 pN for 0.2 s. The prolonged contact for 20 s at 150 pN (below) does not show significant adhesion. Experiments in EDTA with developed cells (C) in contrast show typical force spectra for the CSA molecule. For 0.2 s at 35 pN, one peak at 20 pN becomes prominent from 1334 traces (inset). After contact for 1 s, the spectrum derived from 1088 traces (not shown) raises a second peak around 45 pN, and after 2 s, a third peak at 74 pN appears from 1792 traces (not shown) (Benoit *et al.*, 2000).

of these data, de-adhesion forces were measured in developing cells in which additional cell adhesion proteins are expressed. Cells in the aggregation stage are distinguished from growth-phase cells by EDTA-stable cell adhesion (Beug, Katz, and Gerish, 1973). When 5 mM EDTA was added to these cells and de-adhesion forces were determined after a contact force of 35 ± 5 pN, binding was observed in roughly half of the traces. The collection of traces shown in Fig. 10C illustrates the type of results obtained at various contact times. Often initial forces rose up to several hundred piconewtons, and unbinding occurred in several steps until the last tether connecting the two cells was disrupted at long contacts. In contrast to these multiple de-adhesion events, single steps of deadhesion prevailed after a contact time of 0.2 s.

The last force step, the one that completely separated the cells, was measured in more than 1000 traces after contact times of 2, 1, or 0.2 s (Fig. 10C). When these data were compiled in histograms, a pronounced peak indicating a force quantum of 21 ± 5 pN became apparent. Upon increasing of contact times from 0.2 sec to 2 sec, this peak only negligibly shifted to higher de-adhesion forces (23 pN). The main difference between the histograms resided in the lower contribution of higher forces upon the reduction of contact time. The higher forces contributing to de-adhesion after 2 or 1 s of cell-to-cell contact are interpreted as superimposed multiples of a basic force quantum of 23 pN.

Developmental regulation and EDTA resistance suggest that the measured force quantum of 23 pN is due to the unbinding of csA molecules. However, cells in the aggregation stage differ from growth-phase cells not only in the csA protein but also in several other developmentally regulated cell surface proteins. Therefore, to attribute the peak of 23 pN to the presence of this particular cell adhesion protein, different types of cells in which specifically csA expression was genetically manipulated were employed (Benoit *et al.*, 2000). The csA gene was selectively inactivated by targeted disruption using a transformation vector that recombined into the gene's coding region (Faix *et al.*, 1992). Only 25% of the cells in this csA knock-out strain showed measurable de-adhesion forces as compared to 86% of wild-type cells. Also, cells of a mutant unable to produce csA (Harloff *et al.*, 1989) were transfected with vectors that encode the csA protein under the control of the original promoter. Indeed these "repaired" cells showed adhesion like the wild-type only when developed. Together these results demonstrate that the csA molecule is the primary source of the intercellular adhesion measured by force spectroscopy in the presence of EDTA.

3. Dicussion

The quantized de-adhesion force of 23 pN indicates discrete molecular entities as the unit of csA-mediated cell adhesion. The most likely interpretation of this peak is that one unit reflects the interaction of two csA molecules, one on each cell surface. Nevertheless, since oligomerization may strongly increase the affinity of cell adhesion molecules (Tomschy *et al.*, 1996), we cannot exclude the possibility that defined dimers or oligomers represent the functional unit of csA interactions (Baumgartner *et al.*, 2000; Chen and Moy, 2000).

The measured de-adhesion force of 23 pN for csA is small compared to that of most antibody–antigen or lectin–sugar interactions, which frequently exceeds 50 pN at comparable rupture rates (Dettmann *et al.*, 2000). These moderate intermolecular forces involved in cell adhesion are consistent with the ability of motile cells to glide against each other as they become integrated into a multicellular structure. Moreover, in view of the limited force that the lipid anchor may withstand, much higher molecular unbinding forces would be of no advantage.

Here the separation rate was kept constant at 2.5 μm/s, resulting in force ramps between 100 and 500 pN/s depending on the elasticity of the cells. This rate is on the same order as the protrusion and retraction rates of filopods, the fastest cell surface extensions in *Dictyostelium* cells. With their adhesive ends, the filopods can act as tethers between cells or between cells and other surfaces. Our measurements of separation forces are therefore representative of upper limits to which the cells are exposed by their own motility.

IV. Cell Culture

A. HEC/RL Cell Culture on Coverslips

Measurements on human endometrial cell lines, purchased from the American Type Culture Collection (ATCC, Rockville, MD/USA), i.e., HEC-1-A (short HEC; HTB 112; (Kuramoto *et al.*, 1972)) and RL95-2 (short RL; CRL 1671 (Way *et al.*, 1983)), were performed in JAR medium at 36°C and 5% CO_2. For routine culture, cell lines were grown in plastic flasks in 5% CO_2–95% air at 37°C.

In brief, HEC cells were seeded out in McCoy's 5A medium (Gibco-Life Technology, Eggenstein, Germany) supplemented with 10% fetal calf serum (Gibco); RL cells, in a 1 + 1 mixture of Dulbecco's modification of Eagle's medium and Ham's F12 (Gibco) supplemented with 10% fetal calf serum, 10 mM Hepes (Gibco), and 0.5 μg/ml insulin (Gibco). All media were additionally supplemented with penicillin (100 IU/ml; Gibco) and Streptomycin (100 μg/ml; Gibco). The growth medium was changed every 2 to 3 days, and cells were subcultured by trypsinization (trypsin–EDTA solution; Gibco) when they became confluent. For experiments, cells were harvested by trypsinization from confluent cultures, counted, and adjusted to the desired concentration, i.e., RL95-2 700,000 cells and HEC-1-A 200,000 cells each in 2.0 ml of their respective culture medium (Fig. 2A and 2B). Subsequently, suspended cells were poured out on poly-D-lysine-coated glass coverslips (12 mm in diameter) situated in 4 cm^2 wells. Cells were grown in medium to confluent monolayers and transferred into a Petri dish before used for experiments.

B. JAR Cell Culture on Cantilever

Cantilevers mounted with sephacryl microspheres, as described earlier, were immersed in 0.01% poly-D-lysine for 1 h at room temperature, washed in medium several times, and subsequently incubated with a human JAR choriocarcinoma cell suspension (ATCC: HTB 144 (Patillo *et al.*, 1971)) (200,000 cells/ml RPMI 1640 medium, Gibco, supplemented with 10% fetal calf serum and 0.1% glutamine). After JAR cells had settled, these cantilever–cell combinations were incubated in 5% CO_2–95% air at 37°C. Usually 3 to 4 days after the start of the cultures, cells were grown to confluency and cantilevers were ready to be used for the experiments.

C. *Dictyostelium* Cell Culture

All mutants were derived from the *D. discoideum* AX2-214 strain, here designated as wild-type. Mutant HG1287 was generated by E. Wallraff (Beug, Katz, and Gerish, 1973). In mutant HG1287, csA expression was eliminated by a combination of chemical and UV mutagenesis. In this mutant not only the csA gene but also other genes may have been inactivated by this shot-gun type of mutagenesis. Cells were cultivated in nutrient medium as described (Malchow *et al.*, 1972) in Petri dishes up to a density of 1×10^6 cells/ml. For transformants HTC1 (Barth *et al.*, 1994), CPH (Beug, Katz, and Gerish, 1973), and T10 (Faix *et al.*, 1992), 20 μg/ml of the selection marker G418 was added to stabilize csA expression. Before measurements were taken, cells were washed and resuspended in 17 mM K/Na buffer, pH 6.0, and used either immediately as undeveloped cells or after shaking for about 6 h at 150 rpm as developed cells. The temperature was about 20°C. For the measurement, cells were suspended in 17 mM K/Na phosphate buffer, pH 6.0, and spread on polystyrene Petri dishes, 3.5 cm in diameter, at a density of about 100 cells/mm^2. To chelate Ca^{2+}, 5 mM ethylendiaminotetraacetic acid (EDTA) was added at pH 6.0 in the same buffer. To avoid laser beam scattering of the detection system, nonadherent cells were removed by gently rinsing the dish after 10 min.

V. Final Remarks

The two concepts of either monolayer interactions or single-cell interactions illuminate complementary aspects of the complex cellular adhesion mechanisms. By reducing the complexity, as in the case of measurements between individual *Dictyostelium* cells, processes on the single molecular level are resolved. And the principle of gaining adhesion strength by oligomerization of molecular binding partners can be assumed from these measurements. Insights into the complexity of molecular arrangements, during cell adhesion processes, become possible by the measurements between interacting monolayers.

Bond rupture experiments are performed under nonequilibrium conditions, thus the measured forces are rate dependent. As shown by several groups (Grubmüller *et al.*, 1995; Merkel *et al.*, 1999; Rief *et al.*, 1998), this rate dependence may reveal additional information on the binding potential. For living cells this detailed analysis will be important to relate cell adhesion to the rate of cell movement or shear forces in the blood stream (Chen and Springer, 1999).

The combination of nanophysics with cell biology establishes a mechanical assay that relates qualitatively cooperative molecular processes during contact formation, or even quantitatively the expression of a gene, to the function of its product in cell adhesion. This type of single-molecule force spectroscopy on live cells is directly applicable to a variety of different cell adhesion systems. A wide field of applications of this cell-based molecular assay is predictable, for instance, in investigating mutated cell adhesion proteins or coupling of cell adhesion molecules to the cytoskeleton and also in the evaluation of adhesion-blocking drugs. Furthermore, not only initial steps in the receptor-mediated adhesion of particles to phagocyte surfaces but also interaction of cells with natural and artificial surfaces of medical interest can be measured with this technique.

Acknowledgments

This work became possible only through collaborations with M. Thie, R. Röspel, B. Maranca-Nowak, and U. Trottenberg at the Uni-Klinikum Essen in H.-W. Denker's institute; D. Gabriel, E. Simmeth, and M. Westphal at the MPI-Martinsried in G. Gerisch's institute; M. Grandbois at the University of Missouri-Columbia; and W. Dettmann, A. Wehle, and A. Kardinal in the LMU München at H. E. Gaub's institute. We are also grateful to the Deutsche Forschungsgemeinschaft and the Volkswagenstiftung for funding.

References

Albers, A., Thie, M., Hohn, H.-P., and Denker, H.-W. (1995). Differential expression and localization of integrins and CD44 in the membrane domains of human uterine epithelial cells during the menstrual cycle. *Acta Anatom.* **153,** 12–19.

Barth, A., Müller-Taubenberger, A., Taranto, P., and Gerisch, G. (1994). Replacement of the phospholipid-anchor in the contact site A glycoprotein of *Dictyostelium discoideum* by a transmembrane region does not impede cell adhesion but reduces residence time on the cell surface. *J. Cell Biol.* **124,** 205–215.

Baumgartner, W., Hinterdorfer, P., Ness, W., Raab, A., Vestweber, D., Schindler, H., and Drenckhahn, D. (2000). Cadherin interaction probed by atomic force microscopy. *PNAS* **97,** 4005–4010.

Benoit, M., Gabriel, D., Gerisch, G., and Gaub, H. E. (2000). Discrete molecular interactions in cell adhesion measured by force spectroscopy. *Nature Cell Biol.* **2,** 313–317.

Beug, H., Katz, F. E., and Gerisch, G. (1973). Dynamics of antigenic membrane sites relating to cell aggregation in *Dictyostelium discoideum*. *J. Cell Biol.* **56,** 647–688.

Beug, H., Katz, F. E., Stein, A., and Gerisch, G. (1973). Quantitation of membrane sites in aggregating Dictyostelium cells by use of tritiated univalent antibody. *Proc. Natl. Acad. Sci. U.S.A.* **70,** 3150–3154.

Binnig, G., Quate, C. F., and Gerber, C. (1986). Atomic force microscope. *Phys. Rev. Lett.* **56,** 930–933.

Bruinsma, R., Behrisch, A., and Sackmann, E. (2000). Adhesive switching of membranes: Experiment and theory. *Phys. Rev. E.* **61,** 4253–4267.

Chen, A., and Moy, V. T. (2000). Cross-linking of cell surface receptors enhances cooperativity of molecular adhesion. *Biophys. J.* **78,** 2814–2833.

Chen, S., and Springer, T. A. (1999). An automatic breaking system that stabilizes leukocyte rolling by an increase in selectin bond number with shear. *J. Cell Biol.* **144,** 185–200.

Choquet, D., Felsenfeld, D. P., and Sheetz, M. P. (1997). Extracellular matrix rigidity causes strengthening of integrin-cytoskeleton linkages. *Cell* **88,** 39–48.

Curtis, A. S. G. (1970). Problems and some solutions in the study of cellular aggregation. *Symp. Zool. Soc. London* **25,** 335–352.

Dai, J., and Sheetz, M. P. (1998). Cell membrane mechanics. *In* "Methods Cell Biology," (M. P. Sheetz, ed.), Vol. 55, pp. 157–171. Academic Press, San Diego.

Denker, H.-W. (1994). Endometrial receptivity: cell biological aspects of an unusual epithelium. *Ann. Anat.* **176,** 53–60.

Dettmann, W., Grandbois, M., Andrè, S., Benoit, M., Wehle, A. K., Kaltner, H., Gabius, H.-J., and Gaub, H. E. (2000). Differences in zero-force and force-driven kinetics of ligand dissociation from β-galactoside-specific proteins (plant and animal lectins, immunoglobulin G) monitored by plasmon resonance and dynamic single molecule force microscopy. *Arch. Biochem. Biophys.* **383,** 157–170.

Domke, J., Dannöhl, S., Parak, W. J., Müller, O., Aicher, W. K., and Radmacher, M. (2000). Substrate Dependent Differences in Morphology and Elasticity of Living Osteoblasts Investigated by Atomic Force Microscopy. *Colloids Surf. B Biointerfaces,* **19,** 367–379.

Evans, E. A. (1985). Detailed mechanics of membrane-membrane adhesion and separation II. Discrete kinetically trapped molecular cross-bridges. *Biophys. J.* **48,** 185–192.

Evans, E. (1995). Physical Actions in Biological Adhesion. *In* "Handbook of Biological Physics," (R.a.S., E. Lipowsky, ed.), Vol. 1B, pp. 723–754. Elsevier Science Amsterdam.

Evans, E., and Ritchie, K. (1997). Dynamic strength of molecular adhesion bonds. *Biophys. J.* **72,** 1541–1555.

Faix, J. (1999). Contact site A. *In* Guidebook to the Extracellular Matrix, Anchor, and Adhesion Proteins (T.K.a.R. Vale, ed.), Oxford Univ. Press, London.

Faix, J., Gerisch, G., and Noegel, A. A. (1992). Overexpression of the csA cell adhesion molecule under its own cAMP-regulated promoter impairs morphogenesis in *Dictyostelium. J. Cell Sci.* **102,** 203–214.

Felsenfeld, D. P., Choquet, D., and Sheetz, M. P. (1996). Ligand binding regulates the directed movement of betal integrins on fibroblasts. *Nature* **383,** 438–440.

Florin, E.-L., Moy, V. T., and Gaub, H. E. (1994). Adhesive forces between individual ligand-receptor pairs. *Science* **264,** 415–417.

Fritz, M., Radmacher, M., and Gaub, H. E. (1993). In vitro activation of human platelets triggered and probed by SFM. *Exp. Cell Res.* **205**(1), 187–190.

Gimzewski, J. K., and Joachim, C. (1999). Nanoscale science of single molecules using local probes. *Science* **283,** 1683–1688.

Goldmann, W. H., Galneder, R., Ludwig, M., Kromm, A., and Ezzell, R. (1998). Differences in F9 and 5.51 cell elasticity determined by cell poking and atomic force microscopy. *FEBS Lett.* **424,** 139–142.

Grandbois, M., Beyer, M., Rief, M., Clausen-Schaumann, H., and Gaub, H. E. (1999). How strong is a covalent bond? *Science* **283,** 1727–1730.

Grandbois, M., Dettmann, W., Benoit, M., and Gaub, H. E. (2000). Affinity imaging of red blood cells using an atomic force microscope. *J. Histochem. Cytochem.* **48,** 719–724.

Grubmüller, H., Heymann, B., and Tavan, P. (1995). Ligand binding: molecular mechanics calculation of the streptavidin-biotin rupture force. *Science* **271,** 997–999.

Harloff, C., Gerisch, G., and Noegel, A. A. (1989). Selective elimination of the contact site A protein of *Dictyostelium discoideum* by gene disruption. *Genes Dev.* **3,** 2011–2019.

Hinterdorfer, P., Baumgartner, W., Gruber, H. J., Schilcher, K., and Schindler, H. (1996). Detection and localization of individual antibody-antigen recognition events by atomic force microscopy. *Proc. Natl. Acad. Sci. U.S.A.* **93,** 3477–3481.

Hochmuth, R. M., Shao, J.-Y., Dai, J., and Sheetz, M. P. (1996). Deformation and flow of membrane into tethers extracted from neuronal growth cones. *Biophys. J.* **70,** 358–369.

Hoh, J. H., and Schoenenberger, C.-A. (1994). Surface morphology and mechanical properties of MDCK monolayers by atomic force microscopy. *J. Cell Sci.* **107,** 1105–1114.

Holmberg, M., Wigren, R., Erlandsson, R., and Claesson, P. M. (1997). Interactions between cellulose and colloidal silica in the presence of polyelectrolytes. *Colloids Surf. A Physicochem. Eng. Aspects* **129–130,** 175–183.

John, N., Linke, M., and Denker, H.-W. (1993). Quantitation of human choriocarcinoma spheroid attachment to uterine epithelial cell monolayers. *In Vitro Cell. Dev. Biol.* **29A,** 461–468.

Johnsson, B., Löfas, S., and Lindquist, G. (1991). Immobilization of proteins to a carboxymethyldextran-modified gold surface for biospecific interaction analysis in surface plasmon resonance sensors. *Anal. Biochem.* **198,** 268–277.

Kamboj, R. K., Wong, L. M., Lam, T. Y., and Siu, C.-H. (1988). Mapping of a cell-binding domain in the cell adhesion molecule gp80 of Dictyostelium discoideum. *J. Cell Biol.* **107,** 1835–1843.

Kreis, T., and Vale, R. (eds.) (1999). "Guidebook to the Extracellular Matrix, Anchor, and Adhesion Proteins." Oxford Univ. Press, London.

Kuo, S. C., Hammer, D. A., and Lauffenburger, D. A. (1997). Simulation of detachment of specifically bound particles from surfaces by shear flow. *Biophys. J.* **73,** 517–531.

Kuramoto, H., Tamura, S., and Notake, Y. (1972). Establishment of a cell line of human endometrial adeno-carcinoma in vitro. *Am. J. Obstet. Gynecol.* **114,** 1012–1019.

Malchow, D., Nägele, B., Schwarz, H., and Gerisch, G. (1972). Membrane-bound cyclic AMP phosphodi-esterase in chemotactically responding cells of *Dictyostelium discoideum. Eur. J. Biochem. Eur. J. Biochem.* **28,** 136–142.

Marszalek, P. E., Pang, Y. P., Li, H., Yazal, Y. E., Oberhauser, A. F., and Fernandez, J. M. (1999). Atomic levers control pyranose ring conformations. *Proc. Natl. Acad. Sci. U.S.A.* **96.**

Merkel, R., Nassoy, P., Leung, A., Ritchie, K., and Evans, E. (1999). Energy landscapes of receptor-ligand bonds explored with dynamic force spectroscopy. *Nature* **397,** 50–53.

Müller, K. M., Arndt, K. M., and Plückthun, A. (1988). Model and simulation of multivalent binding to fixed ligands. *Anal. Biochem.* **261,** 149–158.

Müller, D. J., Baumeister, W., and Engel, A. (1999). Controlled unzipping of a bacterial surface layer with an AFM, Nov 9:96 (23), p. 13170–13174. PNAS.

Murray, B. A., Yee, L. D., and Loomis, W. F. (1981). Immunological analysis of glycoprotein (contact sites A) involved in intercellular adhesion of Dictyostelium discoideum. *J. Supramol. Struct. Cell. Biochem.* **17,** 197–211.

Oberhauser, A. F., Marszalek, P. E., Erickson, H. P., and Fernandez, J. M. (1998). The molecular elasticity of the extracellular matrix protein tenascin. *Nature* **393,** 181–185.

Oesterhelt, F., Oesterhelt, D., Pfeiffer, M., Engel, A., Gaub, H. E., and Müller, D. J. (2000). Unfolding pathways of individual Bacteriorhodopsins. *Science* **288,** 143–146.

Patillo, R. A., Ruckert, A., Hussa, R., Bernstein, R., and Delfs, E. (1971). The JAR cell line—Contiuous human multihormone production and controls. *In Vitro* **6,** 398.

Ponte, E., Bracco, E., Faix, J., and Bozzaro, S. (1998). Detection of subtle phenotypes: The case of the cell adhesion molecule csA in Dictyostelium. *Proc. Natl. Acad. Sci. U.S.A.* **95,** 9360–9365.

Radmacher, M. (1997). Measuring the elastic properties of biological samples with the atomic force microscopy. *IEEE Eng. Med. Biol.* **16.**

Radmacher, M., Fritz, M., Kacher, C. M., Cleveland, J. P., and Hansma, P. K. (1996). Measuring the visco-elastic properties of human platelets with the atomic force microscope. *Biophys. J.* **70,** 556–567.

Razatos, A., Ong, Y.-L., Sharma, M. M., and Georgiou, G. (1998). Molecular determinats of bacterial adhesion monitored by AFM. *PNAS* **95,** 11,059–11,064.

Rief, M., Clausen-Schaumann, H., and Gaub, H. E. (1999). Sequence dependent mechanics of single DNA-molecules. *Nature Struct. Biol.* **6,** 346–349.

Rief, M., Fernandez, J. M., and Gaub, H. E. (1998). Elastically coupled two-level systems as a model for biopolymer extensibility. *Phys. Rev. Lett.* **81,** 4764–4767.

Rief, M., Gautel, M., Oesterhelt, F., Fernandez, J. M., and Gaub, H. E. (1997). Reversible unfolding of individual titin Ig-domains by AFM. *Science* **276,** 1109–1112.

Rief, M., Oesterhelt, F., Heymann, B., and Gaub, H. E. (1997). Single molecule force spectroscopy on polysac-charides by AFM. *Science* **275,** 1295–1298.

Sagvolden, G., Giaver, I., Pettersen, E. O., and Feder, J. (1999). Cell adhesion force microscopy. *Proc. Natl. Acad. Sci. U.S.A.* **96,** 471–475.

Smith, B. L., Schäffer, T. E., Viani, M., Thompson, J. B., Frederick, N. A., Kindt, J., Belcher, A., Stucky, G. D., Morse, D. E., and Hansma, P. K. (1999). Molecular mechanistic origin of the toughness of natural adhesives, fibres and composites. *Nature* **399,** 761–763.

Springer, T. A. (1990). Adhesion receptors of the immune system. *Nature* **346,** 425–434.

Stadler, J., Keenan, T. G., Bauer, G., and Gerisch, G. (1989). The contact site A glycoprotein of *Dictyostelium discoideum* carries a phospholipid anchor of a novel type. *EMBO J.* **8,** 371–377.

Strunz, T., Oroszlan, K., Schäfer, R., and Güntherodt, H.-J. (1999). Dynamic force spectroscopy of single DNA molecules. *Proc. Natl. Acad. Sci. U.S.A.* **96,** 11,277–11,282.

Suter, C. M., Errante, L. E., Belotserkovsky, V., and Forscher, P. (1998). The Ig superfamily cell adhesion molecule, apCAM, mediates growth cone steering by substrate-cytoskeletal coupling. *J. Cell Biol.* **141,** 227–240.

Thie, M., Fuchs, P., Butz, S., Sieckmann, F., Hoschützky, H., Kemler, R., and Denker, H.-W. (1996). Adhesiveness of the apical surface of uterine epithelial cells: The role of junctional complex integrity. *Eur. J. Cell Biol.* **70,** 221–232.

Thie, M., Harrach-Ruprecht, B., Sauer, H., Fuchs, P., Albers, A., and Denker, H.-W. (1995). Cell adhesion to the apical pole of epithelium: a function of cell polarity. *Eur. J. Cell Biol.* **66,** 180–191.

Thie, M., Herter, P., Pommerenke, H., Dürr, F., Sieckmann, F., Nebe, B., Rychly, J., and Denker, H.-W. (1997). Adhesiveness of the free surface of a human endometrial monolayer as related to actin cytoskeleton. *Mol. Hum. Reprod.* **3,** 275–283.

Thie, M., Röspel, R., Dettmann, W., Benoit, M., Ludwig, M., Gaub, H. E., and Denker, H.-W. (1998). Interactions between trophoblast and uterine epithelium: Monitoring of adhesive forces. *Hum. Reprod.* **13,** 3211–3219.

Tomschy, A., Fauser, C., Landwehr, R., and Engel, J. (1996). Homophilic adhesion of E-cadherin occurs by a co-operative two-step interaction of N-terminal domains. *EMBO J.* **15,** 3507–3514.

Vestweber, D., and Blanks, J. E. (1999). Mechanisms that regulate the function of the selectins and their ligands. *Physiological Rev.* **79,** 181–213.

Ward, M. D., Dembo, M., and Hammer, D. A. (1994). Kinetics of cell detachment: Peeling of discrete receptor clusters. *Biophys. J.* **67,** 2522–2534.

Ward, M. D., and Hammer, D. A. (1993). A theoretical analysis for the effect of focal contact formation on cell-substrate attachment strength. *Biophys. J.* **64,** 936–959.

Way, D. L., Grosso, D. S., Davis, J. R., Surwit, E. A., and Christian, C. D. (1983). Characterization of a new human endometrial carcinoma (RL95-2) established in tissue culture. *In Vitro* **19,** 147–158.

Willemsen, O. H., Snel, M. M. E., van der Werf, K. O., de Grooth, B. G., Greve, J., Hinterdorfer, P., Gruber, H. J., Schindler, H., van Kooyk, Y., and Figdor, C. G. (1998). Simultaneous height and adhesion imaging of antibody-antigen interactions by atomic force microscopy. *Biophys. J.* **75,** 2220–2228.

Yauch, R. L., Felsenfeld, D. P., Kraeft, S.-K., Chen, L. B., Sheetz, M. P., and Hemler, M. E. (1997). Mutational evidence for control of cell adhesion through integrin diffusion/clustering, independent of ligand binding. *J. Exp. Med.* **186,** 1347–1355.

Zahalak, G. I., McConnaughey, W. B., and Elson, E. L. (1990). Determination of cellular mechanical properties by cell poking, with an application to leukocytes. *J. Biomech. Eng.* **112,** 283–294.

CHAPTER 6

Molecular Recognition Studies Using the Atomic Force Microscope

Peter Hinterdorfer

Institute for Biophysics
University of Linz
A-4040 Linz, Austria

I. Introduction

The potential of the atomic force microscope (AFM) (Binnig *et al.*, 1986) to measure ultralow forces at high lateral resolution has paved the way for molecular recognition studies. The AFM offers particular advantages in biology: measurements can be carried out in both aqueous and physiological environments, and the dynamics of biological processes *in vivo* can be studied. Since structure–function relationships play a key role in bioscience, their simultaneous detection is a promising approach to yielding novel insights into the regulation of cellular and other biological mechanisms. Ligand binding to receptors is one of the most important regulatory elements since it is often the initiating step in reaction pathways and cascades.

Molecular recognition studies provide insight into both detecting specific ligand–receptor interaction forces on the single molecule level and observing molecular recognition of a single ligand–receptor pair. Applications include biotin–avidin (Lee, Kidwell *et al.*, 1994; Florin *et al.*, 1994; Wong *et al.*, 1998), antibody–antigen (Hinterdorfer *et al.*, 1996, 1998; Dammer *et al.*, 1996; Allen *et al.*, 1997; Willemsen *et al.*, 1998; Ros *et al.*, 1998), sense–antisense DNA (Lee, Chrisey *et al.*, 1994; Boland and Ratner, 1995; Strunz *et al.*, 1999), nitrilotriacetate–histidine 6 (NTA–His$_6$) (Conti *et al.*, 2000; Kienberger, Kada *et al.*, 2000; Schmitt *et al.*, 2000), and cellular proteins, either isolated (Dammer *et al.*, 1996; Fritz *et al.*, 1998; Baumgartner, Hinterdorfer, Ness *et al.*, 2000) or in cell membranes (Lehenkari and Horton, 1999; Chen and Moy, 2000; Wielert-Badt *et al.*, 2000). The general strategy is to bind ligands to AFM tips and receptors to probe surfaces (or vice versa), respectively. In a force–distance cycle, the tip is first approached to the surface whereupon receptor–ligand complexes are formed, due to the specific ligand–receptor recognition. During subsequent tip–surface retraction a temporarily increasing force is applied to the ligand–receptor connection until the interaction bond breaks at a critical force (unbinding force).

Such experiments allow for estimation of affinity, rate constants, and structural data of the binding pocket (Hinterdorfer *et al.*, 1996, 1998; Baumgartner, Hinterdorfer, Ness *et al.*, 2000; Kienberger, Kada, Gruber *et al.*, 2000), and comparing these values with those obtained from ensemble-average techniques and binding energies (Moy *et al.*, 1994; Chilkoti *et al.*, 1995) is of particular interest. Several years ago, theoretical findings determined that the unbinding force was dependent on the rate of increasing force (Grubmüller *et al.*, 1996; Evans and Ritchie, 1997; Izraelev *et al.*, 1997) during force–distance cycles. Recent experimental studies confirmed the theoretical findings and revealed a logarithmic dependence of the unbinding force on the loading rate (Merkel *et al.*, 1999; Struntz *et al.*, 1999; Baumgartner, Hinterdorfer, Ness *et al.*, 2000; Kienberger, Kada *et al.*, 2000). These force spectroscopy experiments provide insight into the molecular dynamics of the receptor–ligand recognition process (Baumgartner, Hinterdorfer, Ness *et al.*, 2000) and even render mapping of the interaction potential possible (Markel *et al.*, 1999). Similar experimental strategies were used for studying the elastic properties of polymers by applying external forces (Rief, Oesterhelt *et al.*, 1997; Marzsalek *et al.*, 1998; Oesterhelt *et al.*, 1999; Kienberger, Patushenko *et al.*, 2000) and investigating unfolding–refolding kinetics of filamentous proteins in pull–hold–release cycles (Rief, Gautel *et al.*, 1997; Oberhauser *et al.*, 1998).

Aside from the study of ligand–receptor recognition processes, the localization of receptor binding sites by molecular recognition of a ligand is of particular interest. Simultaneous information for topography and ligand–receptor interaction is obtained by lateral force mapping (Ludwig *et al.*, 1997; Willemsen *et al.*, 1998). Recognition imaging, developed by combing dynamic force microscopy (Han *et al.*, 1996, 1997) with force spectroscopy, allows for the determination of receptor sites with nanometer positional accuracy (Raab *et al.*,1999). This presents new perspectives for nanometer-scale epitope mapping of biomolecules and localizing receptor sites during biological or cellular processes.

In this chapter, the principles of force spectroscopy and recognition imaging are described. Several protocols for anchoring ligands to tips and receptors to probe surfaces are given. Applications of these methodologies to cellular proteins, i.e., (i) the vascular endo-thelian cadherin, a cell–cell adhesion protein, and (ii) the Na^+/D-glucose cotransporter, a nutrient transporting transporter protein, show the potential of molecular recognition force spectroscopy/microscopy in cell biology.

II. Experimental Approach

A. Surface Chemistry

1. Preparation of AFM Tips

The detection of unconstrained ligand–receptor recognition requires a particular link-age design (Hinterdorfer *et al.*, 1996, 1998; Raab *et al.*, 1999; Kienberger, Pastushenko *et al.*, 2000). Covalently coupling the ligand to the tip surface guarantees a sufficiently tight attachment because covalent bonds are about 10 times stronger than typical ligand–receptor bonds (Grandbois *et al.*, 1999). Additionally, the ligand is to be provided with maximal motional freedom around the tip, so that the recognition process is not influ-enced by steric restrictions. Therefore, we developed a strategy for the covalent anchoring of ligands to silicon (Si_3N_4 or SiOH) tips via a flexible crosslinker that enables the li-gand to move and orient freely about the tip and lacks unspecific tip–probe adhesion. It also makes site-directed coupling for a defined orientation of the ligand relative to the receptor possible.

As a crosslinking element, we used poly(ethylene glycol) (PEG), a water–soluble nontoxic polymer with a wide range of applications in surface technology and clinical research. PEG is known to prevent surface adsorption of proteins and lipid structures and appeared therefore ideally suited to our purpose. The flexible crosslinker was syn-thesized in our lab (Haselgrübler *et al.*, 1995) and consisted of a PEG chain of 24 units, corresponding to about an 8-nm extended length. The extension of the crosslinker is comparable to the size of antibodies, which were the most frequently used ligands in our group, and therefore represents a compromise between a sufficient spacing of the ligands from the tip surface and a high lateral and vertical resolution. The crosslinker is heterobifunctional, for the coupling to both the tip surface and the ligands, respectively. An *N*-hydroxysuccinimidyl (NHS) residue on the one end is reactive to amines on the tip, and a 2-pyridyldithiopropionyl (PDP) residue on the other end can be covalently bound to thiols. The ligand density on the tip is adjusted to a value ($\approx 500/\mu m^2$) where only one ligand on the tip is expected to have access to the receptors on the probe. Therefore, single-molecule experiments can be carried out with the described tip sensor design.

The AFM tips are functionalized with ligands, using a thorough cleaning protocol and a three-step binding mechanism. The configuration of the ligand-modified tip is depicted in Fig. 1.

a. Cleaning

Prior to functionalization, the AFM tips (Park, Sunnyvale, CA; MacLevers, Molecular Imaging, Phoenix, AZ) are cleaned in a thorough four-step procedure (Hinterdorfer *et al.*, 1996). The wafers are first defatted in chloroform for 10 min and dried with N_2. They are then incubated in piranha solution (H_2SO_4/H_2O_2, 90/10 (v/v)) for 30 min (except

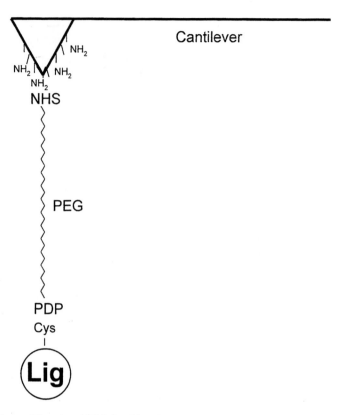

Fig. 1 Linkage of ligands to AFM tips. Ligands are covalently coupled to AFM tips via a heterobi-functional polyethylene glycol (PEG) derivative of 8 nm length. Silicon tips are first functionalized with ethanolamine (NH_2–C_2H_4OH·HCl). Then, the NHS end of the PEG linker is covalently bound to amines on the tip surface before ligands are attached to the PDP end via a free cysteine. Reproduced with permission from Hinterdorfer, P., Kienberger, F., Raab, A., Gruber, H. J., Baumgartner, W., Kada, G., Riener, C., Wielert-Badt, S., Borken, C., and Schindler, H. (2000). Poly(ethylene glycol): An ideal spacer for molecular recognition force microscopy/spectroscopy. *Single Mol.* **1**, 99–103.

MacLevers) and subsequently rinsed with about 100 ml of deionized water before they are dried with N_2. For a final cleaning step and regeneration of the SiOH groups on the tip surface, tips are optionally put in water plasma (Kiss and Gölander, 1990) (Harrick Sci. Corp., Ossining, NY) and immediately used afterwards.

b. Esterification

In the first functionalization step, amines are bound to tip surfaces according to an esterification protocol with slight modifications (Hinterdorfer *et al.*, 1996, 1998). Thirty percent (mol/mol) 2-aminoethanol-Cl is melted in dry dimethyl sulfoxide at 100°C in the presence of 0.3-nm molecular sieve beads. After the solution is allowed to cool down to room temperature, tips are added and incubated for 15 h before they are washed in bare dimethyl sulfoxide and dried with N_2. Such amine-modified tips are stable for weeks when stored in a desiccator.

c. Crosslinker Binding

The crosslinker, $NHS–PEG_{24}–PDP$, is conjugated to amines on AMF tip surfaces via its NHS end. Amine-containing tips are incubated at a concentration of 1–3 mg/ml $NHS–NH–PEG_{24}–PDP$ in $CHCl_3$ containing 0.5% (v/v) triethylamine for 1–3 h at room temperature in an Ar^{2+}-saturated atmosphere. Immediately after washing in $ChCl_3$ and drying with Ar^{2+}, the reaction protocol was followed by the ligand binding step.

d. Ligand Binding

Ligands are bound via free thiols (SH) to the PDP end of the PEG derivative. This type of chemistry is highly advantageous since it is very reactive and renders site-directed coupling possible. However, free thiols are hardly available on native ligands and must therefore be generated.

For this we use three different strategies: (i) Amines of ligands, in particular lysins, are derivatized with N-succinnimidyl-3-(S-acethylthio)propionate (SATP) by incubating the ligands in a ~10-fold molar access of SATP in buffer and by subsequent removal of free SATP by gel exclusion chromatography (Haselgrübler et al., 1995; Hinterdorfer et al., 1996, 1998). Deprotection of the SH groups with NH_2OH leads to reactive groups. Since it is very difficult to react distinct amines with this method, the coupling to the crosslinker is often not specifically site directed. (ii) Half-antibodies are produced by cleaving the two disulfides in the central region of the heavy chain using 2-mercaptoethylamine HCl (Sigma, Vienna, Austria) according to a standard procedure (Pierce, Rockford, IL). The half-antibody is then coupled to the PDP end of the crosslinker via one of the two neighboring cysteines (Raab et al., 1999). (iii) The most elegant method is to mutate a cysteine into the primary sequence of proteins because it allows for a defined sequence-specific coupling of the ligand to the crosslinker.

For all three coupling strategies described earlier, ligands carrying free thiols are reacted to the PDP end of the crosslinker at a concentration of 1–10 μM for 1–3 h in a buffer that represses oxidation (1 mM EDTA in phosphate-buffered saline). Ligand-functionalized tips are stored in buffer in a cold room and retain their functionality over several weeks.

A nice alternative for a most common noncovalent, site-directed high-affinity-binding anchor with large bond strength on the tether has been recently introduced. The binding strength of the NTA–His[6] system, routinely used on chromatographic and biosensor matrices for the binding of recombinant proteins to which a His_6 tag is appended to the primary sequence, was found to be significantly large than typical values of other ligand–receptor systems (Conti et al., 2000; Kienberger, Kada et al., 2000). Therefore, a PEG crosslinker containing an NTA residue, instead of the PDP group, is ideally suited for coupling a recombinant ligand, carrying His_6 in its sequence, to the AMF tip. This general, side-directed, and oriented coupling strategy also allows rigid and fast control of the specificity of ligand–receptor recognition by using Ni^{2+} as a molecular switch of the NTA–His_6 bond.

e. Ligand Density and Functionality

Silicon (Si_3N_4 or SiOH, respectively) substrates (size ≈ 1 cm^2) are treated in parallel with the AFM tips for the determination of the macroscopic ligand density on the surfaces. Three different methods were employed to investigate the number density of the ligands. (i) Antibodies were directly fluorescence labeled prior to their conjugation to surfaces. The substrates were inserted in a wide-field epifluorescence microscope and the fluorescence intensity was measured with sensitive high-resolution fluorescence imaging using a nitrogen-cooled CCD camera. The ligand surface densities were calculated after accurate single fluorophore calibration (Hinterdorfer et al., 1996; Schmidt et al., 1996). (ii) Alternatively, fluorescence-labeled secondary antibodies were ligated to the F_c portion of surface-bound primary antibodies and the F_c density was determined as described in (i). (iii) The ligand site density was determined by an enzyme immunoassay (EIA) similar to that used by Hinterdorfer et al. (1998). Horseradish peroxidase (HRP) antibodies directed to the ligands were bound to the surface, and the enzyme activity was measured in a spectrophotometer. Enzyme densities were calculated after calibration with anti-rabbit–horseradish peroxidase antibody in solution.

The latter two methods provide the advantage in testing the functionality of the ligands on the surface, while the first determines only the total number density. Under our standard conditions, values between 200 and 500/μm^2 are usually obtained with all three protocols. For a typical AFM tip radius of 20 to 50 nm, this value corresponds to about one ligand per effective tip area, which appears to be suited for single-molecule experiments.

2. Probe Surfaces

For the recognition by ligands of the AFM tip, receptors are tightly attached to probe surfaces. Loose receptor fixation could lead to a pull-off of the receptor from the surface by the ligand on the tip, which would consequently block ligand–receptor recognition. The different surface-binding strategies used must be adjusted to the respective properties of the biological samples.

a. Isolated Components

Ideally, water-soluble receptors like either globular antigenic proteins (Hinterdorfer et al., 1996) or extracellular protein chimeras (Baumgartner, Hinterdorfer, Ness et al., 2000) are covalently anchored. When silicon or mica is used as a probe surface, exactly the same surface chemistry is employed for the AFM tips (cf. Section I,A,1). Therefore, the receptor is also provided with motional freedom, which guarantees unconstrained ligand–receptor recognition. The purification step is omitted for mica; instead it is freshly cleaved prior to use. In addition, the number of reactive SiOH groups of the chemically relatively inert mica is optionally increased by water plasma treatment (Kiss and Gölander, 1990) (Harrick Sci. Corp., Ossining, NY).

Some receptor proteins strongly adhere to mica (Raab et al., 1999) via either hydrophobic or electrostatic interaction, in which case it is safe to purely adsorb the receptors from the solution, since the unspecific attachment to the surface is sufficiently strong for recognition force experiments. Electrostatic interaction via Ca^{2+} bridges was also used to adsorb ion channels in a defined orientation to mica (Kada et al., 2000).

Another possibility of binding biomolecules to surfaces is through sulfur–gold chemistry (Dammer *et al.*, 1996; Ros *et al.*, 1998). This strategy has also been used for binding ligands to gold-coated tips. Gold wafers with atomically flat surfaces are perfect probes for AFM because they allow direct anchoring of isolated receptors via free thiols. Receptors on hydrophic chains can be incorporated into self-assembled monolayers (SAM) that form spontaneously on gold and, additionally, can be covalently bound via an SH group on the chain end (Kienberger, Kada *et al.*, 2000). In this way, well-defined surfaces with accurate adjustable lateral densities of reactive sites can be prepared.

b. Membranes and Cells

Various protocols for tight cell anchoring are available. The easiest method for tight cell anchoring is to (i) either grow the cells directly on glass or other surfaces in their cell culture medium (Le Grimellec *et al.*, 1998) or (ii) simply adsorb the cells via adhesive coating like Cell-Tak (Schilcher *et al.*, 1997), gelatin, and poly-lysin. Other hydrophic surfaces like gold or carbon are suitable matrices as well (Wielert-Badt *et al.*, 2000). Covalent binding of cells to surfaces can be accomplished by using PEG crosslinkers similar to those described for tip chemistry, since they react with free thiols on the cell surface (Schilcher *et al.*, 1997). Alternatively, PEG crosslinkers carrying a fatty acid penetrate into the interior of the cell membrane which guarantees a sufficiently strong fixation without interference with membrane proteins (Schilcher *et al.*, 1997). Using glass or mica surfaces, model membranes can be prepared either by vesicle fusion (Kalb *et al.*, 1992) or by the Langmuir–Blodgett technique (Kalb *et al.*, 1992); both result in supported lipid bilayers. With reconstitution techniques, membrane proteins can be embedded into such artificial membranes (Hinterdorfer *et al.*, 1994).

B. Unbinding Force Measurements

1. Force–Distance Cycle

Single-molecule ligand–receptor recognition events are measured in force–distance cycles (Fig. 2a). At a fixed lateral position, a cantilever carrying a ligand is moved toward a probe surface to which receptors are attached and subsequently retracted. The cantilever deflection Δx is measured independent of the tip–surface separation Δz. The force F acting on the cantilever directly relates to the cantilever deflection Δx according to $F = k \, \Delta x$, where k is the cantilever spring constant.

During the tip–surface approach (trace, dashed line) the cantilever deflection remains at zero far away from the surface because there is no detectable tip–surface interaction. At a sufficiently close tip–surface separation, the antibody on the tip has a chance to bind to a receptor on the surface. Upon tip–surface contact ($\Delta z = 0$ nm) a repulsive force develops that increases the harder the tip is pushed into the surface. Subsequent tip–surface retraction (retrace, solid line) leads to relaxation of the repulsive force.

When ligand–receptor binding has occurred, an attractive force develops (unbinding event) in the retrace ($\Delta z = 0$–21 nm) and increases with increasing tip–surface separation. Its shape, determined by the elastic properties of the flexible PEG crosslinker (Kienberger, Pastushenko *et al.*, 2000; Hinterdorfer *et al.*, 2000), shows a nonlinear, parabolic-like characteristic which reflects the increase of the spring constant of the

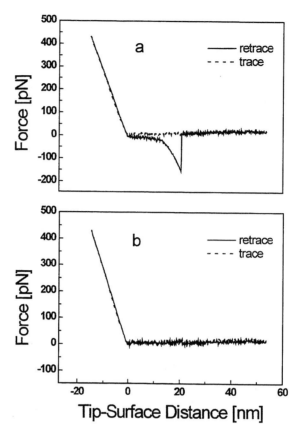

Fig. 2 Single-molecule recognition event. (a) Raw data from a force–distance cycle with a 100-nm z amplitude at 0.9 Hz measured in PBS. The attractive force signal developing in the retrace (0 nm) reflects single-molecule recognition of a receptor on a surface by a ligand on the tip. (b) Force–distance cycle lacking a molecular recognition event. Ligands in solution block receptor binding sites on the surface. Reproduced with permission from Hinterdorfer, P., Kienberger, F., Raab, A., Gruber, H. J., Baumgartner, W., Kada, G., Riener, C., Wielert-Badt, S., Borken, C., and Schindler, H. (2000). Poly(ethylene glycol): An ideal spacer for molecular recognition force microscopy/spectroscopy. *Single Mol.* **1**, 99–103.

crosslinker during extension. Therefore, specific ligand–receptor recognition is easily distinguishable from the linearly shaped, eventually occurring nonspecific tip–surface adhesion signals. The physical connection between tip and surface sustains the increasing force until the ligand–receptor complex dissociates at a certain critical force (unbinding force), and the cantilever finally jumps back to the resting position (at $\Delta z = 21$ nm). The quantitative force measure of the unbinding force of a single ligand–receptor pair is directly given by the force at the moment of unbinding ($\Delta z = 21$ nm).

The specificity of ligand–receptor binding is demonstrated in block experiments (Fig. 2b). Free ligands are injected into a solution so as to block receptor sites on the surfaces. The ligand–receptor recognition signal completely disappears and retrace looks like trace. Apparently, the receptor sites on the surface are blocked by the ligand of the solution, and thus prevent recognition by the ligand on the tip.

2. Unbinding Force Distribution

Hundreds of force–distance cycles are usually recorded to quantify the unbinding force. No deterioration of ligand binding is found, even after storage in buffer for weeks, indicating that the design of the AFM tip sensor is highly stable. Force–distance cycles are stored in digitized form and normalized to a slope of $-k$ in the contact region, where k is the spring constant of the cantilever. Unbinding events are detected using a transition detection algorithm (Baumgartner, Hinterdorfer, and Schindler, 2000) similar to a method for event detection in patch-clamp data. Since full cantilever relaxation is required for reliable height detection, only the last event yielding the unbinding force was used for further analysis.

Distributions of unbinding forces (Fig. 3) are obtained by constructing empirical probability density functions from unbinding force measurements (Hinterdorfer *et al.*, 1996; Baumgartner, Hinterdorfer, and Schindler, 2000b). Single Gaussian functions of unitary area are calculated from the mean and variances of every value of the unbinding force. The Gaussian functions are added up and finally normalized, yielding the empirical probability density function. The advantage of this representation over simple histograms is that the data are weighted by their accuracy, thus yielding a better resolution. Values of unbinding forces give a Gaussian-like distribution (Fig. 3); for example, the maximum is $f \pm \sigma_u = 150 \pm 38$ pN (mean $\pm$ SD). The uncertainty in determining f_u values, given

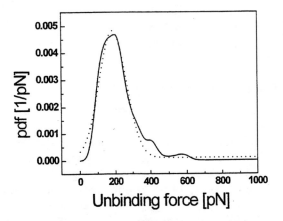

Fig. 3 Distribution of unbinding forces. An empirical probability density function (pdf, solid line) was constructed from about 150 values of unbinding forces (for details see Experimental Approach) obtained in force–distance cycles. Data were fitted with a Gaussian function (dotted line). Reproduced with permission from Raab, A., Han, W., Badt, D., Smith-Gill, S. J., Lindsay, S. M., Schindler, H., and Hinterdorfer, P. (1999). Antibody recognition imaging by force microscopy. *Nature Biotechnol.* **17**, 902–905.

by the thermal noise of the cantilever, was $\sigma_0 \sim 10\,\text{pN}$ for the cantilever used. Therefore, unbinding forces were detectable at a signal-to-noise ratio of $f/\sigma_0 = 15$.

III. Dynamic Force Spectroscopy

A. Principles

1. Bond Lifetime

Ligand–receptor binding is generally a reversible reaction. The average lifetime of a ligand–receptor bond, τ_0, is given by the kinetic offrate k_{off}, according to $\tau_0 = k_{\text{off}}^{-1}$. A force acting on a binding complex essentially reduces its lifetime. At the millisecond time scale of AFM experiments, thermal impulses govern the unbinding process. In the thermal activation model, the lifetime $\tau(f)$ of a bond loaded with a force f is written as $\tau(f) = \tau_{\text{osc}} * \exp((E_b - l * f)/k_B * T)$ (Bell, 1978), where τ_{osc} is the inverse of the natural oscillation frequency, E_b is the energy barrier for dissociation, and l is the effective length of the bond. Consequently, the lifetime $\tau(f)$ under force f compares to the lifetime at zero force, τ_0, according to $\tau(f) = \tau_0 * \exp(-l_r * f/k_B * T)$ (Hinterdorfer et al., 1996).

From unbinding force distributions (cf. Fig. 3), an effective lifetime $\tau(f)$ of the bond under an applied force f can be estimated by the time the cantilever spends in the force window spanned by the standard deviation σ_U of the f_u distribution (Hinterdorfer et al., 1996). The time the force increases from $f - \sigma_U$ to $f + \sigma_U$ is then given by $\tau(f) \approx 2\sigma_U/df/dt$ (Hinterdorfer et al., 1996). In a typical example of a ligand–receptor interaction described in Kienberger, Kada et al. (2000), the lifetime $\tau(f)$ decreased with increasing pulling force f from 17 ms at 150 pN to 2.5 ms at 194 pN. The data were fitted with the Boltzmann ansatz described previously, yielding the exponential lifetime–force relation for the reduction of the lifetime $\tau(f)$ by the applied force f. Data fit also yielded the lifetime at zero force, $\tau_0 = 15$ s, which corresponds to a kinetic offrate of $k_{\text{off}} = 6.7\ 10^{-2}\ \text{s}^{-1}$ (Kienberger, Kada et al., 2000).

2. Unbinding Force versus Loading Rate

Theoretical studies determined that the unbinding force of specific and reversible ligand–receptor bonds is dependent on the rate of the increasing force (Grubmüller et al., 1996, Evans and Ritchie, 1997, Izraelev et al., 1997) during force–distance cycles. In experiments, unbinding forces were found not to assume a unitary value but were rather dependent on both the pulling velocity and the cantilever spring constant (Lee, Kidwell et al., 1994). The theoretical findings were confirmed by experimental studies and revealed a logarithmic dependence of the unbinding force on the loading rate (Merkel et al., 1999; Struntz et al., 1999; Baumgartner, Hinterdorfer, Ness et al., 2000; Kienberger, Kada et al., 2000), which is consistent with the exponential lifetime–force relation described earlier. A force acting on a binding complex reduces the lifetime of the bond due to its input of thermal energy. The input of the mechanical energy during pulling enhances the probability of ligand bond dissociation. During a force–distance cycle, the force increases at a nonlinear rate determined by the force–distance profile of the tether, by which the ligand is coupled to the tip. Finally, the complex dissociates at force f. The main contribution of the thermal activation comes from the part of the force curve which is close to unbinding. Therefore, the f values are dependent on the rate of force increase r; $r = df/dt$ = vertical scan velocity times spring constant, at the end of the recognition signal in the retrace.

In unbinding force distributions, both force f and width σ_U clearly increase with increasing loading rate (Kienberger, Kada et al., 2000). Apparently, at slower loading rates the systems adjusts closer to equilibrium which leads to smaller values of both the force f and its variation σ_U. On a half-logarithmic scale, the unbinding force f rises linear with the loading rate, which is characteristic for a single energy barrier in the thermally activated regime (Merkel et al., 1999).

3. Kinetic Rates, Energies, Binding Pocket

Single-molecule recognition force microscopy studies allow for estimation of kinetic rates (Hinterdorfer et al., 1996, 1998, Baumgartner, Hinterdorfer, Ness et al., 2000; Kienberger, Kada, Gruber et al., 2000), energies (Merkel et al., 1999), and structural parameters of the binding pocket (Hinterdorfer et al., 1996, 1998; Baumgartner, Hinterdorfer, Ness et al., 2000; Kienberger, Kada, Gruber et al., 2000). Quantification of the onrate constant k_{on} for the association of the ligand on the tip to a receptor on the surface requires determination of the interaction time $t_{0.5}$ needed for half-maximal probability of binding. With the knowledge of the effective ligand concentration c_{eff} on the tip available for receptor interaction, k_{on} is given by $k_{on} = t_{0.5}^{-1} c_{eff}^{-1}$. The interaction time $t_{0.5}$ for half-maximal binding can be experimentally determined by measuring the dependence of the binding activity on the ligand–receptor encounter duration (Baumgartner, Hinterdorfer, Ness et al., 2000; Baumgartner, Gruber et al., 2000). The effective concentration c_{eff} is described by the effective volume V_{eff}, and the tip-tethered ligand diffuses about the tip, which yields $c_{eff} = N_A^{-1} V_{eff}^{-1}$, where N_A is the Avogadro number. Therefore, V_{eff} is essentially a half-sphere with a radius of the effective tether length.

The additional estimation of the offrate constant k_{off} as described previously leads to values for the equilibrium dissociation constant K_D, according to $K_D = k_{off}/k_{on}$. The same data fit used to obtain k_{off} also reveals estimates for the energy barrier for dissociation, E_b, and the effective length of the ligand–receptor bond, l (cf. Section III,A,1).

B. Applications to Cellular Proteins

1. Vascular Endothelial (VE) Cadherin

a. Introduction

Vascular endothelial cells form a continuous cellular monolayer that covers the inner surface of blood vessels. This monolayer constitutes the major barrier of the body that separates the blood compartment from the extracellular space of tissues. Cadherin-mediated adhesion between endothelial cellular layers (i) confers mechanical stability against external forces acting on the junctions and (ii) allows fast dynamic cellular remodeling to change the barrier properties of the cell layers, for instance, under inflammatory conditions and in tumor metastasis. The molecular mechanisms regulating intercellular adhesion between endothelial cells are still not understood in detail.

The apparent low affinity between the calcium-dependent VE-cadherins presents an interesting aspect about how the number of adhesive bonds between cells could be regulated by a simple thermodynamic mechanism. Since cadherins are linked with their cytoplasmic domains to the actin filament cytoskeleton, a model was proposed that implies regulation of intercellular binding of cadherins by the degree of cytoskeletal tethering (Baumgartner, Hinterdorfer, Ness *et al.*, 2000). Dissociation of cadherins from the cytoskeleton leads to random lateral diffusion in the cell membrane so that the formation of functional units for *trans*-interaction, i.e., cadherin strand dimers, will follow the principles of a diffusion-limited reaction. Due to diffusion kinetics, short lifetime of bonds, and low affinity, adhesion dimers will rapidly dissociate and become subsequently separated by lateral diffusion, which in turn would result in a reduction of the number of *trans*-interacting cadherins and finally in junctional dissociation. If diffusion of cadherins is restrained by tethering them cytoplasmatically via catenins to the actin filament cytoskeleton, the probability of rebinding after dissociation and thus the number of functional strand dimers will be increased significantly, which will consequently increase the intercellular *trans*-interaction strength.

However, such a transmembrane linkage mechanism would only be effective if both dissociation constant and bond lifetime between *trans*-interacting cadherins are relatively low. Therefore, functional state and binding properties of isolated VE-cadherins were studied by single-molecule atomic force microscopy (Baumgartner, Hinterdorfer, Ness *et al.*, 2000; Baumgartner, Gruber *et al.*, 2000).

b. Conformation

Studies were performed with a chimeric protein consisting of two complete extracellular portions of mouse VE-cadherin appended to the Fc part of human IgG1. The protein was secreted by stably transfected Chinese hamster ovary (CHO) cells and purified and characterized as described by Baumgartner, Hinterdorfer, Ness *et al.* (2000). VE-cadherin-Fc is a *cis*-dimeric protein that migrates in nonreducing sodium dodecyl sulfate (SDS) polyacrylamide (10%) gel electrophoresis at 160–180 kDa. Like other classical cadherins, the external domain of VE-cadherin binds Ca^{2+} and associates into *cis*-dimeric complexes (Fig. 4). The structure of the protein chimera was investigated by single-molecule imaging using dynamic force microscopy (DFM) (Han *et al.*, 1996, 1997) in liquids.

In the presence of $CaCl_2$ (5 mM), VE-cadherin reveals an elongated rod-like morphology of 25–28 nm length and 5–8 nm width (Fig. 4). A globular part of 5–8 nm at one end, most likely reflecting the Fc portion of the molecule, is often visible. In the absence of $CaCl_2$ (5 mM EGTA), molecules show V-shaped or globular structures (Fig. 4). Apparently, the two extracellular domains of the VE-cadherin in the protein chimera are indeed associated with the presence of Ca^{2+} to form an adhesion dimer. After depletion of Ca^{2+}, VE-cadherin dissociates into monomers and eventually collapses into globular structures.

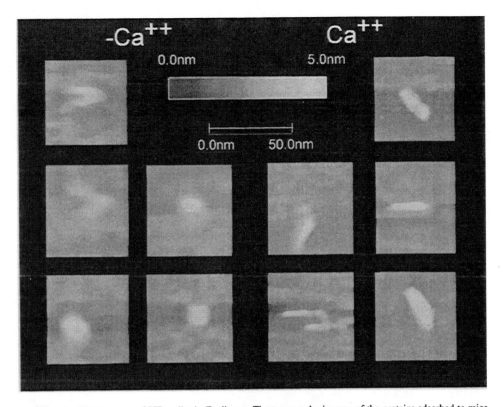

Fig. 4 AFM images of VE-cadherin-Fc dimers. The topography images of the proteins adsorbed to mica were recorded with dynamic force microscopy in isotonic buffer. Cadherin dimers show elongated rod-like structure in the presence of Ca^{2+} and globular to V-shaped morphology in the absence of Ca^{2+}. Reproduced with permission from Baumgartner, W., Hinterdorfer, P., Ness, W., Raab, A., Vestweber, D., Schindler, H., and Drenckhahn, D. (2000). Cadherin interaction probed by atomic force microscopy. *Proc. Natl. Acad. Sci. U.S.A.* **8,** 4005–4010. Copyright (2000) National Academy of Sciences, U.S.A. (See Color Plate.)

c. Binding Strength, Ca^{2+} Dependence, and Trans Association

VE-cadherin-Fc was coupled to both AFM tip and probe surface using the surface chemistry described in Section II,A,1. Recordings, similar to that shown in Fig. 2a, of force–distance cycles showed specific recognition events between tip- and surface-bound VE-cadherin-Fc (Baumgartner, Hinterdorfer, Ness *et al.*, 2000). The specificity of the recognition was proven by the addition of free VE-cadherin-antibody and EGTA in solution, respectively. In both cases, recognition completely disappeared (Baumgartner, Hinterdorfer, Ness *et al.*, 2000) (similar to that in Fig. 2b). The VE-cadherin-antibody blocks *trans*-cadherin–cadherin interaction because it binds to the outermost domain of VE-cadherin whereas EGTA complexes Ca^{2+}, which is required for the formation of the functional strand dimer.

Measuring the binding activity independent of the free Ca^{2+} concentration (Baumgartner, Hinterdorfer, Ness *et al.*, 2000) revealed an apparent K_D of 1.15 mM with a Hill

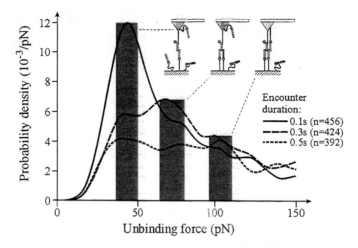

Fig. 5 Unbinding force distribution of *trans*-interacting VE-cadherins. Frequency distribution of unbinding forces between tip- and plate-attached PEG/VE-cadherin-Fc measured at retrace velocities of 800 nm s^{-1} and various tip-to-plate encounter intervals. Data fit to three Gaussian distributions with peak values ($\mu1$–$\mu3$) at about 40, 75, and 120 pN. The dependency of the frequency of $\mu1$–$\mu3$ on encounter duration indicates a diffusion-limited reaction underlying association of tip- and surface-bound cadherins as illustrated. Reproduced with permission from Baumgartner, W., Hinterdorfer, P., Ness, W., Raab, A., Vestweber, D., Schindler, H., and Drenckhahn, D. (2000). Cadherin interaction probed by atomic force microscopy. *Proc. Natl. Acad. Sci. U.S.A.* **8**, 4005–4010. Copyright (2000) National Academy of Sciences, U.S.A.

coefficient of $n_H = 5.04$, indicating high cooperation and steep dependency. Since the K_D measured is close to the extracellular Ca^{2+} concentration and because of the high Hill coefficient obtained, it might be of physiological relevance that a local drop of free Ca^{2+} in the narrow intercellular cleft weakens intercellular adhesion and is therefore involved in facilitating cellular remodeling.

Force–distance cycles were performed in which the tip was allowed to rest on the probe surface at various durations (encounter duration). Accordingly, force distributions using probability density functions were constructed from the unbinding forces (cf. Section II,B,2.) at encounter durations of 0.1, 0.3, and 0.5 s (Fig. 5) (Baumgartner, Hinterdorfer, Ness *et al.*, 2000). In each of the force distribution, three distinct maxima are seen at about 40, 75, and 120 pN, respectively, which suggests that the value of 40 pN corresponds to the adhesive strength quantum between two opposing single-strand dimers. The multiples of this unitary binding force are considered to result from lateral association and load-induced all-or-none cooperative dissociation of three and four interacting strand dimers (Fig. 5). The rather low *trans*-interaction force of a single bond is thus amplified by complexes of cumulative binding strength.

Simultaneous unbinding of two or more independent adhesion dimers is extremely unlikely due to the different coupling sites of the VE-cadherins on both the curved tip and the probe surface (mean distance >10 nm). Lateral oligomerization, however, is possible because of the ability of the tether-linked cadherins to undergo free diffusion and collision within a certain length extension (tether plus cadherin, ~30 nm) (Baumgartner,

Hinterdorfer, Ness *et al.*, 2000). The association of cadherins into complexes with higher-order binding forces is a time-dependent process because the relative size of their peaks in the force distribution increased with increasing encounter duration (Fig. 5). The results suggest that cadherins from opposing cells associate to form complexes and thus increase the intercellular binding strength.

d. Thermodynamic and Structural Parameters

The kinetic onrate constant k_{on} for bond formation between tip- and surface-bound cadherins was determined by the encounter duration needed for half-maximal probability of binding, $t_{0.5}$, and the effective concentration c_{eff} of cadherins available for interaction, $c_{eff} = nN_A^{-1}V_{eff}^{-1}$, according to $k_{on} = t_{0.5}^{-1}nN_A^{-1}V_{eff}^{-1}$, where n is the number of binding partners within the effective volume V_{eff} of a half-sphere with a free equilibrium radius r_{eff} (Baumgartner, Hinterdorfer, Ness *et al.*, 2000; Baumgartner, Gruber *et al.*, 2000) (cf. Section III,A,3). The estimation of r_{eff} in these experiments was based on measures of extended tether lengths from force–distance cycles, yielding $r_{eff} = 12$–22 nm and $V_{eff} = 3.5$–7×10^{-21} L. The average encounter duration for free equilibrium interaction between tip- and surface-bound cadherins during a single force–distance cycle was defined as the time during which the tip–surface distance is $\leq r_{eff}$. Varying the encounter duration resulted in exponential dependency of the probability of recognition events with a half-maximal value at $t_{0.5} = 0.08$ s (Fig. 6). V_{eff} and $t_{0.5}$

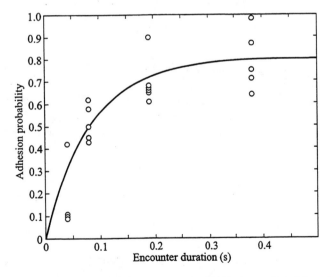

Fig. 6 Binding activity independent of the encounter duration. The probability of force–distance cycles with recognition events was determined independent of the encounter duration between tip- and plate-bound PEG/VE-cadherin-Fc. Each point is the average value calculated from at least 100 force–distance cycles with a 0.5-nm lateral shift between each cycle. Half-maximal adhesion probability is seen at an encounter duration of 0.08 s, allowing calculation of the onrate constant. Reproduced with permission from Baumgartner, W., Gruber, H. J., Hinterdorfer, H., and Drenckhahn, D. (2000). Affinity of trans-interacting VE-cadherin determined by atomic force microscopy. *Single Mol.* **1**, 119–122.

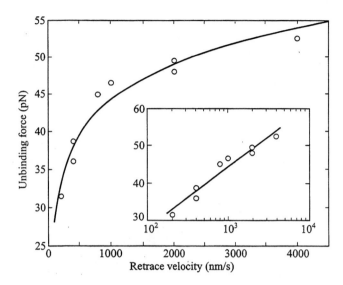

Fig. 7 Dependence of the unbinding force on the pulling velocity. The unbinding force of the first peak (corresponding to $\mu 1$ in Fig. 5) is plotted as a function of the retrace velocity. Each circle represents the average value of at least 300 unbinding events measured at a given retrace velocity. The solid line is a numerical fit of the data to $\tau(f_u) = \tau_0 \exp(-lf/k_B T)$, where τ_0 is the lifetime of unstressed bonds and l is the unbinding width between adhering cadherins. Inset. Data and and fit plotted on a logarithmic scale. Reproduced with permission from Baumgartner, W., Hinterdorfer, P., Ness, W., Raab, A., Vestweber, D., Schindler, H., and Drenckhahn, D. (2000). Cadherin interaction probed by atomic force microscopy. *Proc. Natl. Acad. Sci. U.S.A.* **8**, 4005–4010. Copyright (2000) National Academy of Sciences, U.S.A.

allow the calculation of k_{on} for VE-cadherin-Fc dimer interaction to be $\sim 10^4\ M^{-1}\ s^{-1}$ (Baumgartner, Hinterdorfer, Ness *et al.*, 2000; Baumgartner, Gruber *et al.*, 2000).

Quantifying the dependence of the unbinding force and the bond lifetime on the loading rate yields the kinetic offrate constant k_{off} and the binding pocket bond length 1 (cf. Section III,A,3). Figure 7 shows that the unbinding force increases logarithmically with increasing retract velocity (Baumgartner, Hinterdorfer, Ness *et al.*, 2000), as expected from theory for a single activation barrier (Merkel *et al.*, 1999). The lifetime of bonds at zero force was determined to be $\tau_0 \sim 0.55$ s, yielding an offrate constant $k_{off} = \tau_0^{-1} = 1.8\ s^{-1}$. This allowed calculation of the dissociation constant ($K_D = k_{off}/k_{on}$) which was approximately $2.10^{-4}\ M$. If extreme values for r_{eff} (10–30 nm) were considered, the resulting K_D would lie in the boundaries of 10^{-3}–$10^{-5}\ M$. The effective bond length l was found to be in the range of $l \sim 0.6$ nm.

The rather low values for the interaction force, bond lifetime, and adhesive binding affinity make cadherins ideal candidates for adhesion regulation by cytoskeletal tethering.

2. Na$^+$/D–Glucose Cotransporter (SGLT1)

a. Introduction

The Na$^+$/D-glucose cotransporter (SGLT1) represents one of the prototypes of secondary active transport systems for organic and inorganic solutes that are employed by cells to accumulate nutrients. These transport systems are expected to assume different conformations during their catalytic cycles. Despite their general biological importance, the understanding of the structural events accompanying the transmembrane movement of the transportates is rather limited. The topological assignment of epitopes is still a matter of controversy, and conformational changes have been either deduced intuitively or demonstrated after chemical modification of the molecule (Lin *et al.*, 1999).

MRFM was employed to probe recognition of membrane receptors in functional brush border membrane vesicles (BBMV) (Wielert-Badt *et al.*, 2000). Values for kinetic rates k_{on}, k_{off}, and dissociation constants K_D of various ligands were estimated. Furthermore, the data obtained in this study provide information about membrane-sidedness of the epitopes and structural changes of the binding pocket during interaction with the substrate D-glucose and the inhibitor phlorizin (Wielert-Badt *et al.*, 2000).

b. Membrane–Sidedness and Functionality

BBMV were adsorbed to gold in hypotonic buffer and imaged with the AFM using dynamic force microscopy (Han *et al.*, 1996, 1997; Raab *et al.*, 1999), Single vesicular structures of 200 to 500 nm in diameter and 50 to 150 nm in height were clearly resolved. At lateral positions where BBMV were identified in the topographical imaging mode, the x-y scan was stopped and force–distance cycles were employed to probe the specific recognition of a ligand on the tip to the corresponding receptor in the vesicle membrane. In this way, an AFM tip carrying an anti-γ-glutamyltranspeptidase (γ-GT) showed recognition events with an unbinding force of $f = 131$ pN $\pm$ 44 pN on BBMV membranes. Since the binding epitope for the anti-γ-glutamyltranspeptidase antibody is known to be exclusively located on the luminal side of the cell the orientation of the BBMV on the gold surface must be such that the former luminal side faces the aqueous phase and the AFM tip (Wielert-Badt *et al.*, 2000).

Recognition on BBMV was also obtained ($f = 120 \pm 44$ pN) with AFM tips containing phlorizin, a ligand that acts as competitive inhibitor of glucose binding to SGLT1. Therefore, the phlorizin-binding epitope on SGLT1 is freely accessible to the phlorizin on the AFM tip and consequently located on the luminal side as well. Binding of phlorizin to SGLT in BBMV strongly diminished in the presence of free phlorizin and D-glucose, respectively. As expected, both phlorizin and D-glucose in solution compete with phlorizin on the tip for the binding site at the SGLT and therefore block recognition. This result provides evidence that the Na$^+$/D-glucose cotransporters in the surface of adsorbed BBMV are still functionally active. Values for the equilibrium dissociation constants K_D, estimated from the force measurements as described (cf. Section II,A,3), yield $K_D \sim 0.2$ μM for phlorizin/SGLT binding and compare nicely to values from the literature obtained through ensemble-average experiments (Wielert-Badt *et al.*, 2000).

c. Epitope Mapping

For recognition studies of antibodies to SGLT1 in BBMV, PAN3, an antibody that was raised against a part of a supposedly extra-membrane loop of SGLT1, was coupled to the AFM tip. The binding of PAN3 antibody on the AFM tip to Na$^+$/D-glucose cotransporters in BBMV membrane vesicles was examined in three different situations (Fig. 8). In pure NaCl buffer, SGLT1 is (i) in the nontransporting state, changing to (ii) the transporting

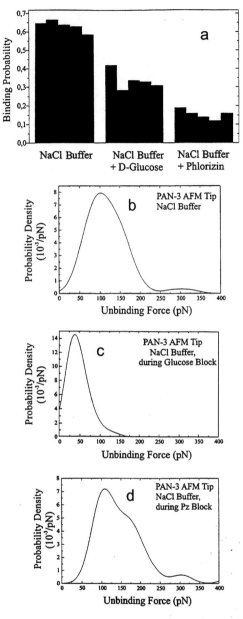

Fig. 8 Influence of phlorizin and glucose on the binding of PAN3 antibody to Na^+/D-glucose cotransporter SGLT1. (a) Binding probability. The probability for binding of PAN3 antibody on the AFM tip to SGLT1 in BBMV is shown for three different situations. The value of 0.62 in pure NaCl buffer (cf. Section III,B) decreased to about 0.3 in the presence of 250 mM D-glucose and 0.15 in the presence 1 mM phlorizin, respectively. Each bar corresponds to 240 force–distance cycles. (b–d) Unbinding force. The maximum of the probability density function of unbinding forces is at 98 pN in pure NaCl buffer (b), at 47 pN in NaCl buffer containing 250 mM D-glucose (c), and at 102 pN in NaCl buffer containing 1 mM phlorizin (d), respectively. Reproduced with permission from Raab, A., Han, W., Badt, D., Smith-Gill, S. J., Lindsay, S. M., Schindler, H., and Hinterdorfer, P. (1999). Antibody recognition imaging by force microscopy. *Nature Biotechnol.* **17,** 902–905.

state in the presence of D-glucose in NaCl buffer and to (iii) the fully blocked state upon addition of phlorizin. Both unbinding force distribution and binding probabiliy, i.e., the probability to detect a recognition event in a force–distance cycle, were determined.

In SGLT–PAN3 binding in pure NaCl, the probability of 0.6 decreased to 0.3 when D-glucose was present in NaCl buffer (Fig. 8). In contrast, the binding probability remained unchanged (0.6) in a buffer containing D-glucose but lacking Na^+ (not shown). This confirmed former findings that D-glucose requires Na^+ for binding to the external surface of SGLT. The presence of D-glucose in NaCl buffer did not only decrease the SGLT–PAN3 binding probability (Fig. 8a) but also had an influence on the unbinding force (Figs. 8b and 8c). While the most probable unbinding force in pure NaCl buffer was about 100 pN (Fig. 8b), this value changed to about 50 pN when D-glucose was added (Fig. 8c). Apparently, D-glucose binding to SGLT influences PAN3 recognition in two ways: it either (i) completely blocks recognition or (ii) reduces the interaction force from 100 to 50 pN.

The addition of phlorizin to NaCl buffer almost completely blocked recognition of SGLT1 by the PAN3 antibody (binding probability 0.25, cf. Fig. 8a). The force distribution of the few remaining recognition events (Fig. 8d) was similar to that obtained in pure NaCl buffer (Fig. 8b), with its maximum at about 100 pN (Fig. 8d). This suggests that the remaining PAN3-binding activity arises from a less than 100% block, most likely a result of the dynamic exchange of phlorizin binding and release from SGLT1.

Altogether, the results lead to the conclusion that we observed at least three different conformations of the PAN3-binding epitope of the SGLT: (i) Binding of PAN3 antibody to SGLT in pure NaCl buffer with the observed interaction force of 100 pN is considered as undisturbed SGLT–PAN3 recognition. (ii) The almost complete block by phlorizin in solution suggests that the binding epitope of SGLT is not accessible to the PAN3 antibody when phlorizin is bound. A similar inaccessibility of the epitope was also observed as one of the two influences of D-glucose binding of SGLT. (iii) Alternatively to the complete block, D-glucose binding to SGLT can also lead to a reduction of the SGLT–PAN3 interaction force from 100 to 50 pN.

The two conformations of the PAN3 recognition epitope of the SGLT1 that are induced by D-glucose binding may well be connected with one or more conformational changes that SGLT1 undergoes during D-glucose transport across the brush border membrane. However, whether the blocked states arising from D-glucose and phlorizin binding, respectively, belong to the same conformational state of SGLT1 remains to be elucidated. (Wielert-Badt *et al.*, 2000).

IV. Recognition Imaging

A. Lateral Force Mapping

For the localization of antigenic sites the probe was laterally scanned during force–distance cycles and the binding probability was determined independent of the lateral position (Hinterdorfer *et al.*, 1996, 1998). Probes contained a low density of antigens (human serum albumin, HSA) with ~100 nm mean distance between single HSA molecules. Force–distance cycles with an AFM tip containing antibodies against HSA were performed at 100 nm amplitude and 3 Hz with a lateral velocity of 0.6 nm/s, resulting in one force–distance cycle per 0.2 nm. Unbinding events occurred singly, and were only detected at certain lateral positions. Data were sampled every 2.6 nm, and the binding probability was calculated for each sampling point.

Binding profiles for single HSA molecules showed a maximum (Hinterdorfer et al., 1996, 1998). The position of the HSA was determined from the position of the maximum with an accuracy of 1.5 nm. Fit of the binding profile with a Gaussian function yielded a width of $r_{eff} = 6$ nm. This value reflects the dynamic reach of the antibody on the tip. Apparently, antigenic sites are detected within 6 nm apart from the center of the AFM tip. The antibody on the tip can diffuse and orient within a half-sphere of a 6-nm radius, provided by the flexible 8-nm-long PEG cross-linker by which the antibody was tethered to the tip. The design of the antibody AFM tip sensor appears apt for a microscopy capable of imaging surface topography and distribution of recognition sites on the single-molecule level simultaneously (Hinterdorfer et al., 1996, 1998).

Simultaneous information on topography and forces was recently obtained by lateral force mapping, i.e., performing an approach–retract cycle in every pixel of the image, on a micrometer-size streptavidin pattern with a biotinylated AFM tip (Ludwig et al., 1997). With a similar configuration height and adhesion force, images were simultaneously obtained with resolution approaching the single-molecule level (Willemsen et al., 1998). The strategies of force mapping, however, either lack high lateral resolution (Ludwig et al., 1997) and/or are much slower in data acquisition (Hinterdorfer et al., 1996, 1998; Willemsen et al., 1998) than topography images, since the frequency of the retract–approach cycles performed in every pixel is limited by hydrodynamic forces in the aqueous solution. In addition, obtaining the force image requires the ligand to be disrupted from the receptor in each retract–approach cycle. For this, the z amplitude of the retract approach cycle must be at least 50 nm, and therefore the ligand on the tip is without access to the receptor on the surface for most of the time during the experiment.

B. Dynamic Recognition Force Microscopy

An imaging method for the mapping of antigenic sites on the surface was recently developed (Raab et al., 1999) by combining molecular recognition (Hinterdorfer et al., 1996) with dynamic force microscopy (DFM) (Han et al., 1996, 1997). This technique provides very gentle tip–surface interactions and the specific interaction of the antibody on the tip with the antigen on the surface can be used to localize antigenic sites, thus recording recognition images. The magnetically coated tip was oscillated by an alternating magnetic field at an amplitude of 5 nm while being scanned along the surface. Since the tether has a length of 6 nm, the antibody on the tip always has a chance of recognition when passing an antigenic site which increases the binding probability enormously.

Antibody–antigen recognition was monitored by the reduction of the oscillation amplitude yielding a lateral resolution of 3 nm. Since the oscillation frequency is more than one hundred times faster than typical frequencies in conventional force mapping, the data acquisition rate is much higher. A recognition image of 500 nm size with a 1-nm pixellation can be obtained in a few minutes which is comparable to measuring times for normal topography images. With this methodology, topography and recognition images can be obtained at the same time and distinct receptor sites in the recognition image can be assigned to structures from the topography image (Raab *et al.,* 1999).

Half-antibodies were used to provide a monovalent ligand on the tip. The antigen, lysozyme, was tightly adsorbed to mica under conditions that yielded a low surface coverage (for details, see Raab *et al.,* 1999). A topographical image of this preparation was first recorded in buffer using a bare AFM tip as a control (Fig. 9c). Single lysozyme molecules were clearly resolved (Fig. 9c). A cross-section analysis (Fig. 9d, profile in black) reveals that the molecules appear to be 8 to 12 nm in diameter and 2.0 to 2.5 nm in height.

Imaging with a half-antibody tethered to the tip under conditions identical to those used to obtain the topographical image gave strikingly different images (Fig. 9e), which differed significantly in both height and diameter compared to the topographical image (Fig. 9c). Cross-section analysis (Fig. 9d, trace in red) reveals a height of 3.0 to 3.5 nm and a diameter of 20 to 25 nm. Profiles obtained from the recognition image appear at least 1 nm higher and 10 nm broader than profiles from the topographical image.

The antibody–antigen recognition process during imaging is depicted in Fig. 9b. Approaching the antigen in a lateral scan from the left, the antibody on the tip binds to the antigen about 10 nm before the tip end is above the antigenic site (Fig. 9b, left tip), due to the flexible tethering provided by the crosslinker. In the bound state, the z oscillation of the cantilever is additionally reduced by the attractive force of the crosslinker–antibody–antigen connection which is acting as a nonlinear spring. Since the AFM detects the z projection of the force, the amount of the attractive force measured increases when the tip moves further to the right and reaches its maximum just above the position of the antigenic site (Fig. 9b, tip in middle).

This amplitude reduction leads to an increasing tip–surface separation induced by the feedback loop of the AFM. Upon further tip movement to the right the z component of the attractive force decreases again resulting in a decreasing tip–surface separation. At lateral distances comparable to the length of the antibody–crosslinker connection the antibody on the tip dissociates from the antibody on the surface and the attractive force goes to zero.

The diameter of cross-section profiles obtained from the recognition image (Fig. 9d, red trace) corresponds to about twice the length of the crosslinker (6 nm) plus antibody (6 nm). Increased heights detected in comparison to profiles of the topographical image (Fig. 9d, black trace) reflect the amplitude reduction owing to antibody–antigen recognition. Cross-section profiles of the recognition image as shown in Fig. 1d (red trace) were fitted with a truncated power law function. Maxima of the profiles indicate the position of the antigenic site. The accuracy of maximum determination which in turn reflects the positional accuracy of determining the position of the antigenic site was 3 nm. The specific nature of the antibody–antigen interaction was tested by injecting free antibody into the liquid cell so as to block the antigenic sites on the surface, and subsequent images showed a reduction of apparent height (Fig. 9f).

The described methodology (Raab *et al.,* 1999) is applicable with any ligand, and therefore it should prove possible to recognize many types of proteins or protein layers

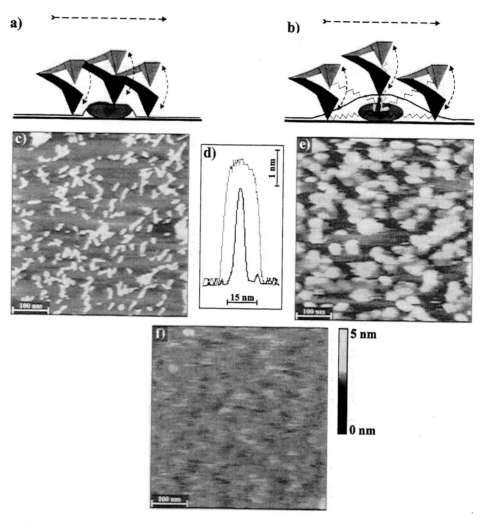

Fig. 9 (a) AFM tip–lysozyme interaction during topography imaging. The red line indicates the height profile obtained from a single lysozyme molecule (shown in green) with the AFM using a bare tip. (b) AFM tip–lysozyme interaction during recognition imaging. Half-antibodies (shown in red) are bound to the AFM tip via a flexible tether (jagged line) for the recognition of lysozyme (shown in green) on the surface. Imaging results in a height profile as indicated (red line). (c) Topography image. Single lysozyme molecules can be clearly resolved. Sometimes small lysozyme aggregates are observed. Image size was 500 nm. False color bar for heights from 0 (dark) to 5 (bright) nm. (d) Height profiles. Cross-section profiles of single lysozyme molecules obtained from the topography (black line) and the recognition (red line) image. (e) Recognition image. The bright dots represent recognition profiles of single lysozyme molecules. Imaging was performed using an AFM tip carrying one half-antibody with access to the antigens on the surface. Conditions were exactly the same as those in (c). (f) Block image. The image was obtained under the same conditions as those in the recognition image (e) in the presence of free antibody in solution. Recognition is blocked as apparent from the lack of recognition profiles. Reproduced with permission from Raab, A., Han, W., Badt, D., Smith-Gill, S. J., Lindsay, S. M., Schindler, H., and Hinterdorfer, P. (1999). Antibody recognition imaging by force microscopy. *Nature Biotechnol.* **17**, 902–905. (See Color Plate.)

and carry out epitope mapping on a nanometer scale. The gentle interaction inherent in magnetically excited dynamic force microscopy should permit an extension of the technique to soft cellular surfaces where dynamic processes may be mapped over time scales from seconds to hours.

Acknowledgments

This work was supported by the Austrian Ministry of Science, project GZ 200.027/2 and by the Austrian Research Funds, projects P12801/2MED. The author acknowledges H. Schindler, F. Kienberger, H. Gruber, and A. Raab (University of Linz), W. Baumgartner and D. Drenckhahn (University of Würzburg), and S. Wielert-Badt and R. Kinne (MPI Dortmund) for their contributions to this chapter. I would like to dedicate this paper to Prof. Hansgeorg Schindler who died on August 28, 2001.

References

Allen, S., Chen, X., Davies, J., Davies, M. C., Dawkes, A. C., Edwards, J. C., Roberts, C. J., Sefton, J., Tendler, S. J. B., and Williams, P. M. (1997). Spatial mapping of specific molecular recognition sites by atomic force microscopy. *Biochemistry* **36,** 7457–7463.

Baumgartner, W., Hinterdorfer, P., Ness, W., Raab, A., Vestweber, D., Schindler, H., and Drenckhahn, D. (2000). Cadherin interaction probed by atomic force microscopy. *Proc. Natl. Acad. Sci. U.S.A.* **8,** 4005–4010.

Baumgartner, W., Hinterdorfer, P., and Schindler, H. (2000). Data analysis of interaction forces measured with the atomic force microscope. *Ultramicroscopy* **82,** 85–95.

Baumgartner, W., Gruber, H. J., Hinterdorfer, H., and Drenckhahn, D. (2000). Affinity of trans-interacting VE-cadherin determined by atomic force microscopy. *Single Mol.* **1,** 119–122.

Binnig, G., Quate, C. F., and Gerber, Ch. (1986). Atomic force microscope. *Phys. Rev. Lett.* **56,** 930–933.

Bell, G. I. (1978). Models for the specific adhesion of cells to cells. *Science* **200,** 618–627.

Boland, T., and Ratner, B. D. (1995). Direct measurement of hydrogen bonding in DNA nucleotide bases by atomic force microscopy. *Proc. Natl. Acad. Sci. U.S.A.* **92,** 5297–5301.

Chen, A., and Moy, V. T. (2000). Cross-linking of cell surface receptors enhances cooperativity of molecular adhesion. *Biophys. J.* **78,** 2814–2820.

Chilkoti, A., Boland, T., Ratner, B., and Stayton, P. S. (1995). The relationship between ligand-binding thermodynamics and protein-ligand interaction forces measured by atomic force microscopy. *Biophys. J.* **69,** 2125–2130.

Conti, M., Falini, G., and Samori, B. (2000). How strong is the coordination bond between a histidine tag and Ni-Nitriloacetate? An experiment of mechanochemistry on single molecules. *Angew. Chem.* **112,** 221–224.

Dammer, U., Hegner, M., Anselmetti, D., Wagner, P., Dreier, M., Huber, W., and Güntherodt, H.-J. (1996). Specific antigen/antibody interactions measured by force microscopy. *Biophys. J.* **70,** 2437–2441.

Evans, E., and Ritchie, K. (1997). Dynamic strength of molecular adhesion bonds. *Biophys. J.* **72,** 1541–1555.

Florin, E. L., Moy, V. T., and Gaub, H. E. (1994). Adhesion forces between individual ligand receptor pairs. *Science* **264,** 415–417.

Fritz, J., Katopidis, A. G., Kolbinger, F., and Anselmetti, D. (1998). Force-mediated kinetics of single P-selectin/ligand complexes observed by atomic force microscopy. *Proc. Natl. Acad. Sci. U.S.A.* **95,** 12,283–12,288.

Grandbois, M., Beyer, M., Rief, M., Clausen-Schaumann, H., and Gaub, H. E. (1999). How strong is a covalent bond. *Science* **283,** 1727–1730.

Grubmüller, H., Heymann, B., and Tavan, P. (1996). Ligand binding: Molecular mechanics calculation of the streptavidin-biotin rupture force. *Science* **271,** 997–999.

Han, W., Lindsay, S. M., and Jing, T. (1996). A magnetically driven oscillating probe microscope for operation in liquid. *Appl. Phys. Lett.* **69,** 1–3.

Han, W., Lindsay, S. M., Dlakic, M., and Harrington, R. E. (1997). Kinked DNA. *Nature* **386,** 563.

Haselgrübler, Th., Amerstorfer, A., Schindler, H., and Gruber, H. J. (1995). Synthesis and applications of a new poly(ethylene glycol) derivative for the crosslinking of amines with thiols. *Bioconjugate Chem.* **6,** 242–248.

Hinterdorfer, P., Baber, G., and Tamm, L. K. (1994). Reconstitution of membrane fusion sites. A total internal reflection fluorescence microscopy study of influenza hemagglutinin-mediated membrane fusion. *J. Biol. Chem.* **269,** 20,360–20,368.

Hinterdorfer, P., Baumgartner, W., Gruber, H. J., Schilcher, K., and Schindler, H. (1996). Detection and localization of individual antibody-antigen recognition events by atomic force microscopy. *Proc. Natl. Acad. Sci. U.S.A.* **93,** 3477–3481.

Hinterdorfer, P., Schilcher, K., Baumgartner, W., Gruber, H. J., and Schindler, H. (1998). A mechanistic study of the dissociation of individual antibody-antigen pairs by atomoic force microscopy. *Nanobiology* **4,** 39–50.

Hinterdorfer, P., Kienberger, F., Raab, A., Gruber, H. J., Baumgartner, W., Kada, G., Riener, C., Wielert-Badt, S., Borken, C., and Schindler, H. (2000). Poly(ethylene glycol): An ideal spacer for molecular recognition force microscopy/spectroscopy. *Single Mol.* **1,** 99–103.

Izraelev, S., Stepaniants, S., Balsera, M., Oono, Y., and Schulten, K. (1997). Molecular dynamics study of unbinding of the avidin-biotin complex. *Biophys. J.* **72,** 1568–1581.

Kada, G., Blaney, L., Jeyakumar, L. H., Kienberger, F., Pastushenko, V. Ph., Fleischer, S., Schindler, H., Lai, F. A., and Hinterdorfer, P. (2001). Recognition force microscopy/spectroscopy of ion channels: Applications to the skeletal muscle Ca^{2+} release channel (RYR1). *Ultramicroscopy* **86,** 129–137.

Kalb, E., Frey, S., and Tamm, L. K. (1992). Formation of supported planar bilayers by fusion of vesicles to supported phospholipid monolayers. *Biochim. Biophys. Acta* **1103,** 307–316.

Kienberger, F., Kada, G., Gruber, H. J., Pastushenko, V. Ph., Riener, C., Trieb, M., Knaus, H.-G., Schindler, H., and Hinterdorfer, P. (2000). Recognition force spectroscopy studies of the NTA-His6 bond. *Single Mol.* **1,** 59–65.

Kienberger, F., Pastushenko, V. Ph., Kada, G., Gruber, H. J., Riener, C., Schindler, H., and Hinterdorfer, P. (2000). Static and dynamical properties of single poly(ethylene glycol) molecules investigated by force spectroscopy. *Single Mol.* **1,** 123–128.

Kiss, E., and Gölander, C.-G. (1990). Chemical derivatization of muscovite mica surfaces. *Coll. Surf.* **49,** 335–342.

Lee, G. U., Kidwell, D. A., and Colton, R. J. (1994). Sensing discrete streptavidin-biotin interactions with atomic force microscopy. *Langmuir* **10,** 354–357.

Lee, G. U., Chrisey, A. C., and Colton, J. C. (1994). Direct measurement of the forces between complementary strands of DNA. *Science* **266,** 771–773.

Legrimellec, C., Lesniewska, E., Giocondi, M. C., Finot, E., Vie, V., and Goudonnet, J. P. (1998). Imaging the surface of living cells by low-force contact-mode atomic force microscopy. *Biophys. J.* **75,** 695–703.

Lehenkari, P. P., and Horton, M. A. (1999). Single integrin molecule adhesion forces in intact cells measured by atomic force microscopy. *Biochem. Biophys. Res. Commun.* **259,** 645–650.

Lin, J.-T., Kormanec, J., Homerova, D., and Kinne, R. K.-H. (1999). Probing transmembrane topology of the high affinity sodium/glucose cotransporter (SGLT1) with histidine tagged mutants. *J. Membr. Biol.* **170,** 243–252.

Ludwig, M., Dettmann, W., and Gaub, H. E. (1997). Atomic force microscopy imaging contrast based on molecuar recognition. *Biophys. J.* **72,** 445–448.

Marzsalek, P. E., Oberhauser, A. F., Pang, Y.-P., and Fernandez, J. M. (1998). Polysaccharide elasticity governed by chair-boat transitions of the glucopyranose ring. *Nature* **396,** 661–664.

Moy, V. T., Florin, E.-L., and Gaub, H. E. (1994). Adhesive forces between ligand and receptor measured by AFM. *Science* **266,** 257–259.

Oberhauser, A. F., Marzsalek, P. E., Erickson, H. P., and Fernandez, J. M. (1998). The molecular elasticity of the extracellular matrix tenascin. *Nature* **393,** 181–185.

Oesterhelt, F., Rief, M., and Gaub, H. E. (1999). Single molecule force spectroscopy by AFM indicates helical structure of poly(ethylene-glycol) in water. *New J. Phys.* **1**, 6.1.–6.11.

Raab, A., Han, W., Badt, D., Smith-Gill, S. J., Lindsay, S. M., Schindler, H., and Hinterdorfer, P. (1999). Antibody recognition imaging by force microscopy. *Nature Biotechnol.* **17**, 902–905.

Rief, M., Oesterhelt, F., Heyman, B., and Gaub, H. E. (1997). Single molecule force spectroscopy on polysaccharides by atomic force microscopy. *Science* **275**, 1295–1297.

Rief, M., Gautel, M., Oesterhelt, F., Fernandez, J. M., and Gaub, H. E. (1997). Reversible unfolding of individual titin immunoglobulin domains by AFM. *Science* **276**, 1109–1112.

Ros, R., Schwesinger, F., Anselmetti, D., Kubon, M., Schäfer, R., Plückthun, A., and Tiefenauer, L. (1998). Antigen binding forces of individually addressed single-chain Fv antibody molecules. *Proc. Natl. Acad. Sci. U.S.A.* **95**, 7402–7405.

Schilcher, K., Hinterdorfer, P., Gruber, H. J., and Schindler, H. (1997). A non-invasive method for the tight anchoring of cells for scanning force microscopy. *Cell. Biol. Int.* **21**, 769–778.

Schmidt, Th., Schütz, G. J., Baumgartner, W., Gruber, H. J., and Schindler, H. (1996). Imaging of single molecule diffusion. *Proc. Natl. Acad. Sci. U.S.A.* **93**, 2926–2929.

Schmitt, L., Ludwig, M., Gaub, H. E., and Tampe, R. (2000). A metal-chelating microscopy tip as a new toolbox for single-molecule experiments by atomic force microscopy. *Biophys. J.* **78**, 3275–3285.

Strunz, T., Oroszlan, K., Schäfer, R., and Güntherodt, H.-G. (1999). Dynamic force spectroscopy of single DNA molecules. *Proc. Natl. Acad. Sci. U.S.A.* **96**, 11,277–11,282.

Wielert-Badt, S., Hinterdorfer, P., Gruber, H. J., Lin, J.-T., Badt, D., Schindler, H., and Kinne, R. K.-H. (2000). Epitope mapping of the Na^+/D-glucose cotransporter in brush border membranes by molecular recognition force microscopy. Submitted for publication.

Willemsen, O. H., Snel, M. M. E., van der Werf, K. O., de Grooth, B. G., Greve, J., Hinterdorfer, P., Gruber, H. J., Schindler, H., van Kyook, Y., and Figdor, C. G. (1998). Simultaneous height and adhesion imaging of antibody antigen interactions by atomic force microscopy. *Biophys. J.* **57**, 2220–2228.

Wong, S. S., Joselevich, E., Woolley, A. T., Cheung, C. L., and Lieber, C. M. (1998). Covalently functionalyzed nanotubes as nanometre-sized probes in chemistry and biology. *Nature* **394**, 52–55.

CHAPTER 7

The Biophysics of Sensory Cells of the Inner Ear Examined by Atomic Force Microscopy and Patch Clamp

Matthias G. Langer[*] and Assen Koitschev[†]

[*] Division of Sensory Biophysics
Universität Tübingen
72076 Tübingen, Germany

[†] Department of Otorhinolaryngology
Universität Tübingen
72076 Tübingen, Germany

METHODS IN CELL BIOLOGY, VOL. 68
Copyright 2002, Elsevier Science (USA). All rights reserved.
0091-679X/02 $35.00

I. Introduction

Atomic force microscopy (AFM) is a well-established technique used to both image the topography of rigid samples such as semiconductors and measure molecular binding forces in the piconewton range (e.g., biotin–avidin, unfolding of titin). Though this technique has already been used in many biological applications (e.g., elasticity measurements), only a few physiologists have attached high importance to the AFM. Scientists doing hearing research are particularly interested in properties of mechanically activated ion channels of sensory cells in the inner ear. For such investigations, AFM combined with conventional physiological methods provides new experimental perspectives. Here, a combination of an AFM and a patch-clamp setup is reported demonstrating the possibility of measuring the force exerted to individual stereocilia of sensory hair cells in the inner ear and the current of the transduction channel simultaneously. Examples are presented exposing the capabilities of this instrument to study the functional properties of sensory hair bundles in the mammalian cochlea. AFM works in vaccum, air, or liquid. The possibility to use AFM in liquid provides a new perspective for studies on living cells under physiological conditions. Applications in cell biology focus on the mechanical properties of subcellular structures such as either the elastic properties of the cytoskeleton (Rotsch and Radmacher, 2000) or the elasticity of secretory granules (Parpura and Fernandez, 1996). Besides pure mechanics, AFM may have also a role in methodological strategies to study the chemical and electrical properties of molecules involved in signal transduction of cells. One approach to examine the specific force interaction between single molecules was to measure the unbinding forces between a ligand molecule (e.g., avidin) attached to an AFM tip and a receptor molecule (e.g., biotin) attached to a substrate (Florin *et al.,* 1994). Recently, the possibility of measuring the structural properties of single molecules anchored in membrane patches was reported (Oesterhelt *et al.,* 2000). Here we present the possibility of using the AFM in combination with a patch clamp as a tool for the examination of the mechanics and the electrical response of cochlear hair cells of the rat. Hair cells are mechanosensory cells in the inner ear responsible for one of the major processes of hearing: the transduction of sound into an electrical signal. A receptive organelle (hair bundle) composed of parallel arranged filamentous structures—the stereocilia—is located on top of these cells. Hair bundles of rats consist of two to three rows of parallel arranged stereocilia showing different length for the different rows while stereocilia within a row are roughly of the same height. The hair bundle is the antenna for mechanical stimuli. Incoming sound deflects this hair bundle, thereby opening mechanosensitive transduction channels located in the lipid membrane of stereocilia (Gillespie, 1995; Markin and Hudspeth, 1995). The molecular structure of the transduction channel and the exact process of transduction channel gating are still unknown. It is supposed that the transduction channel is opened by directly pulling at the so-called tip links connecting adjacent stereocilia of a taller and a shorter row (Pickles *et al.,* 1984). Additionally, it was found by scanning electron microscopy (SEM) that fine filaments—the so-called side links—connect stereocilia of the same row. We decided to use AFM as a local stimulator to measure and exert very small forces to individual stereocilia. Besides examining the pure mechanical properties of stereocilia and transduction channels, it is important to concurrently examine the electrical response of the hair cell. The combination of AFM with other techniques of molecular physiology such as patch clamp (Sakmann and Neher, 1995) might be useful to gain access to the electrical phenomena of transduction channel gating. Patch clamp allows one to both measure the membrane potential of living cells, the current mediated by ion channels embedded in the lipid membrane of cells and to study the kinetics of ion channel gating. A first approach combining AFM and patch clamp was reported by Hörber *et al.* (1995). In these experiments, a patch pipette was attached to a piezo-electric tube scanner. Inside-out and outside-out membrane patches were pulled from *Xenopus* oocytes to allow the scanning of the membrane surface. The idea was to simultaneously measure the topography of the membrane surface with the AFM cantilever and control the membrane potential. However, it was difficult to achieve high spatial resolution with AFM on that area of the membrane surface which was not in direct contact with the pipette glass. Nevertheless, the combination of AFM and patch clamp has been presented as a promising technology for studying mechanical gating of transduction channels in the inner ear (Langer *et al.*, 2000). We present an experimental setup which allows one to simultaneously exert a force to the mechanosensory hair bundles with an AFM tip and measure the electrical response by patch clamp. We took advantage of the capability of AFM to locally apply very small forces (approximately 10 pN) to individual stereocilia. Additionally, the imaging faculty of AFM helps to perform reproducible elasticity measurements at localized stereocilia within a stereociliary bundle while measuring the electrical response of the cell in the whole cell-recording mode. The capability to detect the opening and closing of a single transduction channel strongly depends on the mechanical interaction between adjacent stereocilia. Different experimental strategies are presented demonstrating how AFM and patch clamp were applied to cochlear hair bundles and single stereocilia. However, hair cells are not the only samples of interest for simultaneous application of AFM and patch clamp. Other possibilities and further experiments will be discussed.

II. Morphology and Function of Cochlear Hair Cells

A. The Cochlea

This section gives a highly condensed survey of experimental findings on cochlear mechanics and physiology. The cochlea is the organ of hearing of the mammalian inner ear transducing sound into an electrical signal. The spiral structure is a characteristic feature of the mammalian inner ear and consists of three separate spaces filled with liquid: the scala vestibuli, the scala media, and the scala tympani. Scala vestibuli and scala tympani contain perilymph (high Na^+ and low K^+ concentrations), and the scala media is filled with endolymph (high K^+ and low Na^+ concentrations). The scala media reveals a triangular-shaped cross section imbedding the organ of Corti (Fig. 1). Hair cells are an essential part of the organ of Corti. The organ is located on the top of the basilar membrane showing a characteristic vibration pattern when being stimulated by incoming sound. Incoming sound generates a traveling wave propagating along the

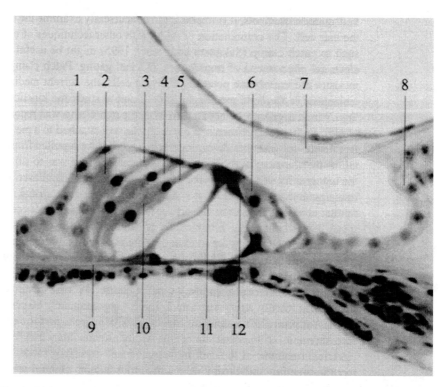

Fig. 1 Light microscopic image of a cross section of the organ of Corti. 1, Hensen cells; 2, outer border cell; 3–5, outer hair cells of different rows; 6, inner hair cell; 7, tectorial membrane; 8, inner sulcus cell; 9, basilar membrane; 10, Deiters cell; 11, outer pillar cell; 12, inner pillar cell. Naturally, the tectorial membrane (7) touches the tallest stereocilia of outer hair cells (3, 4, 5). However, this preparation shows the tectorial membrane removed from the hair bundles, which is an artifact caused during processing for the light microscope.

basilar membrane. For each frequency the membrane exhibits a characteristic vibration pattern with a sharp peak. The position of these peaks strongly depends on the frequency. The performance of the cochlea is equivalent to a Fourier transformation (Von Békésy, 1960). The organ of Corti contains two types of sensory cells: one row of inner hair cells (IHC) and three rows of outer hair cells (OHC) (Fig. 2A). A receptive organelle, composed of so-called stereocilia, is located on top of these cells. The stereocilia differ in length at different positions along the mammalian cochlea reflecting the influence of frequency and are organized in specific patterns, depending on the function of the particular cell population (Furness et al., 1989; Furness and Hackney, 1985; Russell and Richardson, 1987; Lim, 1986; Zine and Romand, 1996). IHC and OHC differ in their morphologic appearance. Hair bundles of IHC consist of two to three straight or slightly curved rows of parallel arranged stereocilia, whereas stereociliary bundles of OHC show a typical V- or W-like pattern (Fig. 2A). Stereocilia are organized in a characteristic manner with respect to their height. The cilia within a single row are the same height, whereas stereocilia of different rows are graded in height, the most within the outer row

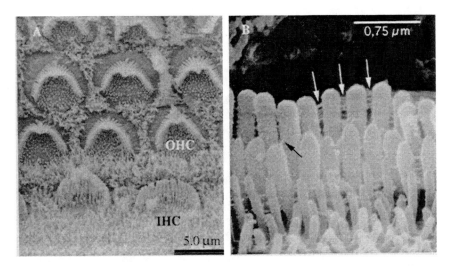

Fig. 2 SEM images of outer and inner hair cells of the organ of Corti obtained from neonatal rats. Before preparation for the SEM, organs were incubated for 1 day in culture medium and for about 2 h in standard extracellular medium during AFM investigation. (A) Arrangement of inner and outer hair bundles 4 days after birth. Stereocilia of OHC are arranged in a V- or W-shape-like pattern (upper three rows) while those of IHC show a more linear shape (labeled as IHC). With proceeding maturation stereocilia of IHC bundles will arrange in lines. (B) IHC bundle of a 3-day-old rat crosslinked by side links (white arrows) and tip links (black arrow). The bundle consists of one row of taller stereocilia and a second row of shorter stereocilia. Smaller filaments at the bottom correspond to microvilli disappearing with increasing age.

being the tallest. Adjacent stereocilia of a hair bundle are connected to one another with fine filaments called links (Fig. 2B). Depending on their geometric arrangement, filaments are divided into two classes: the side links and the tip links (Pickles *et al.*, 1984). In a great number of experiments (Zhao *et al.*, 1996; Marquis and Hudspeth, 1997) it was shown that presence and orientation of links are directly related to the sensory function of hair cells. The tip links, connecting the tips of shorter stereocilia to the lateral wall of adjacent taller stereocilia, are supposed to work like an elastic spring pulling directly at the mechanosensory transduction channels. A force exerted along the major axis of the stereociliary bundle toward the tallest row of stereocilia (excitatory direction) results in a relative displacement between neighboring taller and shorter stereocilia. Tip links are stretched, thereby opening the transduction channels allowing different types of cations such as Ca^{2+} and K^+ to enter the cell through the membrane of its apical surface (Hudspeth and Corey, 1977; Howard *et al.*, 1988). A displacement of stereocilia in the opposite direction (inhibitory direction) slackens the tip links. Those transduction channels still open at the resting position are closed. The transduction channels were found to be located in the apical region of stereocilia (Jaramillo and Hudspeth, 1991; Denk *et al.*, 1995). The second population of links, the side links, laterally connect the stereocilia of the same row (Fig. 2B). The function of these links is not well understood yet, but up to now they are thought to tightly connect the entire hair bundle (Howard and Ashmore, 1986). The cytoskeleton of stereocilia consists of a dense arrangement of parallel oriented actin filaments, called the cuticular plate, forming a core and a root embedded in a platform made of actin on top of the cell (Tilney and Tilney, 1986). At the base, the number of actin filaments is reduced from a few hundred to about a dozen forming a point of flexing. The actin filaments are crosslinked by smaller filaments, probably fimbrin, preventing bending of stereocilia along their length. This construction allows stereocilia to flex around their pivot point located at the bottom where actin filaments penetrate the cuticular plate. A fibrous layer, called the tectorial membrane, overlies the organ of Corti. The tips of the tallest stereocilia of the OHC but not of the IHC touch the bottom side of the tectorial membrane. The origin of stereocilia displacement is a relative displacement between the basilar and the tectorial membrane forming together with the hair cells a layered or sheet-like structure. Bending of the stereocilia results in stretching the tip links and opening of the transduction channel allowing an influx of cations into the hair cell (Gillespie, 1995; Markin and Hudspeth, 1995). The ultrastructure of sensory hair bundles of the mammalian inner ear has been extensively studied by scanning and transmission electron microscopy (SEM and TEM) (Pickles et al., 1984; Lenoir et al., 1987; Russell and Richardson, 1987). These methods are restricted to fixed and dehydrated specimens, however. Structural details of living hair cells have been visualized by light microscopy with a limited spatial resolution due to the limitation of the optical system. Investigation of hair cells with atomic force microscopy (AFM) seems attractive since this method may combine three advantages: the high spatial resolution of scanning probe microscopy (Binnig et al., 1986), the possibility to work under physiological conditions (Hörber et al., 1992), and the opportunity to record mechanical properties of the structures being imaged (Hoh and Schoenenberger, 1994). AFM has already been used for simultaneous imaging and elasticity measurements on living cells under physiological conditions (Hörber et al., 1992; Hoh and Schoenenberger, 1994; Shroff et al., 1995; Radmacher et al., 1996). Therefore, it seems to be an appropriate technique for studying the mechanical characteristics of the hair bundle under physiological conditions.

B. Preparation

Organs of Corti were isolated from 4-day-old rats (Wistar, Interfauna, Germany) according to the methods used by Sobkowicz et al. (1975) and by Russell and Richardson (1987) for mice. They were cut into three segments (basal, medial, apical) and placed with the basilar membrane onto Cell-TAK-coated glass coverslips with a diameter of 10 mm. The sample was transferred into Ø 35-mm Falcon dishes filled with 3 ml culture medium MEM D-VAL with 10% heat-inactivated FCS and 10 mM Hepes buffer, (pH 7.2) supplemented at 37°C and 5% CO_2. Using this preparation, hair bundles are oriented perpendicular to the surface of the supporting glass coverslip. For experiments, specimens were transferred from the incubator to a specimen chamber. Cochlear cultures were bathed at room temperature (22°C) in a solution containing (millimolar concentrations) 144 NaCl, 0.7 NaH_2PO_4, 5.8 KCl, 1.3 $CaCl_2$, 0.9 $MgCl_2$, 5.6 D-glucose, 10 Hepes–NaOH, pH 7.4, osmolarity 304 mOsmol. The solution was exchanged before and after investigation using a conventional gravity-driven perfusion system to keep cells in good condition. At 4 to 6 days after birth the tectorial membrane of rats is already partially developed. In some preparations, the tectorial membrane directly touched the hair bundles of OHC. To guarantee free access of the AFM tip to the hair bundles, the tectorial membrane was removed with a cleaning pipette under the light microscope. Cultures were investigated for a maximum of 2 h.

III. AFM Technology

A. Short Introduction to AFM

The theories of AFM, the technological aspects, imaging techniques, and applications, have already been the topic of numerous publications (e.g., Meyer, 1992). Therefore, only the aspects necessary for understanding the contents of this chapter, not aspects just briefly described in other publications, will be discussed. The AFM experiments described here concentrate on force spectroscopy at the extracellular surface of hair cells. The capability of imaging is only used for correlation of force and position of the sensor tip on the specimen. As described by Binnig et al. (1986), AFM locally measures the force between a small tip and the specimen surface using a free-moving cantilever with the force constant k_L. Scanning the cantilever line for line across the sample while the tip interacts with the surface, we can obtain an image of the sample topography. The vertical deflection of the cantilever can be assigned to each point of the object. The AFM tip is scanned with a piezo-electric tube. The outer electrode is segmented into eight commensurate electrodes electrically connected in a way that the walls at the end of the piezo-electric tube remain perpendicular to the direction of motion (Siegel et al., 1995). The AFM tip consequently moves on a plane surface in contrast to piezo-electric tube scanners with only four outer electrodes scanning on a curved surface. Typically the scan range was between 1 and 8 μm. The force F exerted to the sample depends on the vertical cantilever deflection Δz and the force constant k_L as follows:

$$F = k_L \cdot \Delta z. \tag{1}$$

Typical force constants of commercially available cantilevers vary from 10^{-3} to 10^2 N/m. Different methods for detection of the cantilever deflection such as heterodyne interferometers (Martin et al., 1987), capacitive detection (Neubauer et al., 1990; Göddenhenrich et al., 1990; Miller et al., 1990), and tunneling current detection (Binnig et al., 1986) have been reported. In our instrument, the most common detection method was used: the optical lever method (Meyer and Amer, 1988, Alexander et al., 1989). The light of a laser diode is focused on the gold-coated surface of the AFM cantilever. The reflected light is detected on a position-sensitive four-segmented photo-diode. Deflection of the cantilever beam results in displacement of the laser spot on the photo-diode. The difference in intensity between the pairs of upper and lower segments encodes the vertical cantilever deflection. Torsional forces are detected subtracting the intensity at the left pair of segments from the intensity detected at the right pair of segments. AFM provides a wide range of microscopic techniques, which allow the measurement of different surface and material properties. These techniques can briefly be classified into two different operation modes: the contact and noncontact mode. Both principles, discussing the measurement of a force versus distance curve, can easily be explained (F versus d).

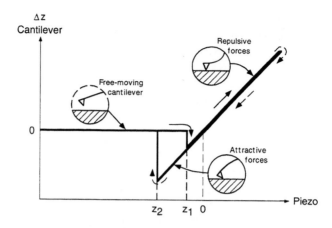

Fig. 3 Principle of force versus distance measurements. The deflection Δz of the cantilever beam is recorded versus elongation of the piezo-electric scanner vertically moving the sample. The force corresponds to the product of cantilever deflection and force constant. Before examination the sample is fully retracted, not interacting with the AFM cantilever (inset: free-moving cantilever). As the scanner extends (arrows), the tip and sample surface attract each other (position z_1) and the cantilever starts to bend downward. This attraction decreases with increasing scanner position. The force goes to zero (equilibrium between attractive and repulsive forces, piezo position 0). As the scanner continues to extend the total force becomes positive (inset: repulsive forces). At the maximum of the curve the scanner stops and retracts. With increasing contraction of the piezo-electric scanner (dashed lines with arrowhead) the cantilever follows this motion beyond zero position and loses contact at position z_2. At this point the tension of the cantilever beam exceeds the attractive forces (see inset), thereby losing contact with the sample surface.

A F versus d is a plot of the deflection of the cantilever versus the extension of the piezo-electric scanner (Fig. 3). On the left side of the curve, the scanner is fully retracted and the cantilever is in its resting position since the tip does not touch the sample. As the scanner moves, the cantilever remains undeflected until it comes close enough to the sample surface. As the atoms of tip and surface are gradually brought together, they first weakly attract each other. This attraction increases until the atoms are so close together that their electron clouds begin to repel each other electrostatically (position z_1). This electrostatic repulsion progressively weakens the attractive force as the interatomic separation continues to decrease. The force approaches zero when the distance between the atoms reaches a couple of angstroms, about the length of a chemical bond (position 0). As the scanner continues to extend, the total van der Waals force becomes positive (repulsive), and the atoms are in contact. This situation corresponds to the contact mode. For the noncontact mode AFM measurements have to be performed in the attractive regime (position z_1).

B. Principle of Hair Cell Measurements

Normally, force versus distance curves are recorded perpendicular to the sample support. In our experiments the cantilever scans in constant height exerting a horizontal force to the mechanosensory structures of outer hair cells. The applied force results in a horizontal displacement of the stereocilia along the axis of symmetry of V-shaped hair bundles. This kind of measurement (Fig. 4A) corresponds to a force versus distance curve performed in the horizontal direction. The feedback electronics normally keeping the measured force at a constant level was switched off during investigation. Every data point in an AFM image represents a spatial convolution (in the general sense, not in the sense of Fourier analysis) of the shape of the tip and the shape of the feature imaged. For most imaging applications this convolution leads to unwanted artifacts. In our measurements, we took advantage of the convolution of AFM tip and rod-like stereocilia geometry permitting us to calculate the stiffness of stereocilia. The van der Waals force curve (Fig. 3) represents just one contribution to the cantilever deflection. Local variations in the form of the F versus d curve indicate variations in the local elastic properties. These local variations in the slope of F versus d were used as a measure for the elasticity of the investigated sample. Langer *et al.* (2000) showed for stereocilia that a steep slope of the force curve indicates a high stiffness of the sample, while a shallow slope indicates a soft sample. For examination, the AFM tip was adapted to the axis of symmetry of the stereociliary bundles of OHC rotating the specimen chamber. A certain hair bundle was visually selected in the light microscope and directly positioned below the AFM tip adjusting the specimen stage. For the vertical fine approach, the specimen chamber was elevated with the help of a piezo-electric stack while scanning the cantilever tip (scan range: 4.5 μm). The approach is stopped as soon as stereocilia of the OHC cause significant deflection of the AFM cantilever (about 20 to 50 nm). For stiffness measurements, the AFM tip successively scans each stereocilium within a hair bundle (Fig. 4A). Only the tallest stereocilia of the hair bundle are touched with the frontal lateral face of the pyramidal tip and deflected in excitatory and inhibitory directions (Langer *et al.*, 2000).

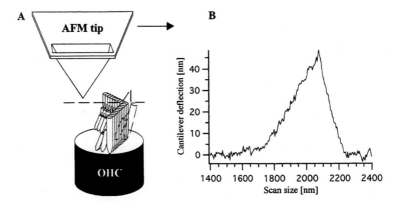

Fig. 4 Principle of stereocilia examination by AFM. (A) The AFM tip scans with the feedback electronics switched off across the hair bundle, thereby deflecting a stereocilium of the tallest row of stereocilia (see dashes stereocilium). Relative displacement between this taller and the adjacent shorter stereocilium (smaller dashes stereocilium) results in a strain of the connecting tip link possibly pulling at a transduction channel. Interaction with the stereocilium results in vertical deflection of the AFM cantilever beam indicated as a broken line for the excitatory scan direction (arrow). (B) Representative example of an AFM curve. The vertical deflection of the AFM cantilever was recorded versus time while scanning an individual stereocilium. The scan frequency was set to 2 Hz. The peak represents a force interaction between AFM tip and stereocilium (between 1700 and 2250 nm).

Figure 4B shows a typical example of an AFM trace. The ordinate displays the vertical deflection of the AFM cantilever while scanning an individual stereocilium. An increase in stiffness as a result, for example, of chemical fixation (Fig. 6 in Langer et al., 2000) leads to an increase in the positive slope of the force curve. This increase in slope reflects that for the same horizontal force F_L exerted by an AFM tip, a stiff stereocilium does not horizontally move as much as a soft stereocilium. Assuming that friction is very low, F_L can be calculated from the measured vertical deflection "a" (Fig. 4B) using

$$F_L = F_N \cdot \tan \alpha = k_{Cant} \cdot a \cdot \tan \alpha, \qquad [2]$$

where α is the angle between the lateral face of the pyramidal-shaped AFM tip and the scan direction, "a" is the vertical deflection of the cantilever beam, and k_{Cant} is the force constant of the AFM cantilever.

It was shown (Langer et al., 2001) that the lateral force constant k_L of a stereocilium is given by

$$k_L = F_L/(c - (a/\tan \alpha)), \qquad [3]$$

where c is the distance the AFM tip moves from the first point of contact to the point where the AFM cantilever reaches deflection "a". The denominator in Eq. [3] corresponds to the horizontal displacement of the tip of the stereocilium. Using the horizontal deflection method described earlier, it is essential to know the angle α between the lateral face of the AFM tip and the scan plane. Therefore, pyramidal-shaped Si_3N_4 tips with a well-defined tip angle of 70° were used. For calculation of the force constant the scan line with maximum peak amplitude was chosen. At this particular scan line the AFM tip is in contact with the center of the stereocilium. Contaminants and lubricants may affect our measurements by inducing adhesion and friction forces that distort the stiffness measurements. Such adhesive and friction forces were detected using a horizontal modulation technique.

The investigated hair bundles were scanned adding a sinusoidal modulation signal (peak-to-peak amplitude: 100 nm; frequency: 98 Hz) to the fast scan signal of the piezoelectric tube scanner of the AFM. This results in an additional horizontal forward and backward movement of the AFM tip at higher frequency. When being in contact with a stereocilium, this technique (Göddenhenrich et al., 1994) allows the detection of frictional and attractive forces. For negligible friction, the AFM tip slides up and down on identical paths while scanning the tip forward and backward at this fast modulation frequency. If frictional forces acting between AFM tip and sample surface increase, the cantilever moves on small loops indicating a speed- and direction-dependent interaction between tip and sample.

C. Basic Requirements for an AFM/Patch-Clamp Setup

Section II reported the unique features of hair cells. Not even their functional properties, as the possibility to transform a mechanical into an electrical signal, manifest not only their extraordinary characteristics but also their high degree of structural organization within the organ of Corti. The orientation of hair bundles of isolated hair cells may vary in a wide range when being attached to a supporting glass coverslip, thereby complicating the detection of the hair bundle orientation in the light microscope and access with the AFM tip. In contrast, cultures of the organ of Corti offer a nice way to hold hair cells in upright position. Accordingly, the instrument used to study the mechanical properties of hair bundles must support the identification of individual hair bundles of about 4 to 5 μm in width within the whole organ of Corti (diameter: about 300 μm). The identification exclusively by AFM at big scan ranges is insufficient. Contact between the AFM tip and the microvilli, situated between the hair cells (see Fig. 2A), promptly results in adsorption of contaminants at the tip surface. Accordingly, both proper selection of individual hair bundles under light microscopic control and precise adjustment of the AFM tip above the selected bundle are inevitable. The examination of stereocilia stiffness and gating of the transduction channel require acquisition of all physical values such as force and current as a function of time. In contrast to most commercial AFM instruments, our setup is able to record the correlation in time of all recorded signals and not only the spatial position of the AFM tip. Therefore, the AFM/patch-clamp hardware was combined with data acquisition hardware allowing the simultaneous generation of up to four stimulation patterns such as the scan signal for AFM and time-correlated acquisition of data such as the transduction current. For maximum flexibility, the data acquisition software provides the possibility of generating user-defined stimulation patterns.

D. AFM Setup

Attachment of the organ of Corti with the basilar membrane to the supporting glass coverslip results in orientation of hair cell bodies and stereocilia perpendicular to the specimen support. The identification of stereociliary bundles requires high optical resolution in both vertical and horizontal directions. Stereocilia are about 1.5 to 3.0 μm in length (postnatal rat; age: 3 days) located on top of a 40- to 50-μm-thick cell layer. To allow stable patch-clamp measurements simultaneously to AFM scans, as soon as the microelectrode has sealed to the lipid membrane of a hair cell, we must prevent relative movement between the cell body and the pipette. Therefore, the AFM tip, rather than the sample, is scanned. Figure 5 shows the three major units of the instrument: optical microscope, AFM, and patch-clamp device. The whole setup is mounted on a regulated air-damped table isolating mechanical vibrations at low frequencies (15–100 Hz). An upright optical microscope is mounted on a custom-made xy-translation stage allowing two-dimensional horizontal adjustment. This translation stage provides an independent positioning of the objective with respect to the separately mounted AFM cantilever and the patch-clamp pipette. Optical imaging is done using a water immersion objective (40×/0.75, Achroplan, Zeiss) with a working distance of 1.92 mm. The outside is coated with a nonconductive material allowing electrical recordings at low noise. Mechanical noise of the experimental setup was reduced by mounting AFM and the patch-clamp head stage on a separate platform. For additional higher mechanical stability eyepiece and CCD-camera were separated from the optical microscope. Cells were investigated in a liquid chamber filled with physiological solution. A micrometer

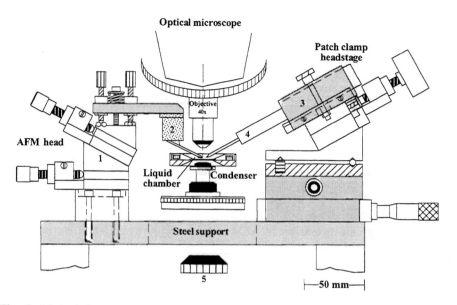

Fig. 5 Mechanical components of the experimental setup. It consists of three major units: the optical microscope, the AFM, and the patch-clamp head stage. 1, xyz-translation device allowing precise positioning of the AFM sensor; 2, electrically shielded piezoelectric tube scanner of the AFM equipped with a cantilever holder made of titanium; 3, patch-clamp headstage mounted on a xyz-translation device; 4, microelectrode holder; 5, halogen light source illuminating the sample from below. The liquid chamber includes a piezo-electric actuator allowing fine approach of the sample to the AFM tip.

screw-driven xy translation device allows horizontal displacement of the specimen chamber in two dimensions. The chamber is additionally adjustable in a vertical direction using a piezo-electric stack providing a resolution of better than 1 nm and a maximum vertical displacement of 20 μm. The patch-clamp setup, the object chamber, and the AFM head are fixed on a separate support, at a height of 190 mm, made of aluminum with a 20-mm-thick U-shaped steel plate on top to prevent mechanical vibrations. The AFM cantilever is mounted on a piezo-electric tube scanner introduced by Binnig and Smith (1986). The maximum horizontal scan size at maximum driving voltage of ± 160 V is 6.5 μm. The maximum elongation is 3.7 μm in the z direction. The AFM cantilever is attached to the end of a rigid titanium beam, which is small enough to position the cantilever between objective and specimen. In the coarse approach, the whole scanning unit is adjustable in the horizontal plane and vertical direction by a xy- and z-translation stage. The light microscope (Axioskop FS I, Zeiss) uses infrared differential interference contrast (DIC) and a water immersion objective providing information from a thin optical plane of the organ of Corti. The combination of an AFM with such optics is essential for the precise vertical approach of the AFM tip to the top of hair bundles of postnatal rats. Due to the limited working distance between objective and specimen, bending of the AFM cantilever had to be detected through the objective (Langer *et al.*, 1997). Detection from below through the condenser of the optical microscope was critical because the laser beam would have to penetrate the optically quite inhomogeneous cell tissue. Moreover, it was necessary to use a collimated instead of a focused laser beam for detection of the AFM cantilever deflection. A collimated laser beam exclusively allows detection of changes in angle rather than in linear movements of the cantilever. Movement of the cantilever along the optical axis of the collimated laser beam does not affect the electrical signal of the photodiode. The reflected beam propagates the same way back to the AFM detector. Only a deflection of the AFM cantilever results in an off-axis angle of the reflected beam shifting the laser spot on the photodiode. Nevertheless, AFM cantilevers had to be controlled in the light microscope for contaminants potentially causing changes in the detector signal.

E. Patch–Clamp Setup

The patch-clamp technique is an electrophysiological method that allows the recording of whole-cell or single-channel currents flowing across biological membranes through ion channels. Using patch clamp we have the possibility to control and manipulate the voltage (voltage clamp) of membrane patches or whole cells such as hair cells. The basic approach to measuring small ionic currents in the picoampere range as the transduction currents in hair cells requires a low-noise-recording technique combined with a precise mechanical positioning of the patch-clamp microelectrode. This paragraph concentrates on the AFM-specific requirements for a patch-clamp setup. For a detailed and general description of the patch-clamp technology, see, e.g., Sakmann and Neher (1995). The electrical connection between microelectrode and preamplifier headstage is kept as small as possible keeping the noise at a low level. Patch-clamp experiments require a precise and drift-free control of the movement of the microelectrode. Therefore, the head stage is attached to the top of a three-dimensional micrometer screw-driven translation stage with a spatial resolution of about 1 μm in each direction. For vertical fine positioning of the patch pipette, the same piezo-electric device as that used for the specimen chamber was implemented into this translation stage. Thus, the vertical movement of the patch pipette and the investigated specimen are synchronized. A Faraday cage made of a metal grate shielding the microelectrode from electrical noise surrounds the whole setup. The piezo-electric tube scanner of the AFM is separately shielded preventing electromagnetic cross-talk between scanner and patch-clamp microelectrode. In voltage-clamp experiments the voltage across the membrane is measured with respect to a bath electrode. Unfortunately, the AFM cantilever holder consists of metal (titanium) which in principle could lead to interfering potentials between the bath electrode and the AFM holder. Therefore, the AFM holder and the bath electrode were electrically connected with a small cable. Supporting cells surrounding the outer hair cells had to be removed using a big cleaning pipette (diameter: 10–15 μm) allowing free access to the cell bodies with the patch-clamp microelectrode. Cleaning pipettes were fabricated of Ø 2 mm borosilicate glass and mounted on a hydraulic-driven three-axis manipulator providing an adjustment range of $10 \times 10 \times 10$ mm. Patch-clamp pipettes were fabricated of Ø 1-mm quartz glass using a laser-based puller. Stimulating voltage steps and intracellular DC potentials were generated using custom-made software.

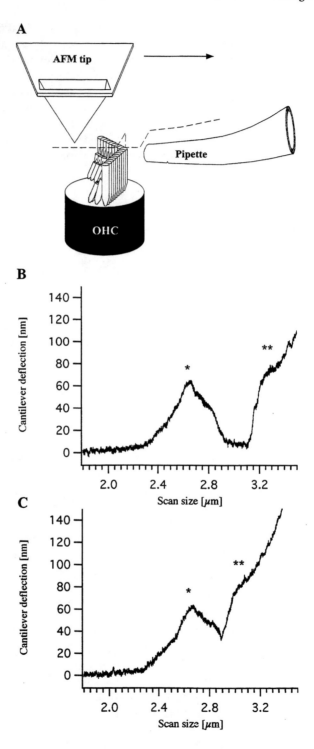

IV. Applications

Although hair bundles and their cross-links have been extensively examined by SEM and TEM many questions about their function remain unanswered. Filamentous links were identified in SEM and TEM, but their elastic and functional properties have not yet been directly measured. Local AFM measurements at individual stereocilia are presented providing information on the mechanical properties of side links and their possible function for hearing. For the further understanding of the mechanoelectrical transduction of stereociliary bundles in the inner ear, two experiments were performed. In the initial experiment, we examined the force transmitted by side-to-side links connecting stereocilia of the same row. The strength of side links determines the magnitude of displacement of adjacent stereocilia not directly interacting with the AFM tip. Results should allow the number of stereocilia displaced when an individual stereocilium is stimulated by AFM to be determined. In a second set of experiments the mechanical properties of stereocilia were correlated to the gating properties of the transduction channel.

A. Effect of Lateral Links on Hair Bundle Mechanics

Force transmission between adjacent stereocilia was examined deflecting one or two stereocilia of a hair bundle with a fine modulating glass fiber tip (diameter: 229 ± 21 nm (mean $\pm$ SD)) while scanning the entire stereociliary bundle with the AFM tip. The force transmitted via the side links was measured at different stereocilia detecting the magnitude of the AFM signal at the modulation frequency with a lock-in amplifier. Organs of Corti were rotated around their vertical axis until the direction of motion of AFM tip and stimulator were aligned parallel to the axis of symmetry of OHC bundles. Stimulation fibers were attached to a piezo-electric tube actuator mounted on a three-axis translation stage tilted by 27°. For parallel alignment of fiber and specimen support, the fiber was appropriately angled close to the tip melting the glass capillary with a heated wire. The fiber was coarsely placed under light microscopic control to the axis of symmetry of the OHC bundle using the three-axis translation stage. The voltage at the piezoelectric stack of the specimen chamber was adjusted for vertical fine positioning of the hair bundle. For the horizontal approach of the stimulating fiber, the piezo-electric tube was axially elongated until the fiber tip touched the top of the stereocilium. The relative arrangement of fiber, AFM cantilever, and OHC bundle during the approach is shown in Fig. 6A. Vertical and horizontal fine positioning of fiber and stereocilium was

Fig. 6 Approaching a glass fiber tip to the top of an individual stereocilium. (A) A fine glass fiber is coarsely approached under the optical microscope toward the major axis of a hair bundle of an OHC. For controlling the approach with nanometer precision, the AFM tip was successively scanned across the hair bundle and the upper edge of the glass fiber (dashed line corresponds to recorded force curve). While scanning, the glass fiber was horizontally moved with a piezo-electric positioning device to the top of an individual stereocilium. (B) AFM trace displaying the interaction with a stereocilium (∗) and the upper edge of the fiber tip (∗∗) during fine approach. The cantilever deflection is plotted versus scan size. The distance between stereocilium (∗) and pipette (∗∗) is about 200 nm. (C) AFM trace recorded after completing the approach. The pipette (∗∗) already touches the stereocilium (∗).

controlled by AFM. The AFM tip scans the upper edge of the fiber tip and the top of a stereocilium during the approach. The resulting AFM detector signal was controlled on the computer live display (Figs. 6B and 6C). After stopping the AFM scan, the fiber was sinusoidally modulated with 22 nm (peak-to-peak) at 357 Hz toward the major axis of the stereociliary bundle. The AFM tip was adjusted in height until interaction with the top of the hair bundle led to a vertical cantilever deflection of about 40 nm. Two hundred line scans were recorded on the entire hair bundle while one to two stereocilia were horizontally displaced with the fiber. The total scan range was 4×4 μm. For calculation of the transmitted force, only line scans representing a force interaction in the excitatory direction were taken into account. A lock-in amplifier, sensitive to signals at 357 Hz, detects the magnitude of the AFM signal with the time constant of the low pass filter set to 5 ms. The magnitude corresponds to the signal amplitude at 357 Hz and does not depend on the phase between the signal and the lock-in reference signal. The lock-in amplifier allows an accurate detection of the force signal at the frequency of interest (357 Hz) with high signal to noise ratio. Lateral force F_L was calculated for each stereocilium of investigated hair cells from the output signal of the lock-in amplifier using Eq. [2]. Figure 7 shows an example of an AFM line scan on a single stereocilium and the fiber (Fig. 7A) and the corresponding magnitude of the cantilever deflection (Fig. 7B) detected with the lock-in amplifier. For each stereocilium the maximum detected force F_L (calculated from the magnitude) was plotted versus stereocilium position (Fig. 8A). A stereocilium or stereocilia being in direct contact with the stimulation fiber were labeled as "0", while adjacent stereocilia being not in direct contact with the fiber were labeled as numbers starting from ± 1. Only that half of the hair bundle was taken into account, where the fiber was in contact with the directly stimulated stereocilium across its entire diameter and the nearest adjacent stereocilium was completely untouched by the fiber. The exact position of stereocilia within the stereociliary bundle and the relative position of the stimulating glass fiber and the stereocilia were detected in the AFM image (Fig. 9). Force interaction between stereocilia and AFM tip led to displacements of stereocilia from 0 nm to 249 ± 41 nm. The relative displacement between the stereocilium stimulated by the glass fiber and the stereocilium displaced by the AFM tip is expected to result in stretching of connecting lateral links. This would allow detection of forces transmitted by lateral links at different degrees of stretching. For a better comparison of the results, forces were normalized with respect to the corresponding maximum force detected at the directly stimulated stereocilium (Fig. 8A). Normalized maximal forces in Fig. 8B rapidly decrease from the directly stimulated to the first adjacent stereocilium. Stereocilia located at positions 1 to 8 reveal only a slight decrease in relative force from 36 to 20%. The background noise level of about 20% is due to acoustical evoked vibrations of the AFM cantilever caused by the oscillating piezoelectric tube and liquid coupling. Absolute forces in Fig. 8A also show a fast decline from stereocilium 0 to 1. Side links connecting the directly stimulated stereocilium with is direct neighbor transmit only a small rate of the maximum force. This result suggests a displacement of only adjacent stereocilia rather than the whole hair bundle. Although it was possible to detect the relative position of fiber tip and stereocilium during the approach, the force exerted to the stereocilium

A

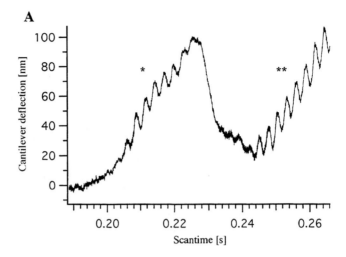

B

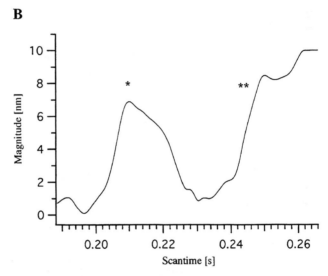

Fig. 7 Principle of force transmission measurement at cochlear hair bundles. A glass fiber is horizontally modulated at 357 Hz, thereby exerting a force to an individual stereocilium. AFM measures the force transmitted from this directly modulated stereocilium to adjacent stereocilia. (A) This trace was recorded on the modulated stereocilium (∗) and the modulating glass fiber (∗∗). The horizontal movement of pipette and stereocilium at 357 Hz induces the characteristic sinusoidal vibration pattern. (B) The AFM signal in (A) is detected at 357 Hz using a lock-in amplifier. This amplifier measures the magnitude of the cantilever deflection at 357 Hz containing no phase-related information. Compared to (A) the noise level is reduced to about 1 nm. The output signal of the lock-in amplifier directly encodes the force transmitted from the modulated to the investigated stereocilium (∗). Reproduced from Langer, M. G., Fink, S., Koitscher, A., Rexhausen, U., Hörber, J. K. H., and Ruppersberg, J. P. (2001). Lateral mechanical coupling of stereocilia in cochlear hair bundles. *Biophys. J.* **80,** 2608–2621, with permission from the Biophysical Society.

A

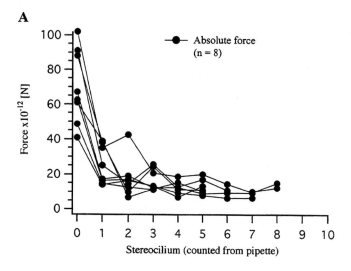

B

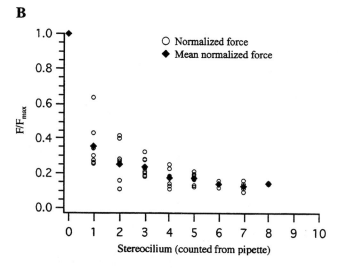

Fig. 8 Force transmitted by side links from stereocilium 0 to adjacent stereocilia of OHC of 4-day-old rats. Data of eight hair bundles (medial region of second and third OHC row) were pooled.(A) Absolute measured lateral forces (F_L, calculated from the magnitude as demonstrated in Fig. 7) are plotted versus stereocilium position. Adjacent stereocilia not in direct contact with the fiber were labeled with numbers starting from 1. Relative arrangement of fiber and hair bundle was controlled in the AFM image. The lateral force transmitted by side links rapidly decreases from the stimulated stereocilium (0) to the nearest adjacent stereocilium (1). In contrast, the forces measured at adjacent stereocilia (from 2 to 8) scatter around 15 pN. Maximum forces measured at stereocilium 0 vary between 40 and 105 pN. (B) Plot of normalized (open circles) and mean normalized forces (filled symbol) versus stereocilium position. The transmitted force shows a maximum decline between stereocilium 0 and 1. From stereocilium 1 to 6 the transmitted force declines slightly (reaching a steady state level at about 15% (positions 6 to 8). Reproduced from Langer, M. G., Fink, S., Koitscher, A., Rexhausen, U., Hörber, J. K. H., and Ruppersberg, J. P. (2001). Lateral mechanical coupling of stereocilia in cochlear hair bundles. *Biophys. J.* **80**, 2608–2621, with permission from the Biophysical Society.

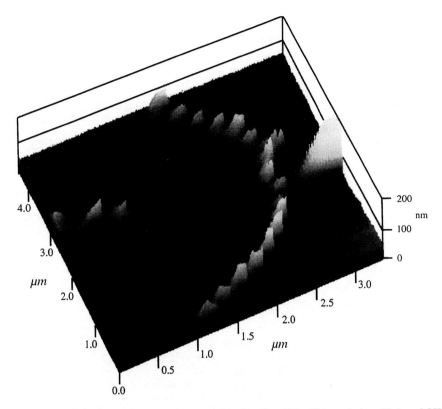

Fig. 9 Three-dimensional surface plot of a stimulating glass fiber (right side) attached to a V-shaped OHC hair bundle (white cones). The overlaid sawtooth-like pattern of the pipette tip is caused by the sinusoidal movement of the driving piezo-electric actuator. The scan size is 3.46 × 4.3 μm. This plot clearly demonstrates the capability of AFM to control and detect the relative position between fiber and stereociliary bundle at high spatial resolution.

could not be measured. Therefore, the maximum force detected at position 0 varies from 41 to 102 pN. We can conclude that forces transmitted by lateral links are much less than those measured for a directly stimulated stereocilium, which is mediated by the elasticity of the stereocilium itself. This result supports the hypothesis of a weak interaction between stereocilia by lateral links. The background amplitude level of about 20% in Fig. 8 makes it difficult to distinguish a weak coupling by lateral links from a lack of coupling. Besides the transmitted force at 357 Hz, AFM curves contain information about the stereocilia stiffness. We used this information for detecting the mechanical effect of the touching glass fiber on stiffness of adjacent stereocilia. If lateral links contribute to the stiffness obtained at individual stereocilia we would expect to see an increase in stiffness for stereocilia adjacent to those touched with a fiber. Stiffness data were calculated according to Eq. [3] only for the excitatory direction, where the AFM tip displaces the stereocilia toward the fiber tip (Fig. 10). Not only the stiffness (filled circles) of the

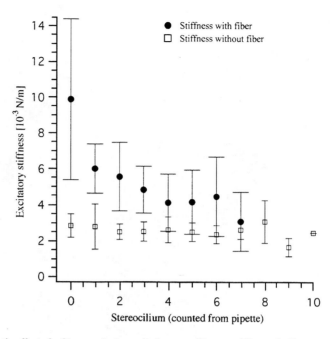

Fig. 10 Elastic effect of a fiber attached to a single stereocilium on stiffness of adjacent stereocilia. Here, we successively measured the stiffness of individual stereocilia within a hair bundle while one of these stereocilia was attached to a glass fiber. Depending on force transmission by side links, stiffness measured at a single stereocilium should also represent more or less the stiffness of adjacent stereocilia. Open squares mark the stiffness of stereocilia investigated without a supporting glass fiber; Filled circles mark the stiffness of stereocilia investigated with a glass fiber attached to a single stereocilium (mean values ± SD). Stiffness of stereocilia investigated without a fiber shows no significant dependence on position. In contrast, measurements with a fiber attached to a stereocilium show an increased stiffness not only for the directly supported but also for adjacent stereocilia (1 to 6), indicating a weak effect of side links on total stiffness. Reproduced from Langer, M. G., Fink, S., Koitscher, A., Rexhausen, U., Hörber, J. K. H., and Ruppersberg, J. P. (2001). Lateral mechanical coupling of stereocilia in cochlear hair bundles. *Biophys. J.* **80**, 2608–2621, with permission from the Biophysical Society.

directly touched stereocilium at position 0 was found to be increased but also the stiffness of stereocilia 1 to 4. Mean stiffness in excitatory direction was $(4.8 \pm 1.8) \times 10^{-3}$ N/m. This is about 1.9 times higher compared to the mean stiffness of stereocilia not touched with a fiber. For better comparison, stiffness data of stereocilia examined without a fiber were plotted as open symbols into the same graph. For position 7, the stereocilium with and the stereocilium without a touching glass fiber show approximately identical stiffness. This demonstrates that lateral links do not tightly couple individual stereocilia of outer hair cells and small independent movements of single stereocilia are possible. Side links between the stereocilia may transmit a force when sufficiently stretched and coupling of stereocilia has a similar stiffness as a single displaced stereocilium. We can conclude that, depending on elongation of side links, it is in principle possible to deflect single stereocilia by AFM at least for the preparation of postnatal rats used here.

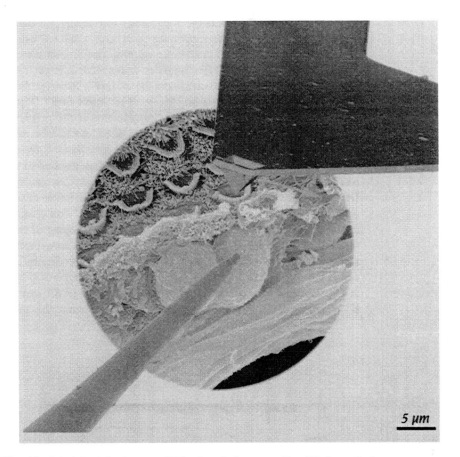

Fig. 11 Principle of simultaneous AFM and patch-clamp recording. This image displays an arrangement of three separate SEM images with identical scale. The center represents a part of the organ of Corti with V-shaped hair bundles on top. The upper right side displays a small part of the cantilever beam with the pyramidal-shaped tip scanning the top of a V-shaped hair bundle of an OHC. The arrow indicates the scan direction. The lower left side shows a glass pipette in contact with the lipid membrane of an OHC allowing the transmembrane currents to be recorded as the transduction current. (See Color Plate.)

B. Gating of the Transduction Channel in OHC

In the experiments described previously, we used AFM only for examination of the elastic properties of hair cells. But how can we benefit from using patch clamp simultaneously with AFM? Many micromechanical measurements have already been performed at entire stereociliary bundles of sensory hair cells using thin glass fibers directly attached to the bundle or fluid jets. The receptor potential or transduction current was measured in response to the displacement of stereocilia. The possibility to study the kinetics of a single transduction channel over the whole range of its open probability requires a technique allowing the stimulation of a single stereocilium. As shown in the previous section, AFM offers the opportunity of exerting a force very locally to an individual stereocilium. After supporting cells were removed using a cleaning pipette, a patch pipette filled with intracellular solution (millimolar concentrations: KCl, 135; $MgCl_2$, 3.5; $CaCl_2$, 0.1; EGTA, 5; Hepes, 5; Na_2ATP, 2.5; pH 7.4) was attached to the lateral wall of an OHC of the outermost row of OHC. Thereby, the glass microelectrode forming a seal on the plasma membrane of an intact OHC (Fig. 11) isolates a small patch. This so-called "cell-attached" configuration is the precursor of the whole-cell configuration where the microelectrode is in direct electrical contact with the inside of the cell. For low-noise measurements of single-ion channels the seal resistance should be typically in the range >1 GΩ. A pulse of suction applied to the pipette breaks the patch creating a hole in the plasma membrane and provides access to the cell interior. During recording, the electrical resistance between the inside of the pipette and the hair cell should be very small. Many voltage-activated K^+-ion channels are embedded in the lipid membrane of outer hair cells. Opening and closing of these channels increase the background noise level during transduction current measurements. The current response of outward rectifying K^+-ion channels was controlled by applying 10-mV steps across the cell membrane (progressively increased from -100 to $+40$ mV) as shown in Fig. 12. The outward currents mainly correspond to K^+ currents of voltage-gated K^+ channels. During transduction current measurements, the holding potential of the

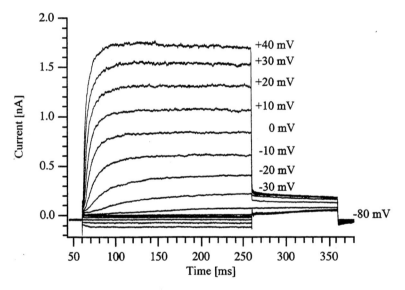

Fig. 12 Activation of voltage-dependent K^+ channels in OHC. This graph displays the electrical response of voltage-dependent K+ channels located in the cell membrane of OHC. Outward currents were activated, applying small voltage steps across the cell membrane. Starting at a holding potential of -80 mV the intracellular potential was changed in 10-mV steps from -100 to $+40$ mV. At voltages below the holding potential (-80 mV) only small inward currents (less than 50 pA) were detected. Starting from -40 mV to a more positive potential, outward-rectifying K^+ currents were detectable. These measurements qualitatively allow testing the leakage and the electrical contact to the intracellular space of examined hair cells.

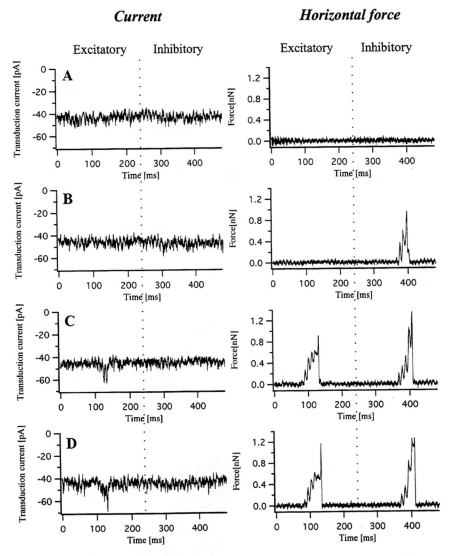

Fig. 13 Simultaneous AFM/patch clamp measurements. The AFM tip scans in the same line while approaching the organ of Corti to the tip. Transduction currents were measured in the whole cell-recording mode keeping the holding potential at −80 mV. (A) Current and vertical force measured before interaction between AFM tip and stereocilium. (B) First contact between the AFM tip and a stereocilium of an OHC. The specimen is continuously moved toward the AFM tip at from 0 to 470 ms. While scanning in excitatory direction the specimen does not yet touch the AFM tip. At about 390 ms the hair bundle is near enough to the tip resulting in a vertical deflection of the AFM cantilever. The cantilever is modulated at about 98 Hz, thereby stimulating the transduction channel several times. (C) The approach was stopped. Now, the stereocilium is displaced in both excitatory and inhibitory directions. At about 130 ms an inward current of 19 pA is detected while displacement in an inhibitory direction does not result in the opening of transduction channels. (D) A second example demonstrating that a transduction current is detectable only for the excitatory but not for the inhibitory direction.

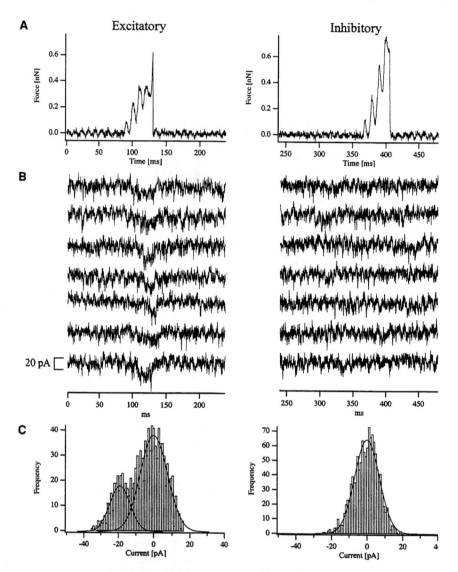

Fig. 14 Examination of the transduction current amplitude. Recorded current and force traces are separately displayed for the excitatory and inhibitory directions of stimulation. (A) These graphs correspond to the force applied to a single stereocilium in excitatory and inhibitory directions. (B) Successive stimulation of the identical stereocilium by AFM resulted in the reproducible opening of transduction channels at between 89 and 131 ms. In contrast, force application in an inhibitory direction did not result in the opening of transduction channels. (C) For detailed analysis, currents recorded during stimulation by AFM (from 89 to 131 ms and from 366 to 407 ms) were pooled and separately displayed for excitatory an inhibitory directions in histograms. The distribution around 0 pA represents the current for the closed state of the channel while the second smaller distribution at around 19 pA represents the current for the open state of transduction channels. Data demonstrate that AFM allows stimulation of only a few transduction channels rather than of all channels of the entire hair bundle.

hair cell was set to -80 mV corresponding to the reversal potential of K^+-ion channels. Currents are expected to be in the range of a few nanoamperes rather than in the range of hundreds of picoamperes. After forming a seal, the AFM tip was moved to the top of the corresponding hair bundle under light microscopic control. The AFM tip successively displaced each stereocilium within a hair bundle as described in Section IV,A. In contrast to force transmission measurements, a sinusoidal voltage was added to the normal AFM scan signal modulating the AFM tip in a horizontal direction with 190 nm at 98 Hz. Thus, the hair bundle was slightly displaced several times while interacting with the lateral face of the AFM tip (Fig. 13). The AFM tip repeatedly scanned in the same line while approaching the hair bundle towards the AFM tip (Fig. 13). As expected, an inward current was not detected until displacement of a stereocilium in excitatory direction (Fig. 13C). In the previous section we demonstrated a weak transmission of force from the directly stimulated stereocilium to adjacent stereocilia implying that only few channels are opened. Are these results of elasticity measurement confirmed by electrophysiological findings? A set of transduction current measurements is displayed in Fig. 14. The tip of an AFM cantilever repeatedly scanned across the same stereocilium of an OHC (from medial turn of a postnatal rat, day 3). Applied horizontal forces ranged up to 0.8 nN (Fig. 14A) resulting in stereocilia displacements of about 350 nm in an excitatory direction and of 250 nm in an inhibitory direction. The AFM tip displaces the stereocilium in the excitatory direction between 89 and 131 ms, thereby opening transduction channels (Fig. 14B). The current amplitude was determined for the period of interaction with the AFM tip plotting histograms of the current (Fig. 14C). The distribution around 0 pA corresponds to the closed state of the channels, while the distribution around 19.1 ± 7.6 pA corresponds to the maximal activation. This current amplitude is twice that expected for a single transduction channel (9.7 ± 1.2 pA; Géléoc et al., 1997). What might underlie this discrepancy? For the detection of single channel currents we have to improve the electrical shielding to reduce the noise level, currently 12.6 pA. Additional high-frequency recordings might be necessary to distinguish the current of a single channel from the total current of several smaller channels. The cut-off frequency of our patch-clamp amplifier was currently limited to 3 kHz allowing the recording of currents at low noise level, but preventing the detection of fast events. However, if we compare the current amplitude of 19.1 ± 7.6 pA in Fig. 14 to the maximum transduction current of about 462 pA measured for entire hair bundles of postnatal mice (Géléoc et al., 1997), the results confirm that AFM is well suited for local stimulation of individual stereocilia.

V. Discussion

In the inner ear the tallest stereocilia of OHC directly contact the tectorial membrane. The process of opening and closing of transduction channels therefore depends on the mechanism of force transmission from the top of stereocilia to the transduction channel. Three structural components are important for understanding the gating process: (i) the complex mechanics of the cytoskeleton of individual stereocilia, (ii) the cross-linkage between adjacent stereocilia, and (iii) the elastic properties of tip links and their connections to the transduction channel. Understanding the functional properties of the functional unit formed by tip link and transduction channels requires separation of these three effects. In this chapter, we show how this problem was addressed using an experimental tool providing local access to individual stereocilia. The results obtained with the presented AFM/patch-clamp setup demonstrate that AFM is an appropriate technique to locally apply forces to individual stereocilia. This simplifies the calculation of stiffness and actual displacement of the mechanosensory structures (iii) of an individual stereocilium. The methods applying force to the entire hair bundle do not allow the discrimination of the contributions of the three interrelated structural components. Presented data confirm that AFM allows local investigation of stereocilia stiffness. At small deflection amplitudes lateral links produce only a weak increase of measured stereocilia stiffness, thereby allowing local stimulation of single stereocilia by AFM. However, it may happen that neighboring stereocilia are significantly displaced by directly pulling at the side links. The transduction current amplitude measured was about twice that expected for a single transduction channel. This might be explained in three different ways: (i) the AFM tip pulled at two serially arranged tip links connecting adjacent stereocilia of different rows, thereby opening one channel each; (ii) the AFM tip pulled at one tip link connected with two transduction channels located at both ends of the tip link; (iii) the AFM tip displaces the directly stimulated stereocilium and one adjacent stereocilium. These questions must be answered in future experiments measuring the effect of side link elasticity on total stiffness as a function of stereocilium displacement. Presently, we can conclude from AFM/patch-clamp measurements that only few transduction channels located very near to the stimulating AFM tip contribute to the total current.

VI. Outlook

The combination of AFM and patch-clamp is not limited to the examination of hair cells; it answers other questions in cell biology, such as, for example, the molecular mechanism of voltage-dependent membrane displacements. Quite often AFM has been used to identify ion channels in the plasma membrane of cells imaging the membrane surface. High-resolution images of proteins in cell membranes were normally obtained on rigid substrates such as, e.g., mica. However, it was difficult to identify ion channels in plasma membranes of intact cells. Even if we identify lump-like structures on the membrane surface appearing similar to the expected structure of the protein, we still must be critical. It is currently impossible to exclude the fact that identified structures correspond to a different type of protein appearing very similar to the protein we would like to localize. In this case it would be very helpful to verify our observation using a second independent technique such as patch-clamp. Patch-clamp could be used as a tool for electrical stimulation of polar molecules in the membrane of whole cells as ion channels while the AFM cantilever locally senses the resulting conformational changes (Mosbacher *et al.*, 1998). In first experiments Mosbacher *et al.* demonstrated that the membrane movement of HEK293 cells became sensitive to the holding potential after transfection with Shaker K^+ channels. The total movement remained in phase with the displacement current of the highly charged transmembrane segment S4 to high frequencies suggesting that Mosbacher *et al.* observed movement of the voltage sensor region rather than changes in the pore gating transition. Using AFM and patch-clamp in combination with chemical blockage, it should be possible to specifically identify ion channels in the plasma membrane and to study their kinetics as well as the force exerted by the ion channel.

Acknowledgments

I would like to thank Peter Ruppersberg for making this project possible and giving me the freedom to independently do my scientific work; J. K. H. Hörber, for his technical support and helpful scientific discussions; Stefan Fink, for his continuous support during experiments and reading this manuscript; and Alfons Rüsch, for helpful discussions. I am grateful to Wolfgang Öffner of the EMBL in Heidelberg for developing the reliable and excellent AFM electronics. This work was financially supported by the Deutsche Forschungsgemeinschaft (Klinische Forschergruppe Hörforschung DFG Nr. Ze 149/6-2 and LA 1227/1-1) and the fortuene program (Projects Nr. 347-2 and Nr. 712-0-0) of the University Clinic Tübingen.

References

Alexander, S., Hellemans, L., Marti, O., Schneir, J., Ellings, V., Hansma, P. K., Longmire, M., and Gurleey, J. (1989). An atomic-resolution atomic-force microscope implemented using an optical lever. *J. Appl. Phys.* **65**, 164–167.

Binnig, G., Gerber, C., and Quate, C. F. (1986). Atomic force microscope. *Phys. Rev. Lett.* **56**(9), 930–933.

Binnig, G., and Smith, D. P. E. (1986). Single-tube three-dimensional scanner for scanning tunneling microscopy. *Rev. Sci. Instrum.* **57**, Issue 8, 1688–1689.

Denk, W., Holt, J. R., Shepherd, G. M., and Corey, D. P. (1995). Calcium imaging of single stereocilia in hair cells: localization of transduction channels at both ends of tip links. *Neuron* **15**(6), 1311–1321.

Florin, E. L., Moy, V. T., and Gaub, H. E. (1994). Adhesion forces between individual ligand-receptor pairs. *Science* 1994 **264**(5157), 415–417.

Furness, D. N., and Hackney, C. M. (1985). Cross links between the stereocilia in the guinea pig cochlea. *Hear. Res.* **18**, 177–188.

Furness, D. N., Richardson, G. P., and Russell, I. J. (1989). Stereociliary bundle morphology in organotypic cultures of the mouse cochlea. *Hear. Res.* **38**(1–2), 95–109.

Géléoc, G. S. G., Lennan, G. W. T., Richardson, G. P., and Kros, C. J. (1997). A quantitive comparison of mechanoelectrical transduction in vestibular and auditory hair cells of neonatal mice. *Proc. Roy. Soc. Lond. B* **264**, 611–621.

Gillespie, P. G. (1995). Molecular machinery of auditory and vestibular transduction. *Curr. Opin. Neurobiol.* **5**(4), 449–455.

Göddenhenrich, T., Lemke, H., and Hartmann, U. (1990). *J. Vac. Sci. Technol.* **A6**, 383.

Göddenhenrich, T., Müller, S., and Heiden, C. (1994). A lateral modulation technique for simultaneous friction and topography measurements with the atomic force microscope. *Rev. Sci. Instrum.* **65**(9), 2870–2873.

Hoh, J. H., and Schoenenberger, C. A. (1994). Surface morphology and mechanical properties of MDCK monolayers by scanning force microscopy. *J. Cell Sci.* **107**(Pt 5), 1105–1114.

Hörber, J. K. H., Häberle, W., Ohnesorge, F., Binnig, G., Liebich, H. G., Czerny, C. P., Mahnel, H., and Mayr, A. (1992). Investigation of living cells in the nanometer regime with the scanning force microscope. *Scan. Microsc.* **6**, 919–930.

Hörber, J. K., Mosbacher, J., Häberle, W., Ruppersberg, J. P., and Sakmann, B. (1995). A look at membrane patches with a scanning force microscope. *Biophys. J.* **68**(5), 1687–1693.

Howard, J., and Ashmore, J. F. (1986). Stiffness of sensory hair bundles in the sacculus of the frog. *Hear. Res.* **23**(1), 93–104.

Howard, J., Roberts, W. M., and Hudspeth, A. J. (1988). Mechanoelectrical transduction by hair cells. *Annu. Rev. Biophys. Biophys. Chem.* **17**, 99–124.

Hudspeth, A. J., and Corey, D. P. (1977). Sensitivity, polarity, and conductance change in the response of vertebrate hair cells to controlled mechanical stimuli. *Proc. Natl. Acad. Sci. U.S.A.* **74**(6), 2407–2411.

Jaramillo, F., and Hudspeth, A. J. (1991). Localization of the hair cell's transduction channels at the hair bundle's top by iontophoretic application of a channel blocker. *Neuron* **7**(3), 409–420.

Langer, M. G., Fink, S., Koitschev, A., Rexhausen, U., Hörber, J. K. H., and Ruppersberg, J. P. (2001). Lateral mechanical coupling of stereocilia in cochlear hair bundles. *Biophys. J.* **80**, 2608–2621.

Langer, M. G., Koitschev, A., Haase, H., Rexhausen, U., Hörber, J. K., and Ruppersberg, J. P. (2000). Mechanical stimulation of individual stereocilia of living cochlear hair cells by atomic force microscopy. *Ultramicroscopy* **82**(1–4), 269–278.

Langer, M. G., Öffner, W., Wittmann, H., Flösser, H., Schaar, H., Häberle, W., Pralle, A., Ruppersberg, J. P., and Hörber, J. K. H. (1997). A scanning force microscope for simultaneous force and patch-clamp measurements on living cell tissues. *Rev. Sci. Instrum.* **68**(6), 2583–2590.

Lenoir, M., Puel, J. L., and Pujol, R. (1987). Stereocilia and tectorial membrane development in the rat cochlea: A SEM study. *Anat. Embryol.* **175**, 477–487.

Lim, D. J. (1986). Functional structure of the organ of Corti: A review. *Hear. Res.* **22**, 117–146.

Markin, V. S., and Hudspeth, A. J. (1995). Gating-spring models of mechanoelectrical transduction by hair cells of the internal ear. *Annu. Rev. Biophys. Biomol. Struct.* **24**, 59–83.

Marquis, R. E., and Hudspeth, A. J. (1997). Effects of extracellular Ca2+ concentration on hair-bundle stiffness and gating-spring integrity in hair cells. *Proc. Natl. Acad. Sci. U.S.A.* **94**(22), 11,923–11,928.

Martin, Y., Willams, C. C., and Wickramashinghe, H. K. (1987). Atomic force microscope-force mapping and profiling on a sub 100-Å scale. *J. Appl. Phys.* **61**, 4723–4729.

Meyer, E. (1992). Atomic force microscopy. *Prog. Surf. Sci. (UK)* **41**, 3–49.

Meyer, G., and Amer, N. (1988). Novel optical approach to atomic force microscopy. *Appl. Phys. Lett.* **53**, 1054 ff.

Miller, G. L., Wagner, E. R., and Sleator, T. (1990). Resonant phase shift technique for the measurement of small changes in grounded capacitors. *Rev. Sci. Instrum.* **61**, 1267.

Mosbacher, J., Langer, M. G., Hörber, J. K. H., and Sachs, F. (1998). Voltage-dependent membrane displacements measured by atomic force microscopy. *J. Gen. Physiol.* **111**(1), 65–74.

Neubauer, G., Cohen, S. R., McClelland, G. M., Horne, D., and Mate, C. M. (1990). Force microscopy with a bidirectional capacitance sensor. *Rev. Sci. Instrum.* **61**, 2296–2308.

Oesterhelt, F., Oesterhelt, D., Pfeiffer, M., Engel, A., Gaub, H. E., and Muller, D. J. (2000). Unfolding pathways of individual bacteriorhodopsins. *Science* **288**(5463), 143–146.

Parpura, V., and Fernandez, J. M. (1996). Atomic force microscopy study of the secretory granule lumen. *Biophys. J.* **71**(5), 2356–2366.

Pickles, J. O., Comis, S. D., and Osborne, M. P. (1984). Cross links between stereocilia in the guinea-pig organ of Corti, and their possible relation to sensory transduction. *Hear. Res.* **15**, 103–112.

Radmacher, M., Fritz, M., Kacher, C. M., Cleveland, J. P., and Hansma, P. K. (1996). Measuring the viscoelastic properties of human platelets with the atomic force microscope. *Biophys. J.* **70**(1), 556–567.

Rotsch, C., and Radmacher, M. (2000). Drug-induced changes of cytoskeletal structure and mechanics in fibroblasts: an atomic force microscopy study. *Biophys. J.* **78**(1), 520–535.

Russell, I. J., and Richardson, G. P. (1987). The morphology and physiology of hair cells in organotypic cultures of the mouse cochlea. *Hear. Res.* **31**, 9–24.

Sakman, B., and Neher, E. (1995). "Single-Channel Recording," Second Edition. Plenum Press, New York/London.

Shroff, S. G., Saner, D. R., and Lal, R. (1995). Dynamic micromechanical properties of cultured rat atrial myocytes measured by atomic force microscopy. *Am. J. Physiol.* **269**(1, Pt 1), 286–92.

Siegel, J., Witt, J., Venturi, N., and Field, S. (1995). Compact large-range cryogenic scanner. *Rev. Sci. Instrum.* **66**(3), 2520–2532.

Sobkowicz, H. M., Bereman, B., and Rose, J. E. (1975). Organotypic development of the organ of Corti in culture. *J. Neurocytol.* **4**(5), 543–572.

Tilney, L. G., and Tilney, M. S. (1986). Functional organization of the cytoskeleton. *Hear. Res.* **22,** 55–77.

Von Békésy, G. (1960). "Experiments in Hearing," p. 745. MacGraw-Hill, New York.

Zhao, Y., Yamoah, E. N., and Gillespie, P. G. (1996). Regeneration of broken tip links and restoration of mechanical transduction in hair cells. *Proc. Natl. Acad. Sci. U.S.A.* **24, 93**(26), 15,469–15,474.

Zine, A., and Romand, R. (1996). Development of the auditory receptors of the rat: A SEM study. *Brain Res.* **721,** 49–58.

CHAPTER 8

Biotechnological Applications of Atomic Force Microscopy

Guillaume Charras,* Petri Lehenkari,† and Mike Horton*

* Bone and Mineral Center
Department of Medicine
The Rayne Institute
University College London
London, WC1E 6JJ
United Kingdom

† Departments of Surgery and Anatomy
University of Oulu
Oulu
Finland

I. Introduction

Ever since the invention of the atomic force microscope in 1986 by Binnig *et al.* (1986), atomic force microscopy (AFM) has been extensively applied to the study of a variety of biological phenomena. Indeed, AFM has several advantages that make it particularly attractive to cell biologists. First, it can yield high-resolution spatial images of live cells under near physiological conditions. Second, due to the ability of the AFM to measure forces, it has become possible to evaluate physical parameters in biological materials, such as material properties and binding forces, which had previously been inaccessible. To date, AFM applications in cell biology can be classified into five broad categories: imaging, material property measurements, binding force measurements, biophysics, and micromanipulation studies.

Several authors have studied the innocuity of AFM imaging on living cells. Using cells incubated with a cytoplasmic fluorescent dye, Parpura *et al.* (1995) showed that standard AFM tips did not induce dye leakage from cells, but sharper tips did. Schaus and Henderson (1997) showed that cells imaged with AFM remained viable for up to 48 h postimaging. However, they also showed that phospholipid membrane components accumulated on the tip during contact imaging. This phenomenon was not observed when force–distance curves where repeatedly taken on cells. You *et al.* (2000) found that continuous contact imaging for up to 2 h in contact mode induced cell retraction. All of these results taken together show that AFM is only minimally disruptive to cells when used for short periods of time.

Imaging is the most straightforward of AFM applications and has served, in particular, to give biology a sense of space by enabling the three-dimensional visualization of biological phenomena. (See the work by Jena and by Le Grimellec and Radmacher in Chapters 2, 3, and 4 in this book.) Imaging has now been extensively applied to a large number of cell types and biological phenomena. Henderson *et al.* (1992) imaged the dynamics of filamentous actin in living glial cells. Lal and co-workers (Shroff *et al.*, 1995) imaged the outgrowth of neurites and witnessed cytoskeletal reorganization. In a particularly valuable study, Kuznetsov *et al.* (1997) examined both the motility and the division of living cells. Nuclear pores and their conformational changes in responses to a variety of compounds were examined by Danker and Oberleithner (2000). Schneider *et al.* (1997) observed the membrane mechanisms involved in exocytosis. Müller (Müller and Engel, 1999) examined the molecular structure of the porin, Ompf, and its rearrangement in response to voltage changes (Engel and Müller, 2000.) (See also Chapter 13 in this work by Müller and Engel.) Many of these results would have be unobtainable using other imaging techniques.

By taking force–distance curves over a whole grid and analyzing each force–distance curve, AFM enables the material properties of cells to be estimated (Radmacher, 1997; Weisenhorn *et al.*, 1993). Although the material properties of cells can be assessed using other techniques such as micropipette aspiration, laser tweezers, or microbead pulling, AFM offers the unique combination of high-degree precision in spatial resolution in material property measurement and the possibility of obtaining measurements from cells spread on substratum. This latter application is the subject of another chapter of this book. (See Chapter 3 in this work by Le Grimellec and co-workers and Chapter 4 in this work by Radmacher).

Whereas many measurements of binding forces between ligand and receptor adsorbed to mica have been reported (Florin *et al.*, 1994; Hinterdorfer *et al.*, 1996) (see also Chapters 6 and 14–16 in this work), there have been few attempts to apply this technique to living cells. Recently, Lehenkari and Horton (1999) were the first to measure the binding forces between integrin receptors in intact cells and Arg–Gly–Asp (RGD) amino acid sequence-containing extracellular matrix protein ligands. Using a modification of their technique, Lehenkari *et al.* (2000) reported the first binding map of functional receptors on living cells. In a similar study, Grandbois *et al.* (2000) showed that it was possible to differentiate red blood cells of different blood groups within a mixed population by using affinity imaging with the blood group A specific lectin from *Helix pomatia*.

Thanks to its capacity to measure cellular profiles and cellular material properties at high resolution, AFM can be applied to biomechanical and biophysical problems. Davies (1997) used high-resolution images of endothelial cells and computational fluid dynamics to calculate the shear stresses on cells due to fluid flow. Sato *et al.* (2000) examined the changes in material properties of bovine endothelial cells after exposure to fluid shear stress. Charras and Horton (2002a) utilized the material properties and topographies of live osteoblasts acquired using AFM as an input into finite element modeling software to calculate the cellular strains resulting from a variety of mechanical stimuli. Several studies have taken advantage of the possibility of acquiring three-dimensional images of cells to study real-time changes in cell volume. Schneider *et al.* (1997) measured the changes in cell volume in endothelial cells upon exposure to aldosterone. Quist *et al.* (2000) used AFM to investigate the modulation of cell volume by extracellular calcium levels and showed that these were mediated through connexin protein containing gap junctions.

In the last category of applications, AFM has been used as an ultraprecise micromanipulator. Domke *et al.* (1999) used the AFM to map the mechanical pulse of cultured cardiomyocytes. Thie *et al.* (1998) examined the adhesive forces between trophoblasts and uterine epithelium using whole cells instead of isolated molecules to functionalize the tips. The adhesion forces recorded between cells were around 3 nN, which is an order of magnitude higher than the molecule–ligand adhesion forces. Recently, Sagvolden *et al.* (1999) developed a new use of AFM to quantify the adhesion forces of cells to a substrate. Charras and colleagues (Charras *et al.*, 2000; Charras and Horton, 2002b), who used AFM to stimulate cells mechanically, measured cellular material properties while monitoring the changes in intracellular calcium resulting from stimulation and showed that cells exhibiting changes in intracellular calcium had been submitted to a higher strain.

In summary, over the years, AFM has shown that it has the potential to become a crucial instrument in cell biology; however, to realize its full potential in this field, a certain number of problems needed to be solved.

1. The AFM had to be interfaced to an optical microscope to be able to choose the cell to be examined.

2. The cells had to be maintained in a near-physiological state during examination, and the culture conditions had to be easily changeable during imaging.

3. As certain cell types are very tall and certain substrates very uneven, the z range of the AFM had to be extended beyond the range that is commercially available for these applications.

4. Phase-contrast and fluorescent imaging of the cells examined using AFM during or postexperimentation had to be possible.

5. A fast and robust method for tip derivatization and tip modification had to be found. (See also Chapter 6 by Hinterdorfer in this work.)

In this chapter, we shall provide examples of methodologies to address these issues with cell biology in mind. We shall detail several postprocessing and experimental protocols. To illustrate the use of these solutions and methodologies, we will show applications taken from our own research. Finally, we shall detail the problems that remain to be solved before large-scale application of AFM in cell biology. We shall also give our views on possible uses of the AFM in both biotechnological and pharmaceutical industries.

II. Methods

A. Microscope–AFM Interface

In all of our own work that we cite in this review, we have exclusively used a Thermomicroscopes (Topometrix) Explorer AFM. While the principles that we expound remain general, it is entirely possible that some of the specific issues that we have encountered or their solutions are "machine specific" and this should be taken into consideration by the reader.

For use with AFM, the obvious choice of an optical microscope is one of "inverted" design (Fig. 1). The microscope should be chosen preferably with several side ports to easily integrate frame grabbing and confocal microscopy capabilities. Because the microscope is a structure with large acoustic and thermal dimensions, it has the potential to be a major source of vibrations within the system. We took several steps to eliminate these. A commercially available air-floated table (TMS) was used upon which to install the microscope. The microscope itself was vibrationally isolated from the table using several layers of standard bubble wrap. Finally, the microscope–AFM interface was designed to make the AFM and the microscope thermally and mechanically united (Lehenkari *et al.*, 2000).

The microscope–AFM interface has to satisfy a certain number of criteria so that examination conditions are as close as possible to the physiological ones. Cells have to

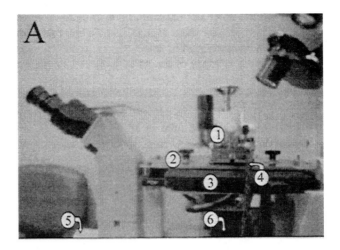

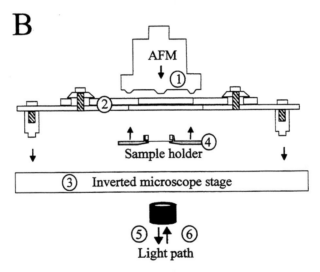

Fig. 1 The hybrid AFM–confocal microscope. Photograph (A) of the experimental setup and a cross-sectional diagram (B) of the AFM-inverted microscope interface and optical path. The Thermomicroscopes Explorer AFM (1) is fitted to the inverted microscope interface (2) that allows alignment of the tip into the desired position, independent of the movement of the microscope stage (3) and the sample holder (4), thus allowing the placement of the desired object into the AFM imaging point. This configuration allows simultaneous light [phase-contrast and CCD capture, (5)] and epifluoresecence confocal imaging [Bio-Rad side port attachment, not shown (6)]. Note that the AFM/AFM holder/sample holder combination (2 + 4) is capable of vertical movement diminishing any interfering "noise" in the closed loop of the AFM-inverted microscope. Reprinted from *Ultramicroscopy* **82**, Lehenkari, P. P., Charras, G. T., Nykänen, A., and Horton, M. A., Adapting atomic force microscopy for cell biology, pp. 289–295. Copyright (2000), with permission from Elsevier Science.

be cultured under normal conditions on glass or plastic coverslips, as transparency is needed to visualize them. It must also be possible to move the sample independently of the AFM scanner head to select the cell to be examined and ensure that the whole cell is within the *xy* range of the AFM scanner. The interface that we designed consisted of two parts: the sample holder and the AFM holder.

The sample holder was designed to fit within the circular area at the center of the microscope platen. In its center, a short, tapered tube was machined. The coverslip was fitted into the central tube in a sandwiched layer consisting of a rubber O ring, the coverslip, and another O ring, and finally, water tightness was ensured by screwing a tapered, hard plastic ring within the tube.

The AFM holder consisted of two parts: first, a base plate that firmly fits to the microscope stage. This base plate had a large circular aperture in the center to enable access to the sample. A smaller plate could slide horizontally above the base plate. Several kinematic mounting holes were machined in the top plate surface so that the AFM mounts could be fit stably and located with precision. A circular aperture was cut in its center to give access to the sample. Once the AFM and the optics were aligned, the top plate could be immobilized with respect to the base plate by tightening two screws.

Medium exchange could be effected using a simple system that fitted above the sample holder and beneath the AFM. This system consisted of two syringes linked to the sample holder via small plastic tubes. One syringe was used to empty the chamber, whereas the other contained the replacement medium.

Temperature could be controlled simply by fitting an external heating coil to the underside of the microscope platen along with a temperature controller. When a CO_2 environmental chamber is not available, great care must taken to properly buffer the culture medium to avoid abrupt pH changes.

B. Integrating AFM with Confocal Microscopy and/or Frame Grabbing

Taking images of the examined cell prior to and after AFM analysis may provide information that is important to an experiment. This can be achieved simply by adding an "off-the-shelf" CCD camera to the side port of the microscope and linking it to a PC equipped with a frame-grabbing card. A simple single or series of phase contrast optical images of the cell can be obtained by this method. Moreover, if a CCD camera of sufficient sensitivity is used, fluorescence images can also be obtained (but see the following discussion of confocal microscopy).

Alternatively, a confocal microscope can be integrated via the side port of the inverted microscope. This enables the monitoring of cellular reaction to AFM experimentation using fluorescent reporter systems such as, for example, the calcium-sensitive dye Fluo-3; fluorescent enzyme substrate activity indicators, as for caspases activated in apoptosis; or cells transfected with green fluorescent protein (GFP) reporter gene constructs. There are a number of manufacturers that provide such add-on confocal laser scanning capabilities, and great care must be taken in choosing the one that best suits the particular AFM system used. Indeed, a number of mechanical parts within the confocal laser scan head move during imaging. Scanning mirrors in the Bio-Rad confocal microscope are controlled by galvanometers and interference is potentially produced, but these have not proved to be problematic at the relatively low-resolution AFM being applied. Other confocal microscopes may produce noise, but they have not been tested. Other moving parts (dichroic filters, confocal apertures) are preset prior to imaging. The remaining components are separated from the confocal scan head and unlikely to produce electrical/mechanical/thermal inteference. In our system, we found the Bio-Rad Radiance 2000 system to be particularly well suited as many of the moving parts are located in a separate box and therefore introduce no additional perturbation to the system.

C. Confocal Microscope Settings for Use with AFM

During confocal microscopy measurements, great care must taken to set the dichroic wheel to separate red and green channels so that the laser light used in the AFM light lever detection system can be eliminated. In our setup, decoupling at 560 nm worked well. A band pass filter should also be used on the green channel to further eliminate red laser bleed through. In our experimental setup, we used a 530- to 560-nm band pass filter. The AFM and optical imaging z positions should also be as far apart as possible to reduce stray light cross-talk. Furthermore, with certain cantilevers, there is a large reflection of the confocal laser beam from the underside of the cantilever. If the fluorescent signal is strong enough, this can be reduced by applying a polarizer to the emitted signal, as the reflected light will be polarized in the same way as the light emitted by the laser source.

D. Increased z-Range Scanner

Cells can range anywhere between 100 nm (at the flat edge of a spread cell) and 50 μm (at the top of the cell body) in height. Therefore, the maximum z range in commercially available AFMs is not suited to the imaging of cells under certain conditions. The first solution that springs to mind is to manufacture longer piezo-electric ceramics for the scan heads. However, because ideal piezo-electric crystals have a linear response to the feeding current, increasing the size of the piezo-electric ceramic results in a loss of accuracy and linearity. To overcome this problem, we designed an AFM scanner head in which two standard specification piezo-electric ceramics are electrically and mechanically coupled in series but physically located parallel to the scan head. This allows the doubling of the z range of the scanner without a large loss of accuracy (Fig. 2).

E. Tip Modifications and Derivatization

Several methods exist for tip derivatization and the subject is detailed in other chapters (See Chapter 6 in this work by Hinterdorfer.) However, we have found a particularly simple way of derivatizing AFM cantilevers. A solution of 10 mg/ml polyethylene glycol (MW: 8000, Sigma Chemicals) was prepared in PBS. A drop of this solution was deposited on a coverslip, and using the optical feedback camera of our AFM, the tip was dipped in the PEG solution for 30 s at room temperature. Unbound PEG was then washed off with PBS. On a fresh coverslip, a drop of the ligand to be used (concentration >1 mg/ml) was deposited, and the tip was again dipped in the solution for 30 s. Unbound ligand was washed off with PBS. One downfall of this simple passive chemisorption

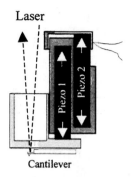

Laser

Cantilever

Fig. 2 Schematic illustration of the double z-range scanner head. The two piezo-ceramics function in series and are located and offset in parallel inside a rigid cylinder that allows movement only in the z-direction over a range of up to 20 μm. Reprinted from *Ultramicroscopy* **82,** Lehenkari, P. P., Charras, G. T., Nykänen, A., and Horton, M. A., Adapting atomic force microscopy for cell biology, pp. 289–295. Copyright (2000), with permission from Elsevier Science.

method is that only a limited number of binding measurements can be effected before the ligand is desorbed. Therefore, the prolonged force–distance cycling required for affinity mapping is not possible using this derivatization technique.

To remedy this, glass beads (Sigma) with diameters varying from 10 to 40 μm were glued onto the cantilevers and derivatized as described previously. The bead-modified tips are coated with a greater number of ligand molecules, and therefore affinity mapping is possible due to the large number of ligand molecules present; hence, sufficient ligand density is maintained on the glass bead tip in relation to the cell membrane contact area. However, the estimation of the number of receptors in any location on the cell is not possible, as the number of ligand molecules on the tip is unknown. The glass beads were glued onto the cantilevers by dipping the cantilevers into UV-activatable glue (GlassBond, Loctite), and particular care was taken to avoid depositing glue on the top side of the cantilever. To ascertain that no glue had been deposited on the top of the cantilever, the AFM gain was checked before and after coating. In case of failure, the cantilevers can be cleaned by dipping them sequentially in 50% ethanol and acetone. Dry beads were then settled onto a coverslip on the sample holder, and the AFM was placed on top of the optical microscope. The cantilever was aligned with the chosen bead and slowly lowered toward the bead. After touch down, the AFM was lifted off the stage, the cantilever was taken off and exposed to daylight to crosslink the glue. Note that cantilevers must be calibrated prior to bead gluing if any numerical values are needed.

III. Analysis

A. Binding Force Measurements on Intact Cells

Most receptors, such as integrin cell adhesion receptors, require optimal cellular surroundings and organization within the plasma membrane in addition to association with lipids and accessory molecules to achieve their correct *in vivo* function and mechanical binding properties. Therefore, it is imperative to investigate the binding properties of a given receptor within its physiological cellular environment.

The actual analysis of the force–distance curve will not be detailed here, as it is the subject of other chapters of this book, which we invite you to refer to. (See Chapters 6 and 14–16 in this work.)

Binding force measurements on cells may be facilitated by brief fixation (30 to 60 s) in 2.5% paraformaldehyde in sucrose buffer. However, great care must be taken not to overfix the cells because binding disappears due to the destruction of the receptor. Moreover, this process is likely to be dependent upon the exact system under study– integrins appear robust in this context, whereas multicomponent receptors or enzyme complexes are likely to be less so. Furthermore, imaging the cells at low temperature may help to reduce both their motility and the lateral movement of molecules within the lipid bilayer.

As there is often a high receptor concentration at the cell surface, there may be several unbinding events within the force–distance curve. Integrins may be present at millions of copies per cell, whereas growth factor receptors may be expressed at several orders of magnitude less—this clearly may have large effects upon the fraction of force–distance cycles that involve a receptor–ligand interaction. Each of these steps should be measured, and a multipeak Gaussian fitting effected on the total binding force distribution in order to detect possible multiple adhesions and calculate minimal binding that may approach single-molecule events. Studies of isolated macromolecule binding (for example, streptavidin–biotin) have shown a cantilever pull-off rate dependency for the value of the minimum single-molecule binding force obtained (Merkel *et al.,* 1999). While such a detailed analysis involving thousands of force–distance measurements is possible and appropriate, it is not feasible with cells which are delicate (cell damage on repeated tip contact is likely) and motile (a really slow pull-off would be impossible).

B. Binding Map Analysis

Binding map analysis is particularly useful for visualizing the location of receptors to a given ligand when no antibody to this receptor exists or when the receptor has not yet been identified. AFM may thus represent a viable alternative to single-cell autoradiography that uses radioisotopically labeled probes and is methodologically complex and slow. Furthermore, it may help to determine whether a receptor that is detected through immunostaining is actually functional in a given cell type.

Prior to analysis on a cell surface, a dry run should be effected on an area with no cells to evaluate the extent of nonspecific binding. If nonspecific binding is present, steps must be taken to eliminate it. Also, taking a phase-contrast image or a high-resolution AFM image prior to binding map analysis will greatly help to visually localize the points where receptor–ligand adhesion takes place. It is recommended not to take the image after the analysis, as the cell may have reacted to the ligand that is being studied, such as through an alteration in shape or rate of movement. A sensor response curve should be

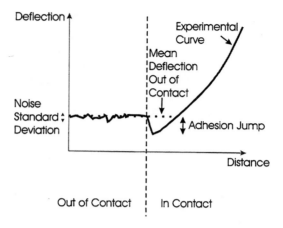

Fig. 3 Typical binding curve for an affinity map experiment. The out-of-contact part of the curve is used to determine the mean value of the deflection and the standard deviation of the noise. The size of the adhesion jump is compared to the size of the noise standard deviation, and if it is more than threefold greater there is a 99.7% chance of the jump being due to true adhesion.

taken on a hard material (e.g., the culture substrate) to obtain the conversion factor between current (nA) and force (nN).

To effect a binding map analysis, one must first acquire force–distance curves in each point of a grid superimposed to the cell surface. Thereafter, each force–distance curve must be analyzed separately to search for an adhesion event. To do so, we suggest a very simple method in which only the retraction curve is needed (Fig. 3). First, one must evaluate a length of the retraction curves where the AFM tip is no longer in contact with the cell in all of the force–distance curves. Second, the mean and the standard deviation of the population of points where the tip is out of contact should be calculated. The tail-end part of the curve can be approximated by a horizontal straight line. We assume that the distribution of tip deflections when the tip is out of contact follows a normal distribution. The experimental curve values can then be examined one by one from the last point that was taken into account in the fitting of the tail end of the curve. A criterion for an adhesion event can be derived using the standard deviation. For a normal distribution, 99.7% of the observations lie within 3 SD of the mean. Therefore, if a point of the curve is further than 3 SD away from the mean of the tail end of the curve, adhesion is taking place with a 0.3% possibility of error. This measurement can be effected without prior flash fixing of the cells.

C. Material Property Analysis

1. Conical Tips

As this is the subject of another chapter of this book and we invite you to refer to it. (See Chapters 3 and 4 in this work by Le Grimellec and co-workers and Radmacher.)

2. Spherical Tips

This calculation needs prior determination of the radius of the sphere and cantilever spring constant. The radius of the sphere can be measured simply by using a digitized image of the sphere/cantilever system.

The material properties can be evaluated using the indentation of a Hertzian half-space with a spherical indentor (Johnson, 1985; Weisenhorn et al., 1993). The indentation resulting from the force F applied by spherical indentor onto the surface of a cell with the elastic modulus E is

$$\delta_{\text{spherical}} = \left(\frac{3}{4}\frac{F(1-\nu^2)}{E}\frac{1}{\sqrt{R}}\right)^{2/3},$$

where R is the radius of the spherical tip and ν is the local Poisson ratio. The loading force F can be obtained from the spring constant of the cantilever k and the deflection $d(F = kd)$.

Let z be the distance traveled by the AFM head and $z0$ be the height at contact. One can express the change in height as a function of the deflection d, the initial deflection $d0$ and the indention δ: $z - z0 = d - d0 + \delta$. The equation of the force–distance curve after contact for the spherical indentor can be rewritten as (adapted from Radmacher et al. (1996))

$$z - z0 = d - d0 + \left(\frac{3}{4}\frac{k(1-\nu^2)}{E}\frac{1}{\sqrt{R}}\right)^{2/3}(d - d0)^{2/3}.$$

A good fit of this curve can be obtained by taking two points of the curve and solving for E and $z0$ (Radmacher et al., 1996). The equation for the spherical indentor can be rewritten as a third-degree equation and solved exactly using Cardano's method (Arnaudies and Fraysse, 1989) or iteratively using numerical methods. The elastic modulus obtained is a measure of the compound local elasticity over the area of indentation.

D. Induced Strain Calculation

Because cells are responsive to mechanical stimuli, it can be informative to calculate the strains induced in the cellular material by AFM indentation to attempt to correlate cellular reaction to applied strain. The induced strain can only be calculated simply for spherical tips. For conical tips, numerical methods such as finite element modeling must be used.

From the elastic modulus and the force applied, one can calculate the strains applied on the surface of the cell by a spherical indentor using the Hertzian theory of contact. The total force P applied by a spherical indentor with a radius R to a surface with the elastic modulus E (determined from the force–distance curve) creates an indentation with a radius a (Johnson, 1985); for example

$$a = \left(\frac{3}{4}\frac{PR(1-\nu^2)}{E}\right)^{1/3}.$$

The relationship between the maximum pressure $p0$ applied and the total load P applied by the AFM is (Johnson, 1985)

$$P = \frac{2}{3}\pi p_0 a^2 \Leftrightarrow p_0 = \frac{3}{2}\frac{P}{\pi a^2}.$$

Thereafter, one can determine the radial displacements $\overline{u_r}$ on the surface of the elastic half-plane (Johnson, 1985) and from those the radial strains $\overline{\varepsilon_{rr}}$ at the surface can be calculated as

$$\overline{\varepsilon_{rr}}(r) = \frac{\partial \overline{u_r}(r)}{\partial r} = \frac{(1-2v)(1+v)}{3E}\frac{a^2}{r^2}p_0\left\{1 - \left(1 - \frac{r^2}{a^2}\right)^{3/2}\right\}$$

$$- \frac{(1-2v)(1+v)}{E}p_0\left(1 - \frac{r^2}{a^2}\right)^{1/2}, r \leq a$$

$$\overline{\varepsilon_{rr}}(r) = \frac{\partial \overline{u_r}(r)}{\partial r} = \frac{(1-2v)(1+v)}{3E}\frac{a^2}{r^2}p_0, r > a.$$

The strain distribution under the area of indentation has both a compressive and a tensile component. Both the tensile and compressive components of the radial surface strain can be calculated for each cell.

Calculation of the radial tangential or vertical strains within the cell thickness can only be performed using finite element modeling techniques. However, a detailed description of this technique is beyond the scope of this chapter and more details can be found in Charras *et al.* (2000) and Charras and Horton (2002b).

E. Interfacing AFM Measurements with Finite Element Modeling Techniques

The simultaneous acquisition of cell topography and a map of the local material properties of the cellular material make AFM an ideal tool for interfacing with finite element modeling techniques. After digital extraction of the cell from the image, topography can be utilized to create the mesh of the cellular structure, and the material properties can be grouped into several collectors and used to describe the local material properties of the cell in the finite element model. As this application is very software dependent, one needs to adapt the methodology depending on the AFM, postprocessing analysis, and FEM software. Further details cannot be given here, but the reader is referred to Barbee *et al.* (1995).

IV. Application Examples

A. Measurement of Adhesion Force of $\alpha_v\beta_3$ Integrin on Osteoclasts

Adhesion forces between an antibody, F11, to rat $\alpha_v\beta_3$, or a linear RGD (Arg–Gly–Asp) containing ligand (Lehenkari *et al.*, 1999), and $\alpha_v\beta_3$ receptors on intact bone cells (osteoclasts and osteoblasts) were measured (Fig. 4). Several unbinding events could be observed in many cases. Multipeak Gaussian fitting revealed that they were integer multiples of each other. Further, ligands which had predicted higher affinities for the receptor gave greater binding forces, and the amino acid sequence/pH/divalent cation

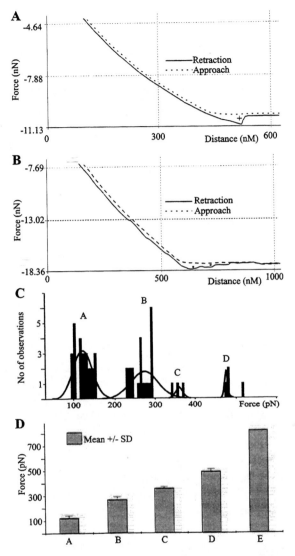

Fig. 4 Measuring interaction forces between ligands and cell-surface receptors by AFM. Interaction forces were evaluated between F11 antibody molecules on the AFM cantilever tip and surface-expressed $\alpha_v\beta_3$ integrin dimers on freshly isolated osteoclasts. (A) Analysis at high retraction speeds (50 μm/s) revealed a typically large single release of multiple molecular interactions (the "jump" in the AFM retraction curve seen at +). (B) Reduction of the retraction speed to 1 μm/s enabled a "stepwise" release of one or more binding events to be visualized (each marked, *) and quantitated. (C) Individual release forces were plotted as a histogram and accumulate around certain values; these can be further analyzed by Gaussian curve fitting [as described in Lehenkari and Horton (1999)]. (D) Results from (C) were tabulated, and since (A–D) show particular multiples of a certain integral force, dividing the values of each group by the number of placements of the group within the histogram, these results could be combined. The strategy employed here revealed that the binding force between F11 and osteoclast $\alpha_v\beta_3$ is 127 $\pm$ 16 pN (mean $\pm$ SD). Adapted from Lehenkari, P. P., and Horton, M. A. (1999). Single integrin molecule adhesion forces in intact cells measured by atomic force microscopy. *Biochem. Biophys. Res. Commun.* **259,** 645–650, with permission from Academic Press.

dependency of the receptor–ligand interaction examined by AFM was similar to that observed under bulk measurement conditions using other techniques (Lehenkari and Horton, 1999).

B. Functional Receptor Mapping of Receptors on Live Osteoclasts

VIP receptor mapping was carried out on live osteoclasts and analyzed with custom written software as discussed earlier (Lundberg *et al.*, 2000). VIP receptors appeared to be localized mainly to the periphery and midzone of the osteoclast cell membrane (Fig. 5).

C. Simultaneous AFM and Confocal Imaging of Live Cells

Mouse B16 melanoma cells were transfected with a green fluorescent protein (GFP) reporter construct linked to F-Actin (Ballestrem *et al.*, 1998). This cell was imaged with the confocal microscope prior to AFM contact imaging. The shape of the cell is easily recognizable and several features can be distinguished in both images. F-actin fibers and focal adhesion structures are easily identified as can the nucleus in both images (Fig. 6).

D. Mechanical Stimulation of Live Osteoblasts

Virtually all cell types have been reported to adapt to their mechanical environment (Donahue *et al.*, 1995). Among these, bone cells are particularly interesting, as bone is a dynamic material that continually adapts its structure in response to mechanical usage (Rubin and Lanyon, 1985). To assess cellular responses, we chose to monitor intracellular calcium intensity changes as it is one of the very early and easily measurable responses to mechanical stimulation. In this application, primary rat osteoblasts were stimulated mechanically using $200\mu m$ V-shaped cantilevers with glass beads glued onto the tip. The force applied was varied between 1 and 30 nN. Cells reacted either instantaneously after the cantilever came into contact with the cell or when the cantilever was lifted off the cell (Fig. 7).

E. Biomechanics

Cells respond to many different mechanical stimuli. Among these, fluid flow has often been used to mechanically stimulate cells. Indeed, endothelial cells submitted to fluid flow for 24 h align in the direction of the flow. Using computational fluid dynamics in conjunction with acquisition of the cellular profile by AFM, Davies (1997) showed that this change in alignment served to reduce the shear stresses on the cell surface of the cells. In our application, we show the flow velocity over a live osteoblast. The AFM topography image was acquired in contact mode, converted into a finite element mesh, and the laminar flow of a viscous incompressible fluid was simulated over the cell surface (shown here only in the center of the cell) (Fig. 8).

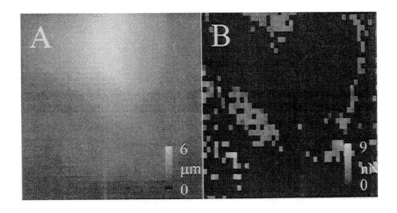

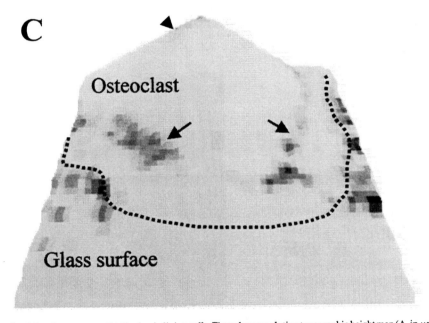

Fig. 5 Mapping receptor distribution in living cells. First a low-resolution topographic height map (A, in μm) was created using the "ball" tip cantilever (hence, the low resolution of the image). The interaction forces between vasoactive intestinal peptide (VIP) and its cellular receptor were analyzed by taking repeated force–distance measurements across the entire surface of freshly isolated rat osteoclasts. Glucagon, a negative control peptide, failed to show any binding under the same measurement conditions. This enabled the distribution of the VIP–receptor binding forces on the cells to be evaluated (B, in nN). The two were then merged into an image of the binding forces displayed on a pseudo-three-dimensional height image of a cell (C). The dashed line shows the outline of the osteoclast; the arrows show the VIP binding sites in the midzone of the cell and was also located at the top of the osteoclast (arrowhead); some nonspecific binding was seen at the glass surface with test and control peptides. Adapted from Lundberg, P., Lie, A., Bjurholm, A., Lehenkari, P., Horton, M., Lerner, U. H., and Ransjo, M. (2000). Vasoactive intestinal peptide (VIP) regulates osteoclastic activity via specific binding sites on both osteoclasts and osteoblasts. *Bone* **27**, 803–810, with permission from Elsevier Science.

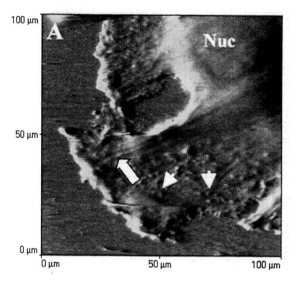

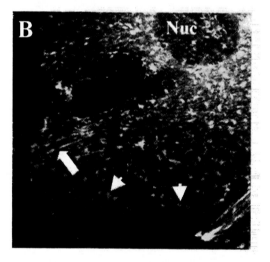

Fig. 6 Combined topography AFM imaging with confocal fluorescence microscopy using cells expressing recombinant green fluorescent protein (GFP)-actin. (A) is a topographic AFM image of a melanoma cell showing intracellular cytoskeletal elements (arrow), demonstrated by their height profile in contact mode scanning under different applied forces. (B) is a fluorescence image of the same cellular region as (A) showing the distribution of fluorescent GFP-actin (i.e., F-actin fibers) analyzed using the FITC channel of the linked confocal microscope. A representative region rich in actin fibers is arrowed to show the coincident distribution by both imaging techniques; actin-rich patches ("focal adhesion complexes") are also seen and examples are marked in both images (arrowheads). The nucleus of the cell (Nuc) is identifiable in both images. (The images are sized at 100×100 μm.) Adapted from Horton *et al.* (2000), with permission. (See Color Plate.)

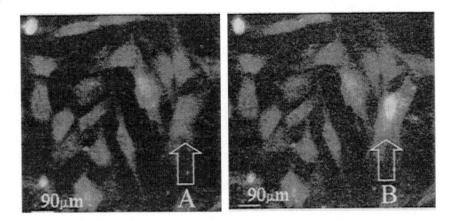

Intracellular Calcium Intensity

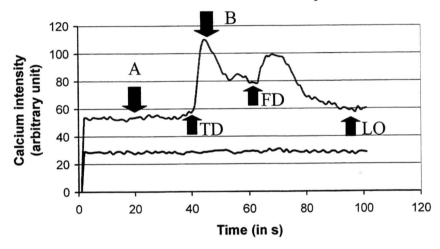

Fig. 7 (A) Osteoblasts loaded with Fluo-3 prior to indentation. The cell about to be indented is indicated by the white arrow. (B) Osteoblasts after indentation. The indented cell (indicated by the white arrow) has increased its intracellular calcium concentration. Time course of the calcium intensity within the indented cell is shown graphically. TD (touch down) indicates the time when the AFM cantilever contacts the cell. FD (force–distance) indicates the time when a force–distance curve is taken on the cell. LO (lift off) indicates the time when the AFM cantilever is lifted from contact with the cell surface. Reprinted from *Ultramicroscopy* **86,** Charras, G., Lehenkari, P., and Horton, M. Atomic force microscopy can be used to mechanically stimulate osteoblasts and evaluate cellular strain distributions. pp. 85–95. Copyright (2000), with permission from Elsevier Science. (See Color Plate.)

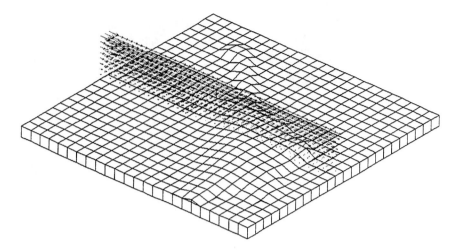

Fig. 8 Computational fluid dynamics (CFD) model of a primary osteoblast submitted to laminar flow. The cellular profile was acquired, extracted from its substrate, digitally plated on a flat surface, and then transformed into a mesh suitable for CFD. The flow speed and direction are represented by the red arrows for a region around the center of the cell. (See Color Plate.)

V. Future Directions and Improvements

A. Problems To Be Solved

In summary, AFM has the potential of becoming a routine piece of equipment, rather than an object of curiosity, in biological laboratories. However, a certain number of problems remain to be addressed by manufacturers in order for this to happen.

1. Tips need to be designed that would be less deleterious to the cells being examined, and enable the determination of material properties with a higher degree of precision.

2. To reliably examine biological phenomena in real time, an increased scanning speed applying less force would be desirable. New methods to control applied force during noncontact imaging need to be developed to improve resolution on soft materials such as cell surfaces (see Chapter 16 in this book by Humphris and Miles).

3. The replacement of the red laser, commonly found in AFMs, by an infrared laser would free the red channel when the AFM is coupled with a confocal microscope.

4. Tips less reflective on the underside would reduce confocal laser reflection when combined with optical microscopy.

5. To realize the full potential of integration with the whole range of biological examination techniques, the design of the AFM needs to leave easy access to the sample. This would enable easy simultaneous AFM and micromanipulation, microinjection or electrophysiology.

6. For chemical force AFM, a robust, easy and reliable way of functionalizing tips needs to be devised.

B. Pharmaceutical Applications and Future Directions

Of the many applications that we have described, chemical force microscopy and affinity mapping have the potential to be used industrially in the evaluation process of candidate pharmaceuticals. The potency and structure–activity relationship of new agonists/antagonists may be evaluated by measuring the binding force of compounds to target receptors. More importantly, the specificity of the compound could be tested by using a range of other cell types or cells in which the receptor of interest has been knocked out or genetically modified. The antagonistic or agonistic properties of a drug may be tested in conjunction with confocal microscopy and fluorescent dyes sensitive to proteins known to be involved in early responses to agonist binding, such as inositol triphosphate up-regulation in response to G-protein activation or the induction of apoptosis. Binding map analysis is particularly useful for evaluating not only the location of receptors to which there exist no antibodies but also the functionality of receptors present within the cell membrane. With development, an AFM-based method could replace the standard technique of receptor autoradiography for such studies.

Adhesion measurement of whole cells as described by Sagvolden et al. (1999) and Thie et al. (1998) may enable the evaluation of cellular responses to new materials, for example, in cardiovascular grafts. Indeed, one of the main problems of cardiovascular grafts is that these need to be replaced within a few years as the cells attach to them have a modified phenotype and form new atherosclerotic plaques or fibrotic strictures, hence reducing their functionality. Evaluation of new orthopedic implants could also be carried out to select those that promote adhesion of osteoblasts over other cell types to induce osteointegration of the new material. In either case, conducting a series of adhesion measurements on candidate graft materials would enable objective selection of the best suited material for the specific purpose: materials that promote adhesion for orthopedic implants and materials that do not promote adhesion for cardiovascular applications.

AFM may be of particular interest in the field of biomechanics. Indeed, cell biomechanics has been hindered mainly by lack of a precise tool enabling the verification of the hypotheses formulated. AFM may help comprehend how cells react to strain, how they adapt to life in strained environments, or how the mechanical and the biochemical pathways interact as has been hypothesized in several theories [for example, percolation (Forgacs, 1995) and tensegrity (Ingber, 1997)]. Furthermore, in conjunction with finite element modeling, AFM may help to answer some of the more intriguing questions posed by biology. For example, how do erythrocytes manage to pass through capillaries whose diameter is smaller than their own?

There is currently a lack of suitable methods to analyze the three-dimensional structure of membrane glycoproteins at high resolution in their native context and configuration. The pioneering work of Müller (Engel and Müller, 2000; Müller et al., 1995; Müller and Engel, 1999) (see Chapter 13 in this work by Müller and Engel) used proteins of bacterial purple membranes which are naturally present as tightly packed two-dimensional arrays of high-purity crystals (such as bacteriorhodopsin and Ompf). This makes equivalent methods for molecules present in the membranes of eukaryotic cells particularly attractive, especially if high-resolution "soft" imaging techniques can be developed. Here, though, membrane glycoproteins are typically present at much lower densities and below levels that would be expected to form crystalloid features. By performing such experiments on eukaryotic cells, essentially one may be able to gain a definitive insight into the structure and function of, for example, ion channels, receptor complexes, or nuclear pores. This would help draw out rational strategies to devise new specific drugs to one particular part of a cell physiological mechanism.

In summary, through its capacity to quantify a number of biological phenomena in engineering terms, AFM may bring certain fields of biology into the era of solution engineering and exploitation of biological properties to reach a well-defined goal.

References

Arnaudies, J. M., and Fraysse, H. (1989). Equations algebriques. Equations de degre 3. *In* "Cours de Mathematiques," pp. 434–442. Dunod Universite, Paris, France.

Ballestrem, C., Wehrle-Haller, B., and Imhof, B. A. (1998). Actin dynamics in living mammalian cells. *J. Cell Sci.* **111**, 1649–1658.

Barbee, K. A., Mundel, T., Lal, R., and Davies, P. F. (1995). Subcellular distribution of shear stress at the surface of flow-aligned and nonaligned endothelial monolayers. *Am. J. Physiol.* **268**, H1765–H1672.

Binnig, G., Quate, C. F., and Gerber, C. (1986). Atomic force microscope. *Phys. Rev. Lett.* **56**, 930.

Charras, G. T., and Horton, M. A. (2002A). Determination of cellular strains by combined atomic force microscopy and finite element modelling. *Biophys. J.*, in press.

Charras, G. T., and Horton, M. A. (2002B). Single cell mechanotransduction and its modulation analyzed by atomic force microscope indentation. *Biophys. J.*, in press.

Charras, G., Lehenkari, P., and Horton, M. (2000). Atomic force microscopy can be used to mechanically stimulate osteoblasts and evaluate cellular strain distributions. *Ultramicroscopy* **86**, 85–95.

Danker, T., and Oberleithner, H. (2000). Nuclear pore function viewed with atomic force microscopy. *Pfluegers Arch.* **439**, 671–681.

Domke, J., Parak, W. J., George, M., Gaub, H. E., and Radmacher, M. (1999). Mapping the mechanical pulse of single cardiomyocytes with the atomic force microscope. *Eur. Biophys. J.* **28**, 179–186.

Donahue, H. J., McLeod, K. J., Rubin, C. T., Andersen, J., Grine, E. A., Hertzberg, E. L., and Brink, P. R. (1995). Cell-to-cell communication in osteoblastic networks: Cell line-dependent hormonal regulation of gap junction function. *J. Bone Miner. Res.* **10**, 881–889.

Engel, A., and Müller, D. J. (2000). Observing single biomolecules at work with the atomic force microscope. *Nat. Struct. Biol.* **7**, 715–718.

Florin, E. L., Moy, V. T., and Gaub, H. E. (1994). Adhesion forces between individual ligand-receptor pairs. *Science* **264**, 415–417.

Forgacs, G. (1995). On the possible role of cytoskeletal filamentous networks in intracellular signaling: an approach based on percolation. *J. Cell Sci.* **108**, 2131–2143.

Grandbois, M., Dettmann, W., Benoit, M., and Gaub, H. E. (2000). Affinity imaging of red blood cells using an atomic force microscope. *J. Histochem. Cytochem.* **48**, 719–724.

Henderson, E., Haydon, P. G., and Sakaguchi, D. S. (1992). Actin filament dynamics in living glial cells imaged by atomic force microscopy. *Science* **257**, 1944–1946.

Hinterdorfer, P., Baumgartner, W., Gruber, H. J., Schilcher, K., and Schindler, H. (1996). Detection and localization of individual antibody-antigen recognition events by atomic force microscopy. *Proc. Natl. Acad. Sci. U.S.A.* **93**, 3477–3481.

Horton, M. A., Charras, G., Ballestrem, C., and Lehenkari, P. (2000). Integration of atomic force and confocal microscopy. *Single Mols.* **1**, 135–137.

Ingber, D. E. (1997). Tensegrity: The architectural basis of cellular mechanotransduction. *Annu. Rev. Physiol.* **59**, 575–599.

Johnson, K. L. (1985). *In* "Contact Mechanics." Cambridge Univ. Press, Cambridge, UK.

Kuznetsov, Y. G., Malkin, A. J., and McPherson, A. (1997). Atomic force microscopy studies of living cells: visualization of motility, division, aggregation, transformation, and apoptosis. *J. Struct. Biol.* **120**, 180–191.

Lehenkari, P. P., Charras, G. T., and Horton, M. A. (1999). New technologies in scanning probe microscopy for the understanding of molecular interactions in cells. Expert Reviews in Molecular Medecine @ http://www-ermm.cbcu.cam.ac.uk.

Lehenkari, P. P., Charras, G. T., Nykänen, A., and Horton, M. A. (2000). Adapting atomic force microscopy for cell biology. *Ultramicroscopy* **82**, 289–295.

Lehenkari, P. P., and Horton, M. A. (1999). Single integrin molecule adhesion forces in intact cells measured by atomic force microscopy. *Biochem. Biophys. Res. Commun.* **259**, 645–650.

Lundberg, P., Lie, A., Bjurholm, A., Lehenkari, P., Horton, M., Lerner, U. H., and Ransjo, M. (2000). Vasoactive intestinal peptide (VIP) regulates osteoclastic activity via specific binding sites on both osteoclasts and osteoblasts. *Bone* **27**, 803–810.

Merkel, R., Nassoy, P., Leung, A., Ritchie, K., and Evans, E. (1999). Energy landscapes of receptor-ligand bonds explored with dynamic force spectroscopy. *Nature* **397**, 50–53.

Müller, D. J., and Engel, A. (1999). Voltage and pH-induced channel closure of porin OmpF visualized by atomic force microscopy. *J. Mol. Biol.* **285**, 1347–1351.

Müller, D. J., Schabert, F. A., Buldt, G., and Engel, A. (1995). Imaging purple membranes in aqueous solutions at sub-nanometer resolution by atomic force microscopy. *Biophys. J.* **68**, 1681–1686.

Parpura, V., Doyle, R. T., Basarsky, T. A., Henderson, E., and Haydon, P. G. (1995). Dynamic imaging of purified individual synaptic vesicles. *Neuroimage* **2**, 3–7.

Quist, A. P., Rhee, S. K., Lin, H., and Lal, R. (2000). Physiological role of gap-junctional hemichannels. Extracellular calcium-dependent isosmotic volume regulation. *J. Cell Biol.* **148**, 1063–1074.

Radmacher, M. (1997). Measuring the elastic properties of biological samples with the AFM. *IEEE Eng. Med. Biol. Mag.* **16**, 47–57.

Radmacher, M., Fritz, M., Kacher, C. M., Cleveland, J. P., and Hansma, P. K. (1996). Measuring the viscoelastic properties of human platelets with the atomic force microscope. *Biophys. J.* **70**, 556–567.

Rubin, C. T., and Lanyon, L. E. (1985). Regulation of bone mass by mechanical strain magnitude. *Calcif. Tissue Int.* **37**, 411–417.

Sagvolden, G., Giaever, I., Pettersen, E. O., and Feder, J. (1999). Cell adhesion force microscopy. *Proc. Natl. Acad. Sci. U.S.A.* **96**, 471–476.

Sato, M., Nagayama, K., Kataoka, N., Sasaki, M., and Hane, K. (2000). Local mechanical properties measured by atomic force microscopy for cultured bovine endothelial cells exposed to shear stress. *J. Biomech.* **33**, 127–135.

Schaus, S. S., and Henderson, E. R. (1997). Cell viability and probe-cell membrane interactions of XR1 glial cells imaged by atomic force microscopy. *Biophys. J.* **73**, 1205–1214.

Schneider, S. W., Yano, Y., Sumpio, B. E., Jena, B. P., Geibel, J. P., Gekle, M., and Oberleithner, H. (1997). Rapid aldosterone-induced cell volume increase of endothelial cells measured by the atomic force microscope. *Cell Biol. Int.* **21**, 759–768.

Shroff, S. G., Saner, D. R., and Lal, R. (1995). Dynamic micromechanical properties of cultured rat atrial myocytes measured by atomic force microscopy. *Am. J. Physiol.* **269**, C286–C292.

Thie, M., Rospel, R., Dettmann, W., Benoit, M., Ludwig, M., Gaub, H. E., and Denker, H. W. (1998). Interactions between trophoblast and uterine epithelium: Monitoring of adhesive forces. *Hum. Reprod.* **13**, 3211–3219.

Weisenhorn, A. L., Khorsandi, M., Kasas, S., Gotzos, V., and Butt, H. J. (1993). Deformation and height anomaly of soft surfaces studied with an AFM. *Nanotechnology* **4**, 106–113.

You, H. X., Lau, J. M., Zhang, S., and Yu, L. (2000). Atomic force microscopy imaging of living cells: A preliminary study of the disruptive effect of the cantilever tip on cell morphology. *Ultramicroscopy* **82**, 297–305.

CHAPTER 9

Cellular Membranes Studied by Photonic Force Microscopy

Arnd Pralle and Ernst-Ludwig Florin

Cell Biology and Biophysics Program
European Molecular Biology Laboratory
D-69117 Heidelberg, Germany

I. Introduction

Since the formulation of the fluid mosaic model for cellular membranes of cells by Singer and Nicolson (1972), it has been recognized that the membranes are rather well-structured interfaces whose structure is important for their diverse functions (reviews: Vaz and Almeida, 1993; Jacobson *et al.*, 1995). For a quantitative understanding and subsequent modeling of membrane-bound processes, such as the lateral interaction of membrane receptors, not only the lateral structure but also the interaction forces between various membrane components and their mobility have to be known. The photonic force microscope (PFM) reviewed here provides a novel tool to quantify those important parameters on the plasma membrane of intact cells at superior spatial and temporal resolutions.

The first biophysical characterizations of intact cell membranes at the sub-light-microscopic level were pursued using techniques such as scanning force microscopy (SFM) and single-particle tracking (SPT). The membrane of living cells was imaged at superior resolution by conventional force microscopy (Häberle *et al.*, 1991; Grimellec *et al.*, 1994). The viscoelasticity, the bending modulus of the membrane (Evans and La Celle, 1975; Evans, 1983), and the elasticity of the membrane cytoskeleton (Radmacher *et al.*, 1992) were determined using related techniques. The lateral heterogeneity of the plasma membrane was shown by SPT studying the diffusion of individual membrane proteins (Edidin *et al.*, 1991; Kusumi *et al.*, 1993; Zhang *et al.*, 1993). However, quantitative models of membrane processes such as lateral interaction between proteins in signal transduction require knowledge of the biophysical membrane properties near the molecular scales. Here, the conventional SFM lacks dynamics and sensitivity in force, while traditional SPT lacks spatial and temporal resolutions.

The recently developed PFM allows the measuring of a number of physical properties of the plasma membrane at improved resolution. The PFM employs a laser trap as force transducer with sensitivity in the sub-piconewton range. Various position sensors record the force acting on the probe by measuring the three-dimensional displacement of the probe from its resting position. Under appropriate conditions, the temporal and spatial resolutions suffice for studying molecular diffusion and mechanics at the scale of a few molecules. It is possible either to introduce molecular specificity to the sensor or even to use a single molecule as a sensor itself.

The PFM has been applied to image the membrane of developing neurons (Florin *et al.*, 1997) and to determine the elasticity of their plasma membrane. The viscosity of the plasma membrane and the rate of diffusion of single-membrane proteins were determined at exceeding temporal and spatial resolutions (Pralle *et al.*, 2000). Here, we describe the design principles and the operation of the PFM. The various operation modes and the data analysis are demonstrated on the bases of applications of the PFM in cell biology.

II. Photonic Force Microscopy

A. Imaging and Characterizing the Plasma Membrane

The small forces in the PFM are well suited to image the plasma membrane of cells, especially in regions with weak structural support by the cytoskeleton or limited adhesion to the substrate, like, e.g., new branches in developing neurons. A scanning probe image of the outer surface of such a small neurite from a cultured rat hippocampal neuron is shown in Fig. 1a, and the corresponding differential interference contrast microscope images (DIC) are shown in Figs. 1b and 1c. Two-dimensional images are formed by laterally scanning a trapped latex bead across biological samples while recording the bead's deflection from its resting position. Under the experimental conditions, the

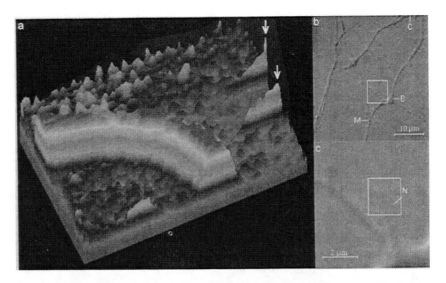

Fig. 1 (a) A PFM scan of a small neurite (N) branching (B) from a major neurite (M) of a growing hippocampal neuron. (b, c) Different scale DIC overviews including the scan area. The PFM scan measures the neurite to be 400 nm high and 300 nm wide. Adapted from (1997) *J. Struc. Biol.* **119**, Florin *et al.* Photonic force microscope based on optic tweezers and two-photon excitation for biological application, pp. 202–211, (1997), with permission from Elsevier Science. (See Color Plate.)

maximal imaging force applied by the probe is well below 5 pN, and the lateral force is at maximum threefold higher than the axial force. These low forces minimize mechanical deformations on soft biological samples. The softness of neuronal membranes and the steep structures of these cells have limited measurements of their mechanical properties. The image of the plasma membrane of the neurite in Fig. 1 was acquired in a constant height mode. The constant height mode of the PFM is limited to flat surfaces with corrugations smaller than the trapping range along the optical axis, which is about 0.5 μm. However, the PFM can be used with a feedback circuit in any conventional force microscopy mode.

A very useful approach to study cells with their steep edges and trenches is a PFM tapping mode functioning much like the force volume scans in SFM (Radmacher *et al.*, 1996). In each image point, the laser focus holding the probe particle approaches the surface and is then retracted a fixed distance. This way, the probe can either climb up and down steep slopes or enter deep trenches. Figures 2c and 2d show a tapping mode scan of a branching neurite. At 1.2 μm high, the image of the structure exemplifies how this mode extends the z range of the PFM, while the sensor still moves into the trench between the two branches. The force can be reduced to fractions of piconewtons using the tapping mode. In the second example of the PFM tapping mode, a line profile of a scan over the surface of a fibroblast near its nucleus provides an example of a tilted surface imaged with high aspect ratio (Fig. 2e). The plasma membrane of single fibroblasts cultured on coverslips grows almost vertically out of the surface to cover the large spherical nucleus, while the remainder of these cells is mostly flat. The PFM can resolve these steep membrane

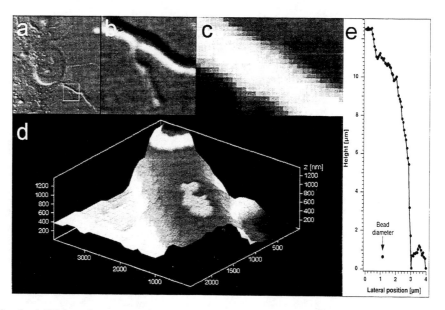

Fig. 2 A PFM tapping mode image of another neurite of a hippocampal neuron is shown in (c) and (d). Being up to 1200 nm high, these structures could only be resolved in the PFM tapping mode. Different scale DIC overviews including the scan area are presented in (a) and (b). Part (e) displays a line scan taken from an extreme example of the tapping mode in which the probe climbed down the side of a fibroblast cell near the nucleus. (See Color Plate.)

structures because the probe is a submicrometer sphere trapped in three dimensions without mechanical connection to the microscope actuator or sample holder (Fig. 5). Hence, the PFM is especially well suited for biological applications, because it overcomes the restriction to well-oriented surfaces of objects as long as they are transparent to the wavelength of the laser. PFM scans yield topographical and elasticity information of highly corrugated and very soft surfaces with a resolution limited by the probe size and the thermal position fluctuations (see discussion on resolution in the following).

B. Quantification of Molecular Interactions

The observation of the motion of membrane-embedded molecules attached to beads by videomicroscopy, known as single-particle tracking, has provided insights into the protein diffusion in the membrane and the lateral organization of the plasma membrane (De Brabander *et al.*, 1991; Cherry, 1992; Kusumi *et al.*, 1993). However, videomicroscopy studies are limited to two dimensions and provide normally only a low temporal resolution, 25–33 fps. Using a sensitive camera with shorter integration time, the frame rates can be increased, and Tomishige and Kusumi (1999) have achieved a 4.5-kHz bandwidth. The spatial resolution is coupled to the temporal resolution because a membrane molecule with a typical diffusion coefficient of $D = 1 \times 10^{-10}$ cm^2/s moves in an area of about 800 nm^2 between two subsequent video images (25 fps). The same molecule diffuses only in an area of 0.4 nm^2 during the 20-μs interval between two PFM

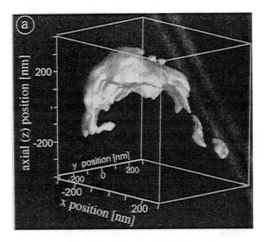
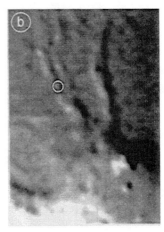

Fig. 3 Three-dimensional particle tracking of a probe attached to a Thy1.1 molecule diffusing on the neurite membrane. The position of the probe was measured every 20 μs over a period of 42 s resulting in 2.1 × 10⁶ data points. (a) Surface plot of the volume in which the molecule–bead complex is found 85% of the measurement time. (b) DIC image of the bead (marked by the circle) on the neurite. Adapted from Pralle, A., Prummer, M., Florin, E.-L., Stelzer, E. H. K., and Hörber, J. K. H. Three-dimensional position tracking for optical tweezers by forward scattered light. *Micro. Res. Technol.* **44**, 378–386, Copyright © (1999, John Wiley & Sons). Reprinted by permission of Wiley-Liss, Inc., a subsidiary of John Wiley & Sons, Inc.

position measurements, which is about the area occupied by a single lipid molecule. In addition, the PFM allows the determination of the position of the bead, hence, of the molecular tracer in three dimensions, ensuring that the actually three-dimensional path of diffusion on the corrugated cell surface is measured and not only a two-dimensional projection as in video microscopy.

Figure 3 shows the result of a three-dimensional tracking experiment of a Thy1.1 molecule diffusing on a neurite membrane. For clarity, the information is presented graphically as a surface representation of the volume in which the sphere has been diffusing during 85% of the time. The tubular outline of the neurite is clearly visible as empty space around which the molecule is diffusing. The diffusion coefficients can be determined.

The local rate of diffusion obtained from the three-dimensional single molecule tracking in the plasma membrane allow study of the dynamics of a single-membrane molecule at a scale not attainable previously. Using this approach, it became possible to measure the viscous drag imposed on membrane components by the lipid bilayer directly separating it from effects of obstacles like cytoskeleton-anchored proteins.

The viscous drag of an individual membrane protein in the plasma membrane of living cells is measured by observing the thermal position fluctuations of an attached sphere. The damping of the motion is dominated by the viscous drag on the membrane domain of the protein because of the 1000-fold higher viscosity of the lipid bilayer compared to that of the aqueous medium. For a local diffusion measurement, an antibody-coated sphere is captured in the solution and placed onto the cell membrane, while maintaining the interaction force between the bead and the membrane below 0.1 pN. The viscous drag on

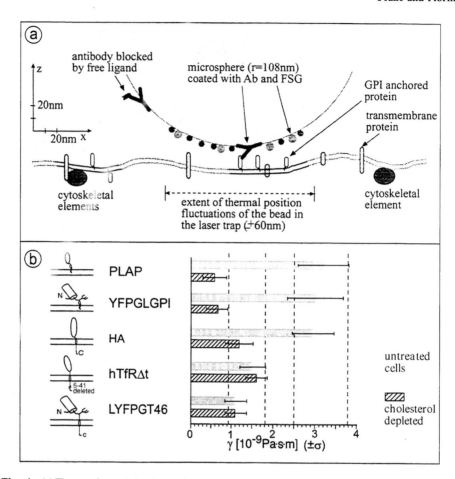

Fig. 4 (a) The experimental situation during a local membrane protein diffusion measurement is shown in this scaled model: the sphere ($r = 108$ nm) is bound via an adsorbed antibody to the membrane protein, in this case a GPI-anchored protein that is part of a raftdomain. To achieve single-molecule binding conditions, the surface of the sphere not covered by antibodies is blocked with fish skin gelatin, and the majority of the antibodies are blocked with soluble antigen. The darker sections in the lipid bilayer symbolize raft domains. The extent of the thermal position fluctuations is kept well below the average distance between diffusion barriers. (b) Summary of the result of local diffusion measurements of single proteins in the membrane of intact cells. The proteins diffuse in intact cells not according to the size of their membrane anchor but according to the type of membrane anchor (shaded bars): the viscous drag of the GPI-anchored proteins (PLAP and YFPGLGPI) is larger than that of the nonraft transmembrane proteins (hTfRΔt and LYFPGT46). In cells in which the rafts have been disrupted by cholesterol extraction the proteins diffuse according to the size of their membrane anchor (crossed bars); the viscous drag of the GPI-anchored proteins is smaller than that of the transmembrane proteins. If the rafts are left intact, the raft proteins diffuse with the raft, which leads to a higher viscous drag than that of the transmembrane proteins in the remainder of the membrane. Reproduced from Pralle *et al. The Journal of Cell Biology*, 2000, **148**, 997–1007 by copyright permission of The Rockefeller University Press.

the same sphere is first recorded in the bulk solution, then near the membrane, and finally after binding to the membrane protein. The comparison of these three measurements allows separating the influence of the sphere diffusing unbound near the membrane from the binding of the bead to the membrane protein. The continuous observation with high temporal resolution resolves directly individual binding events ensuring single-molecule measurements (Pralle *et al.*, 2000).

A scaled scheme of a local diffusion measurement clarifies the advantages gained by using the PFM (Fig. 4a): Confining the motion of a particle with the laser trap to an area small compared to the spacing between immobile membrane components minimizes the effects of immobile obstacles. The measured diffusion coefficient is mainly determined by the interaction of the membrane region of the protein with the lipids in the membrane. Figure 4b summarizes the results of local protein diffusion measurements aimed at elucidating the existence and nature of lipid rafts, special lipid microdomains rich in sphingolipids, and cholesterol. These measurements showed that the local protein diffusion in the membrane of intact cells depends on both the membrane anchor of the protein and the local lipid composition, in agreement with measurements on artificial membrane systems and the theory of protein diffusion in a membrane. The measurements showed also that proteins thought to reside in lipid rafts diffuse independently of their type of membrane anchor slower than transmembrane proteins outside these domains (Fig. 4b). The results provided evidence of lipid rafts in intact fibroblasts and an estimate of their size and stability (Pralle *et al.*, 2000).

III. Experimental Considerations

The PFM is based on a single-beam gradient laser trap (Ashkin *et al.*, 1986, 1990) combined with high-resolution three-dimensional position manipulators and sensors. The laser trap offers a novel type of cantilever to build a scanning force microscope: a sub-micrometer-sized dielectric particle held and scanned by the gradient forces of a strongly focused laser beam. Depending on the properties of the optically trapped particle and the readout, various scanning microscopes have been proposed: a scanning optical near-field microscope (Malmquist and Hertz, 1992; Hertz *et al.*, 1995), a local pH sensor (Sasaki *et al.*, 1996), and a simple optical force microscope (Ghislain and Webb, 1993). We built the first PFM with a detection scheme based on two-photon fluorescence intensity to image soft three-dimensional biological samples (Florin *et al.*, 1997). Also, the PFM was modified to be the first microscope to exploit the thermal position fluctuations of the position sensor to analyze the environment of individual molecules (Pralle *et al.*, 1999, 2000).

Figure 5 shows a scaled comparison of the force transducer and sensor of a conventional SFM and those of the PFM. The much smaller PFM probe has an at least 100-fold smaller hydrodynamic resistance, a smaller spring constant ($10^{-6} - 10^{-4}$ N/m), a virtually diminishing mass, and its motion is completely over-damped in water. These characteristics allow faster and less disturbing measurements. The absent mechanical connection to the outside allows imaging of deep narrow trenches and even inside of

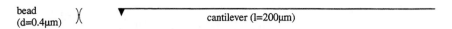

Fig. 5 A scaled comparison of the probes of a classical SFM and those of the PFM. A mechanical cantilever ($l = 200$ μm) with a pyramidal tip and a 0.2-μm sphere trapped in the laser beam are displayed.

transparent three-dimensional objects. The accessible forces range between 0.1 and 20 pN and depend not only on the laser intensity and wavelength but also on the properties of the focus and probe particle (see following). In some experiments, the low trapping forces can be disadvantageous as adhesive forces between the probe, and the sample might become dominating. However, in cells, many macromolecular complexes are held together by interaction potentials not much larger than the thermal energy, a range in which the PFM has proven to be a powerful tool.

A. Design Principles

The basic setup of a PFM follows the principle of single-beam gradient traps, which are mostly implemented in inverted microscopes. A guide for the design of laser traps can be found in Mehta *et al.* (1998). Here, we focus on the points that are special for the PFM, i.e., the combination of three-dimensional manipulators with high-resolution three-dimensional particle tracking and the stability of the setup.

1. Optics of the Laser Trap

The combination of an optical trap with three-dimensional manipulation capabilities and high-resolution position detectors requires special design principles. Figure 6 displays the optical paths in our PFM. To manipulate objects in lateral directions, we use either a piezo-driven scanning mirror or a scanning stage. The probe is positioned along the optical axis by the piezo-driven objective mount.

a. Fluorescence Detection

The position of the trapped particle is measured using either its emitted fluorescence light or by the forward scattered laser light (see Section III,B). Since the fluorescently labeled probe is essentially a point source, the emitted light can be detected with high signal-to-noise ratio in a confocal manner. The fluorescence detection unit consists of a pair of lenses, a pinhole, filters, and a PMT. Because we use a two-photon excitation process, the pinhole is not really necessary. It is still used, because it efficiently rejects the DIC illumination light allowing for simultaneous scanning and observation of the sample by DIC microscopy. The filters are designed to reject the remaining DIC illumination and trapping laser light.

The dichroic mirror 1 (*DM1*) splits the trapping laser from the fluorescent light. To maintain flexibility, it transmits the entire visible spectrum. The second dichroic mirror (*DM2*) couples the trapping laser into the optical path of the microscope. It reflects light between 500 and 610 nm and in the near-infrared. All other wavelengths are accessible

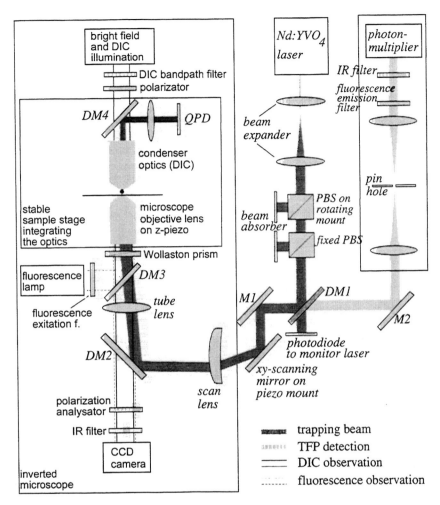

Fig. 6 Optical paths in the PFM built around an inverted microscope with DIC equipment whose wavelength of 700 nm is chosen to reduce photon damage. The DIC and fluorescence observation light paths are drawn outlined while the specific PFM light paths are solid and labeled in italic. The IR laser trapping beam is widened using a beam expander formed by two lenses. By diverting part of the beam with a polarizing beam splitter (PBS) the laser intensity sent to the microscope can be adjusted. The beam is redirected via the scanning mirror and scanning lens and coupled into the optical path of the microscope via dicroic mirror *DM2*. An oil immersion objective lens mounted on a piezo focuses the beam. The forward-scattered laser light is collected by the condenser lens and projected by *DM4* and a lens onto the QPD for the position detection. The TPF is uncoupled from the laser path by *DM1* and detected confocally by a photomultiplier.

to DIC microscopy. Access to longer wavelength is important for *in vivo* observation of neurons, because they are known to suffer under strong illumination in the visible spectrum (unpublished data).

b. DIC Optics

The DIC optics in the PFM differs from the commercial optics in several aspects. The accessible wavelengths are restricted for reasons discussed earlier. The analyzer is normally positioned between the objective and the tube lens. To increase the detection efficiency for fluorescence light, we positioned the analyzer between *DM2* and the camera. This slightly degrades the DIC quality, because of the convergent rays at that position. The dichroic mirror *DM4* separates the DIC illumination from the trapping laser and collimates it on a QPD. This mirror causes a lateral displacement of the DIC illumination. The QPD should be positioned around the back-focal plane of the condenser lens (Gittes and Schmidt, 1998).

c. Stability

The entire microscope and optics are mounted on a vibration-isolated table. However, to achieve a resolution of a few angstroms requires additionally a compact design of the components steering the trapping laser with the probe particle, holding the sample, and detecting the relative position of sample and probe. Our design is arranged around a main plate in which the sample holder containing an xy-piezo stage is directly integrated. The objective lens focusing the trapping laser is mounted with a piezo drive at the bottom of the main plate, omitting any coarse focusing. The light-collecting condenser and the QPD with its optics are mounted on a tripod on the main plate. This design minimizes the distances and number of movable adjustments between the trap-generating objective lens and the position sensor while providing enough flexibility for the experiments.

The power and pointing stability of the trapping laser are extremely critical, because the laser determines the relative sample-probe separation and simultaneously serves as a light source for the position detection. Microradian shifts in the laser pointing translate into nanometer movements of the laser focus in the object plane. Laser intensity fluctuations are indistinguishable from relative z-position changes of the probe.

Laser traps used in biology, usually designed with a diode-pumped solid-state infrared (IR) laser, are a good compromise between availability and stability and provide wavelengths which are very slightly absorbed by water and biomolecules but are easily detectable. Our system uses the 1064-nm line of a $Nd:YVO_4$ laser (T20-B10-106Q, Spectra Physics, Germany).

d. Scan and Feedback Control and Data Recording

A personal computer is used to generate the scan pattern, to control the PFM, and to record the data. The actual requirements for the recording depend on the type of scan performed. In general, three channels for the position of the scanner, plus another three for the position of the probe, and eventually the probe's fluorescence need to be recorded. Custom-designed amplifiers are used to optimize the amplification and offset of all signals for the data acquisition board. The fluctuation analysis of the thermal motion requires an acquisition rate of at least 50 kHz to provide reasonable statistics.

Ideally the data are recorded at a constant rate of several 100 kHz over the length of an experiment, which might be several minutes. We use data acquisition boards from National Instruments (www.ni.com) and a digital signal processor (DSP) board. The DSP is used to create the scan pattern and to implement the tapping mode and a feedback in constant force mode. The boards can be controlled using software packages like LabView (National Instruments, www.ni.com), the visual programming suite (Microsoft, www.microsoft.com), and IgorPro (Wavemetrics, www.wavemetrics.com). The three-dimensional reconstruction of position histograms and energy landscapes are represented with AVS (Advanced Visual Systems Inc., www.avs.com).

B. Details of the Position Detection

Because of the harmonic nature of the trapping potential for small displacements, the force acting on the trapped bead can be deduced from the displacement of the probe within the trap. Due to the three-dimensional trapping volume, the position of the probe should ideally be measured in all three dimensions. The PFM contains two position detectors: first, a device measuring the two-photon fluorescence intensity emitted by a fluorescently labeled bead to determine the axial displacement (Florin et al., 1997); and second, a quadrant photodiode (QPD) placed in the back-focal plane (BFP) of the condenser measuring the interference of the laser light scattered by the probe with unscattered light (Finer et al., 1994; Allersma et al., 1998; Pralle et al., 1999). The position sensor based on fluorescence is predominantly sensitive along the optical axis; however, the actual signal is a convolution of the three-dimensional displacement. The fluorescence intensity is largely independent of any distortion by the sample or the scan. It is therefore the preferred position signal for larger scans on cells. However, one has to integrate over a few milliseconds to obtain reasonable photon statistics for the laser intensities ($\sim$100 mW in the focal plane) and bead size ($r = 0.1 \, \mu m$) used. The maximum sensitivity of the fluorescence intensity is about 15% change in fluorescence intensity per 100 nm (Florin et al., 1997). The upper part of Fig. 7 shows the fluorescence signal of a lateral and an axial scan through a 0.2-μm sphere. Because the minimum of the trapping potential along the optical axis is behind the geometric focus, the two-photon fluorescence signal can be used in most applications as an axial position sensor. The exact location of the trapping minimum along the optical axis depends on the properties of the laser focus and the probe.

The lower part of Fig. 7 displays the corresponding signals for the interference of the scattered light with the unscattered laser light measured by the QPD. The detection principle based on the interference of the light scattered at the probe with unscattered light has been used to measure one-dimensional lateral displacements and forces in single-molecule experiments (Finer et al., 1994; Svoboda and Block, 1994). Gittes and Schmidt (1998) presented a first explanation for the one-dimensional position detection. We have shown that a QPD placed in the BFP indeed yields information about the displacement of the probe in all three dimensions (Pralle et al., 1999). Because the QPD diode detects the light of the trapping laser, this position sensor is not limited by the light intensity and can be very fast (up to 1 MHz, only limited by the capacity of the QPD, the amplifier, and the recording device). However, any scattering caused by

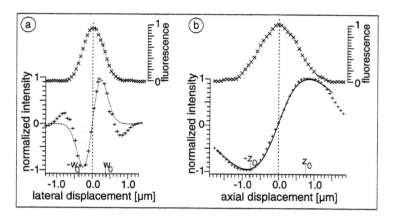

Fig. 7 Upper half: Line profiles of the TPF intensities measured when moving the focus across a 0.6-μm-diameter bead immobilized on a coverslip perpendicular to (a) and along (b) the optical axis. Lower half: Corresponding signal form the QPD measured along the x axis (a) and along the optical axis (b). The fits are taken from the theoretical two-dimensional description for the QPD detection for a focus with a beam-waist radius of $w_0 = 0.33$ μm and a Rayleigh length of $z_0 = 0.76$ μm.

inhomogeneities in the sample may distort the signal. This detection method is best suited for small scans, and is ideal for the detection of the Brownian motion of the probe without moving the laser relative to the sample (see scan modes in the following). The measured intensity changes are described as interference of the light scattered on the sphere with the noninteracting transmitted light. The lateral position of the sphere with respect to the optical axis is measured as the difference between the left and the right halves of the QPD and the difference between the upper and the lower halves. Displacements along the z axis will instead affect the total light intensity detected by the QPD. The detector response is modeled as interference of the scattered light from the trapped particle, described as Rayleigh scatterer, and unscattered laser light. Figures 7a and 7b display one-dimensional profiles taken perpendicular to (a), and along (b) the optical axis with the theoretical description is overlaid. The lateral sensitivity of the interference signal is about threefold higher than the axial sensitivity because the focal volume is elongated along the optical axis. The signals are linear for small displacements. However, all three signals are coupled for larger displacements, and it is advisable to record the three-dimensional position response on a known object to calculate a correction matrix that compensates the coupling between the signals.

C. The Probe

1. Probe Material and Size

Polystyrene or silica spheres, colloid gold particles, or elongated glass particles can be used for PFM experiments. More sophisticated probes may be available in the near future. Probe material, size, and shape selection must be optimized for each type of experiment. Maximizing trapping force and detection contrast needs to be combined with best lateral resolution, with minimizing the surface interactions, or with specific surface modifications.

The factors influencing the trapping force and the surface modifications necessary for specific targeting of the probe limit the choice of probe materials to mostly polymer, silica, and gold particles. The trapping forces depend mainly on the probe size, the ratio of the probe's index of refraction n to that of the suspending medium, and on the focal dimensions. For particles which are small compared to the wavelength of the trapping light, the gradient force is proportional to the particle radius r and the polarizability α (Visscher and Brakenhoff, 1992). Hence, polystyrene (latex) spheres ($n = 1.57$) are more strongly trapped than silica spheres ($n = 1.45$). In diffraction-limited laser traps the focal dimensions are on the order of the wavelength.

The two-photon fluorescence detection provides the best signal-to-noise ratio for spheres with diameters comparable to the focus, as larger spheres average the intensity profile of the laser beam and smaller ones contain fewer fluorophores. The same is true for the detection relying on the interference of the scattered light as spheres just smaller than the focal volume scatter the most amount of light, optimizing the interference signal detected by the quadrant photo-diode.

A sphere as sensor provides the advantage of isotropic resolution independent of the orientation of the sensor to the surface in space. An asymmetrical probe with a pointed end orientated along the optical axis provides increased lateral resolution on flat surfaces (Stout and Webb, 1998). The contact area of the probe and the sample determine the lateral resolution achievable in imaging. Hence, in the case of spherical probes, the one with the smallest diameter providing still reasonable signal-to-noise ratio and the desired trapping strength should be chosen for a particular PFM experiment. We use mostly 0.2-μm polystyrene spheres (average measured radius: 108 nm $\pm$ 3.8 nm). Polystyrene spheres are produced by several manufactures in a broad range of diameters from 20 nm to several micrometers: molecular probes (www.probes.com), Bangs (www.bangslabs.com), Seradyn (www.seradyn.com), and others. Spheres with various surface functional groups to bind proteins covalently and a selection of fluorophore or dye labels are available. The spheres used in the examples shown here were orange-fluorescent (530 nm ex./560 nm em.) carboxyl-modified latex (CML) from molecular probes. These are the preferred probe particles in our group because their fluorophores are excitable by a two-photon process by the IR-trapping laser, and a wide range of sizes with the same surface functional group is available. Polystyrene spheres tend to exhibit stronger nonspecific interactions with surfaces and often have tendrils of polymer hanging off the surface, which can tether them to a substrate (Dabros *et al.*, 1994; Zocchi, 1996). Also, for some molecules, like the molecular motors, kinesin and myosin, silica spheres seem to provide a favorable surface for adsorption (Svoboda and Block, 1994). Plain silica spheres can be obtained from Bangs (www.bangslabs.com) and others.

2. Probe Surface Modifications

To probe the properties or the environment of a specific protein, that protein needs to be attached to the sphere acting as sensor. The protein, the antibody, and the ligand are adsorbed to the sphere and may be additionally covalently crosslinked to the sphere. Adsorption interferes less with the activity of the adsorbed proteins than covalent attachment and is usually reliable and stable enough. However, the conditions have to be optimized for each protein and sphere type, because good adsorption depends strongly on the electrostatic and hydrophobic interactions between the surfaces. Small ligands might need to be attached covalently to the spheres using a spacer to provide enough distance from the surface to preserve their activity. Because noncoated CML spheres are highly charged, they should be coated with a protein blocking unspecific interactions with the cell surface, such as fish skin gelatin (FSG), bovine serum albumin (BSA), or casein (Sigma, www.sigma-aldrich.com). A basic adaptable protocol for antibody adsorption on CML spheres can be found in Sako and Kusumi (1995) and the guideline of the manufactures. Before coating, the spheres are washed three times in 0.2 M boric acid buffer (adjusted to pH 9 using 1 M NaOH) to wash out any solvents left in the spheres from the manufacturing process. A 1% solution of the 0.2-μm CML spheres is incubated with 1 mg/ml antibody in a 50 mM MES buffer, pH 6, for 30 min at room temperature. This is a typical antibody–sphere–surface ratio, however the exact concentration of the coating protein must be optimized for the total surface area and the surface charge density of the spheres in each experiment.

The Brownian motion of the small polystyrene spheres ($r \leq 100$ nm) is sufficient for mixing; however, larger ones or silica spheres should be incubated on a wheel. After coupling, the spheres are incubated for 30 min with 10 mg/ml FSG and washed twice in 10 mg/ml FSG in PBS; another wash is performed immediately before the experiment. Spheres prepared by adsorbing the ligand remain active for many days when stored at 4–8°C. For increased long-term stability the proteins may be covalently coupled to the spheres. Typically the carboxyl beads are crosslinked to the amino groups of the adsorbed protein using ethylcarbodiimed (EDAC, Sigma). We find that crosslinking can reduce the activity of the coated proteins and can increase the likelihood of unspecific adsorption to the cellular membrane.

To optimize the binding procedure, the amount of protein bound to the surface of the spheres should be measured with an assay like the BCA assay from Pierce (www.piercenet.com) or the NanoOrange Protein Quantification from Molecular Probes (www.probes.com). The BCA test is less sensitive; however, it is more compatible with fluorescent spheres, as the spheres can be removed before the measurement because the resulting BCA-Cu$^+$ complex is stable.

To optimize the spheres for single-membrane protein binding, the specific ligand or antibody can be coadsorbed with a similar unspecific protein. Alternatively, after coating the spheres completely with the ligand, a small amount of free receptor without membrane anchor is added to block all but the desired number of binding sites. For each experiment, the conditions providing single-molecule events should be tested by statistical analysis. The binding times and fraction of the beads binding during the observation interval should be measured and compared to a Poisson distribution scaled by $1/n$, where n is the average number of active binding sites on each sphere. The number of active binding sites needed depends on the size of the contact area of the size.

D. Calibration of the Force Sensor

Because the trap potential and the position detector response depend directly on the properties of the sphere and laser focus, it is necessary to calibrate the laser trap and the position sensor with each sphere used for an experiment at a location near the actual measurement.

The trapping potential $V(r)$ can be determined by measuring the position distribution of the trapped particle (Florin *et al.*, 1998). The Boltzmann probability density $P(r)dr$ to find a thermally excited particle in a potential $V(r)$ at position r in the interval $[r, r + dr]$ is $P(r) = c^* \exp[-V(r)/k_B T]$, with c chosen to normalize $\int P(r)dr = 1$. Conversely, the trapping potential can be determined by the probability distribution as $V(r) = k_B T^* \ln(P(r)) + k_B T^* \ln(c)$, where c is an offset. This method allows profiling of the trapping potential even below the thermal energy with temporal and spatial resolutions given by the strength of the potential and the bead size, while requiring only minimal knowledge about the system, i.e., the temperature. For a harmonic trapping potential a stiffness $\kappa = 2V(r)/r^2$ can be defined.

The local detector sensitivity β is determined from the thermal position fluctuations using the Stokes drag γ of the sphere. The motion of a Brownian particle in a harmonic potential is characterized by an exponentially decaying position autocorrelation function $<r(0)^*r(t)> = <r^2>e^{-t/\tau}$ with the mean square amplitude $<r^2> = k_B T/\kappa$ and the correlation time $\tau = \gamma/\kappa$. Thus, the local viscous drag γ and the diffusion coefficient $D = k_B T/\gamma$ of a sphere in a harmonic potential are calculated from the measured correlation time τ of the motion and the stiffness κ of the potential (Pralle *et al.*, 1998). To determine the local detector sensitivity β the autocorrelation time of the positions τ and the spring constant of the trap are calculated from the raw data, yielding an uncalibrated spring constant $\acute{\kappa}$ (in units Nm/V^2 instead on N/m). Because $\gamma = \kappa\tau$ and $\kappa = \acute{\kappa}\beta^2$, the sensitivity β is determined from $\beta^2 = 6\pi\eta r/\acute{\kappa}\tau$, which is valid for a sphere in a harmonic potential as long as the position fluctuations remain within the linear response range of the detector and the calibration is performed at least 10 times the radius of the bead away from the surface.

In experiments determining the local diffusion in the cell membrane, the lateral spring constant of the laser trap was adjusted to about $\kappa \approx 1\mu$N/m for a sphere of 0.2 μm in diameter. The sample chamber was maintained at $36 \pm 1°$C leading to lateral rms position fluctuations of ± 60 nm.

E. Resolution of the PFM

The *resolution* of the PFM needs to be discussed for the particular experiment. The main characteristics of any microscopy are spatial and temporal resolutions. In addition, force microscopes need to optimize the force sensitivity, which is a combination of the precision of the position measurement of the deflection of the force sensor and the compliance of the force sensor. A force sensor with compliance close to the compliance of the sample provides optimal force conditions. Since the compliance of the PFM can be tuned by adjusting the laser power, forces from 1 to 100 pN can be measured with subpiconewton resolution.

At these small forces, thermal motion becomes an important factor in the position measurement of the sensor, hence influencing the spatial and force resolutions. The thermal motion of the interaction area of the sensor with the sample during the measurement interval reduces the achievable spatial resolution. Hence, the spatial resolution is coupled with the temporal resolution. Usually, measurements are performed slower than the position autocorrelation time of the sensor. In the PFM, the situation depends on the experiment and the position sensor used: the two-photon fluorescence is slower due to the low light intensities, while the QPD detecting the interference signal provides position measurements much faster than the autocorrelation time allowing novel methods of data analysis (see scan modes). To image the surface topography of cells, the two-photon fluorescence intensity signal is used, as it is less susceptible to distortions by light scattering inside the cell. Under these conditions, the spatial resolution depends on the amplitude of the Brownian motion and the contact area of the sensor with the surface.

The position sensing based on the interference pattern of the scattered light yields the current position of the probe more precisely because it provides subnanometer resolution at a bandwidth sufficiently broader than the typical autocorrelation time of the Brownian motion in the optical trap. In this case, the topographic resolution is solely dependent on the interaction area of the sensor with the environment. One way to reduce the contact area of the sensor would be by using an asymmetrical probe. Another way would be to keep the sphere outside of the interaction area but rigidly connected to a single-protein molecule, which serves as sensor for its environment. An example for the latter approach is the local diffusion measurement of single molecules in the cell membrane (Pralle *et al.*, 2000).

F. PFM Recording Modes

While some PFM scanning modes are similar to conventional SFM modes, the laser trap has some unique features allowing additional scan modes. The absence of any mechanical lever allows scanning of any three-dimensional shape through space. Either the sample can be scanned using an x-y-z piezo stage, or the trapping laser can be moved. The choice depends on both the area and the shape of the scan. While the latter provides higher scan speeds, it is prone to introduce focus variations in larger scans. A novel scanning alternative unique to the PFM is the use of the Brownian motion of the probe to sample small volumes inside the trapping volume.

1. Contact Mode

In the constant-height mode, the sphere trapped by the laser beam is brought into contact with the surface and then moved over the surface along an area of scan lines (Fig. 8a). The two-photon intensity is recorded to detect the axial displacement of the sphere out of its resting position in the trap to measure the topography of the surface. At the beginning of the scan, the sphere trapped in solution away from the surface is approached to the surface by moving the piezo-mounted microscope objective away from the sample chamber. A drop in the two-photon fluorescence intensity indicates the contact with the surface. Due to the weak axial spring constant of the trap in comparison to the lateral one, a protrusion of the surface displaces the bead predominantly along

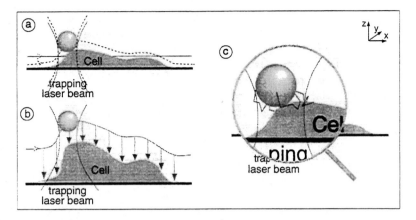

Fig. 8 Illustration of the various recording modes of the PFM: (a) to image a surface in the contact mode, the focus holding the probe particle is scanned over the surface laterally. This can be done either by maintaining a constant distance to the support (solid line) or by using a feedback, moving the focus up and down (dashed line) to maintain a constant force between probe and sample. (b) In the PFM tapping mode the focus is approached to the sample in each image point, and upon contact is retracted a predefined distance. (c) Three-dimensional SPT of a sphere bound to a diffusing membrane particle: the laser trap is held steady, and the Brownian motion of the diffusing particle is used to record the interaction with the environment.

the optical axis, i.e., vertically ($+z$) away from the surface. The displacement results in a further decrease of the two-photon fluorescence intensity. An image of the surface topography is acquired by recording the fluorescence intensity while raster scanning an area. If the bead is displaced too far away from the focus, it escapes the trapping potential. Therefore, the height of the object has to be smaller than the trapping range of the laser trap, which is about 0.8 μm.

The vertical working range is substantially extended by using a feedback circuit that drives the piezo-mounted objective lens up or down maintaining a constant fluorescence intensity and constant position of the probe in the laser trap, thus creating a constant force mode. Because of the large mass of the objective lens, the response time of the feedback is limited.

While these scan modes rely on the two-photon fluorescence intensity as a measure for the axial displacement of the probe in the trap, it is advisable to simultaneously record the signals from the quadrant detector as well. These signals provide information about the three-dimensional displacement of the probe and, taken together, help to reveal possible scan artifacts in the normal topographic image.

2. Tapping Mode

In the tapping mode, the sphere trapped by the laser beam is brought repeatedly into contact with the surface (Fig. 8b). The PFM tapping mode can be compared to the force–scan volumes acquired by conventional SFM (Radmacher *et al.*, 1996). In each point of an image, the surface is approached while recording the two-photon fluorescence intensity. When the fluorescence intensity decreases below a preset set fraction of the intensity measured for the free sphere, the sphere is retracted a fixed distance and moved to the next point. The tapping mode enables the measurement of virtually vertical slopes. Because the contact times and forces are reduced, the spheres are less often lost due to nonspecific adhesion to the cell surface. The vertical range in the tapping mode is limited either by the working range of the driving piezo, i.e., 100 μm, or by spherical aberration effects, which restrict the range of stable trapping for larger distances from the coverslip surface.

The tapping mode feedback is implemented via a DSP board and by a computer that also displays the image and individual force scans. A reference fluorescence intensity for the free sphere is measured in each point to avoid image distortion due to bleaching of the sphere and laser intensity variations in the sample plane. The height of the endpoint of each force scan depends on the imaged topography. The probe is retracted at constant distances from the last contact with the surface, enabling the PFM to climb up the extremely steep edge, without the need for extremely long and time-consuming force scans. The elasticity of the surface is computed from the slope of the two-photon fluorescence intensity decrease. Again, using the QPD to detect the forward-scattered light, the lateral displacement of the sphere upon contact can be recorded simultaneously.

3. Fast Three-Dimensional Single-Particle Tracking

To measure the local environment of single-membrane proteins, no active scanning is necessary, but the thermal position fluctuations of a sphere in a weak trapping potential. The rms thermal position noise in a trapping potential of 2 μN/m is $\approx$45 nm. Measuring this motion precisely using the forward scattered light allows recording of the three-dimensional diffusion on the cell surface with high temporal resolution. The free trapping potential is plotted to visualize the volume accessible to the bead. Any deviations thereof are due to interacting potentials or obstructions such as a surface of stable object or immobile membrane components. The local viscous drag can be determined by analyzing the motion along the track.

G. Sample Preparation

It is essential to prepare the sample surfaces as cleanly as possible to minimize the nonspecific interaction between the probe and the sample and to avoid collecting small biological particles like vesicles in the laser trap. The cellular samples are prepared as follows: Baby-hamster kidney (BHK-21) cells are grown, according to standard cell culture procedures, in a tissue culture flask with supplemented Glasgow(G)-MEM and passaged every 2–3 days. The hippocampal neurons are extracted from 18-day-old rat embryos, plated on poly-L-lysine-coated coverslips in a dish that was preincubated with glia cells and grown at 37°C and 5% CO_2 in N_2 culture medium (Goslin and Banker, 1991). Circular glass coverslips (11 mm) are used as substrate for the cells. These are cleaned and sterilized (either autoclaved or washed in acetone/ethanol and dried in sterile air). The cells, BHK fibroblasts or hippocampal neurons, are plated at low density on the coverslips and allowed to grow 3–5 days. At this stage, the early development of the major processes and the growth cone morphology of the neurons can be studied.

For imaging of living cells, the cells are washed and imaged in filtered culture medium the same. In the case of the neuronal cells, it is advisable to use the culture medium from the dish in which the cells had been growing to maintain the exact composition of the medium during the experiment. For the experiments on fixed cells, the cells are washed twice in PBS, fixed in 1% glutaraldehyde for 10 min at room temperature, washed three times in PBS, and incubated for 10 min in 50 mM NH$_4$Cl to block any free aldehyde groups. Cells for live imaging are washed again in PBS containing 10 mg/ml FSG. The scanning experiments are carried out in culture medium for living cells and in PBS for fixed cells. In both cases, 10 mg/ml FSG is added to the solution and the microscope stage is heated to 35°C. All solutions should be filtered through 0.1-μm SuporeAcrodisc filters (Gelman Sciences, www.pall.com/gelman).

To study nonendogenous membrane proteins, the cells can be either transfected 12–14 h prior to the experiment using a trasporter such as lipofectamine (Gibco, www.lifetech.com) or infected using a retrovirus-based system, like the adenovirus. In any case it is useful to cotransfect the cells with a cytoplasmic green fluorescent protein (E-GFP) to facilitate the search for successfully transfected cells. The virus system provides the advantages of being very reproducible, yielding high ratios of infected cells and disturbing the composition of the plasma membrane the least.

References

Allersma, M. W., Gittes, F., deCastro, M. J., Stewart, R. J., and Schmidt, C. F. (1998). Two-dimensional tracking of ncd motility by back focal plane interferometry. *Biophys. J.* **74**(2), 1074–1085.

Ashkin, A., Dziedzic, J. M., Bjorkholm, J. E., and Chu, S. (1986). Observations of a single-beam gradient force optical trap for dielectric particles. *Opt. Lett.* **11**, 288–290.

Ashkin, A., Schütze, K., Dziedzic, J. M., Euteneuer, U., and Schliwa, M. (1990). Force generation of organelle transport measured in vivo by an infrared laser trap. *Nature* **428**, 346–348.

Cherry, R. J. (1992). Keeping track of cell surface receptors. *Trends Cell Biol.* **2**, 242–244.

Dabros, T., Warszynski, P., and van den Ven, T. G. M. (1994). Motion of latex spheres tethered to a surface. *J. Colloids Surf. B* **4**, 327–334.

Dai, J., and Sheetz, M. P. (1995). Mechanical properties of neuronal growth cone membranes studied by tether formation with laser optical tweezers. *Biophys. J.* **68**(3), 988–996.

De Brabander, M., Nuydens, R., Ishihara, A., Holifield, B., Jacobson, K., and Geerts, H. (1991). Lateral diffusion and retrograde movements of individual cell surface components on single motile cells observed with nanovid microscopy. *J. Cell Biol.* **112**, 111–124.

Edidin, M., Kuo, S. C., and Sheetz, M. P. (1991). Lateral movements of membrane glycoproteins restricted by dynamic cytoplasmic barriers. *Science* **254**, 1379–1382.

Evans, E. A. (1983). Bending elastic modulus of red blood cell membrane derived from buckling instability in micropipette aspiration tests. *Biophys. J.* **43**(1), 27–30.

Evans, E. A., and La Celle, P. L. (1975). Intrinsic material properties of the erythrocyte membrane indicated by mechanical analysis of deformation. *Blood* **45**(1), 29–43.

Finer, J. T., Simmons, R. M., and Spudich, J. A. (1994). Single myosin molecule mechanics: piconewton forces and nanometre steps. *Nature* **368**, 113–118.

Florin, E.-L., Pralle, A., Hörber, J. K. H., and Stelzer, E. H. K. (1997). Photonic Force Microscope based on optical tweezers and two-photon excitation for biological Applications. *J. Struct. Biol.* **119**, 202–211.

Florin, E. L., Pralle, A., Stelzer, E. H. K., and Hoerber, J. K. H. (1998). Photonic force microscope calibration by thermal noise analysis. *Appl. Phys. A* **66**, S75–S78.

Ghislain, L. P., and Webb, W. W. (1993). Scanning-force microscope based on an optical trap. *Opt. Lett.* **18**, 1678–1680.

Gittes, F., and Schmidt, C. F. (1998). Interference model for back focal plane displacement detection in optical tweezers. *Opt. Lett.* **23**, 7–9.

Goslin, K., and Banker, G. (1991). "Culturing Nerve Cells." M.I.T. Press Cambridge, MA.

Grimellec, C. L., Lesniewska, E., Cachia, C., Schreiber, J. P., Fornel, F. d., and Goudonnet, J. P. (1994). Imaging of the membrane surface of MDCK cells by atomic force microscopy. *Biophys. J.* **67**, 36–41.

Häberle, W., Hörber, J. K. H., and Binnig, G. (1991). Force microscopy on living cells. *J. Vac. Sci. Technol.* **B9 2**, 1210–1213.

Hertz, H. M., Malmqvist, L., Rosengren, L., and Ljungberg, K. (1995). Optically trapped non-linear particles as for scanning near-field optical microscopy. *Ultramicroscopy* **57**, 309–312.

Jacobson, K., Sheets, E. D., and Simons, R. (1995). Revisiting the fluid mosaic model of membranes. *Science* **268**, 1441–1442.

Kusumi, A., Sako, Y., and Yamamoto, M. (1993). Confined lateral diffusion of membrane receptors as studied by single particle tracking (nanovid microscopy). Effects of calcium-induced differentiation in cultured epithelial cells. *Biophys. J.* **65**(5), 2021–2040.

Malmquist, L., and Hertz, H. M. (1992). Trapped particle optical microscopy. *Opt. Commun.* **94**, 19–24.

Mehta, A. D., Finer, J. T., and Spudich, J. A. (1998). Reflections of a lucid dreamer: Optical trap design considerations. *In* "Methods in Cell Biology" (M. Sheetz, ed.), Vol. 55, pp. 47–69. Academic Press, San Diego.

Pralle, A., Florin, E. L., Stelzer, E. H. K., and Hoerber, J. K. H. (1998). Local viscosity probed by photonic force microscopy. *Appl. Phys. A* **66**, S71–73.

Pralle, A., Keller, P., Florin, E.-L., Simons, K., and Hörber, J. K. H. (2000). Sphingolipid—Cholesterol rafts diffuse as small entities in the plasma membrane of mammalian cell. *J Cell Biol.* **148**, 997–1007.

Pralle, A., Prummer, M., Florin, E.-L., Stelzer, E. H. K., and Hörber, J. K. H. (1999). Three-dimensional position tracking for optical tweezers by forward scattered light. *Micro. Res. Technol.* **44**, 378–386.

Radmacher, M., Fritz, M., Kacher, C. M., Cleveland, J. P., and Hansma, P. K. (1996). Measuring the viscoelastic properties of human platelets with the atomic force microscope. *Biophys. J.* **70**, 556–567.

Radmacher, M., Tillmann, R. W., Fritz, M., and Gaub, H. E. (1992). From molecules to cells: Imaging soft samples with the atomic force microscope. *Science* **257**, 1900–1905.

Sako, Y., and Kusumi, A. (1995). Barriers for lateral diffusion of transferrin receptor in the plasma membrane as characterized by receptor dragging by laser tweezers: fence versus tether. *J. Cell Biol.* **129**(6), 1559–1574.

Sasaki, K., Shi, Z. Y., Kopelman, R., and Masuhara, H. (1996). Three-dimensional pH microprobing with an optically-manipulated fluorescent particle. *Chem. Lett.* **2**, 141–142.

Singer, S. J., and Nicolson, G. L. (1972). The fluid mosaic model of the structure of cellmembranes. *Science* **175**(23), 720–731.

Stout, A. L., and Webb, W. W. (1998). Optical force microscopy. *In* "Methods in Cell Biology" (M. Sheetz, ed.), Vol. 55, pp. 47–69. Academic Press, San Diego.

Svoboda, K., and Block, S. M. (1994). Biological applications of optical forces. *Annu. Rev. Biophys. Biomol. Struct.* **23**, 247–285.

Tomishige, M., and Kusumi, A. (1999). Compartmentalization of the erythrocyte membrane by the membrane skeleton: inter compartmental hop diffusion of bond 3. *Mol. Biol. Cell* **10**(8), 2475–2479.

Vaz, W. L. C., and Almeida, P. F. F. (1993). Phase topology and percolation in multiphase bilayers: Is the biological membrane a domain mosaic. *Curr. Opin. Struct. Biol.* **3**, 482–488.

Visscher, K., and Brakenhoff, G. J. (1992). Theoretical study of optically induced forces on spherical particles in a single beam trap. I. Rayleigh scatterers (cellular micromanipulator). *Optik* **89**, 174–180.

Zhang, F., Lee, G. M., and Jacobson, K. (1993). Protein lateral mobility as a reflection of membrane microstructure. *Bioessays* **15**(9), 579–588.

Zocchi, G. (1996). Mechanical measurement of the unfolding of a protein. *Europhys. Lett.* **35**, 633–638.

CHAPTER 10

Methods for Biological Probe Microscopy in Aqueous Fluids

Johannes H. Kindt, John C. Sitko, Lia I. Pietrasanta,*
Emin Oroudjev, Nathan Becker, Mario B. Viani,†
and Helen G. Hansma

Department of Physics
University of California
Santa Barbara, California 93106

*Current address: Laboratorio de Electrónica Cuántica, Departamento de Física, Pabellón I - Ciudad Universitaria, C1428EHA Buenos Aires, Argentina.
†Current address: Asylum Research, Santa Barbara, CA 93117.

I. Introduction

It is easier and often faster to image biological samples in air than in aqueous fluid. But imaging in aqueous fluids is almost always preferable if one wants to see biomaterials in near-physiological environments. When imaging in fluid, one sees biomaterials not only under conditions where their structures are native but also under conditions where the biomaterials retain their biological activity. This activity can be monitored and, using advanced fluid handling techniques, investigated under changing environmental conditions.

It has been said that atomic force microscopy (AFM) is unnatural because the atomic force microscope (AFM) looks at biomaterials on surfaces instead of in test tubes. The development of biological AFM has also been handicapped by "test tube biology," because *in vitro* biological systems have been developed to work in test tubes, while the AFM looks at biological systems on surfaces. But living systems are filled with surfaces, especially membranes. Therefore surfaces are arguably more relevant biologically than test tubes. In fact, AFM may be a leader in a new field, Surface Biology, which will grow into a major research area in the new century.

This article covers methodology for using AFMs and other probe microscopes with which the authors are familiar. These are Digital Instruments scanning probe microscopes (SPMs) and the Asylum Research Molecular Force Probe (MFP). The MFP is a new instrument optimized for molecular pulling experiments of the type shown in Fig. 1.

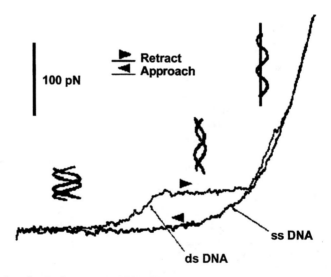

Fig. 1 A single molecule of overstretched DNA. This graph shows a force measurement of a single tethered molecule of Lambda Digest DNA showing the B–S and the melting transition. Arrowheads indicate pulling direction as follows: DNA stretch is ▶ and DNA relaxation is ◀. During the extension of the molecule (red trace), the DNA first goes through the B–S transition (the plateau) and then melts to single-stranded DNA (ss-DNA) at a higher force. During relaxation of the molecule (blue trace), the DNA does not reanneal, so the curve is a simple freely jointed chain, indicative of ss-DNA. The traces were made at a pulling speed of 1 mm/s. Data courtesy of H. Clausen-Schaumann and R. Krautbauer, Gaub Lab, LMU-München. Data were obtained with a cantilever from Park Scientific Microlevers on a Molecular Force Probe from Asylum Research (http://www.asylumresearch.com). (See Color Plate.)

II. Substrates/Surfaces

"Substrates" in this context are the surfaces that biomaterials are placed on for AFM imaging. Common substrates for biological AFM are mica and glass. Glass is flat enough for imaging cells but is generally too rough for easy visualization of DNA, especially under fluid.

Biomaterials such as DNA and proteins are usually imaged on mica, which has a root-mean-square roughness of only 0.06 ± 0.01 nm (Hansma and Laney, 1996). Silylated mica and other treated micas such as Ni(II)-mica (Bezanilla *et al.*, 1994; Hansma and Laney, 1996) and Mg(II)-mica are also used. AP-mica is the most common of the silylated mica substrates (Bezanilla *et al.*, 1995; Lyubchenko *et al.*, 1992); its RMS roughness of 0.09 ± 0.01 nm is only slightly rougher than mica.

The biomaterials of interest need to adhere at least weakly to the substrate if they are to be imaged well in aqueous fluid.

III. Basic Methods for Atomic Force Microscopy in Aqueous Fluids

A. Imaging without an O Ring

This is the default method for many SPMs, and it is an optional method when using the MultiMode SPM.

Given the importance of biological imaging in fluid, one wants to be able to image in fluid as simply as possible. One thing that makes biological imaging easier, when using the Digital Instruments MultiMode AFM, is to leave out the O ring. One can usually image for about an hour under a drop of fluid before evaporation becomes a problem. There are at least two ways to set up samples for imaging in fluid. Often one can simply place a drop of 30–35 μL, containing the biomaterial of interest, on the cantilever in the fluid cell and then quickly turn over the fluid cell and insert it into the AFM over the substrate. Of course one wants to be sure beforehand that the cantilever will not crash onto the substrate, so one may want to do a "coarse approach" with the dry cantilever + fluid cell + substrate in the AFM before adding the sample solution.

If one wants to image for longer than an hour, one will want to add a few microliters (μL) of water or buffer to the fluid cell periodically. One can do this with a syringe or a microliter pipetter, inserted into the space between the fluid cell and the sample. Or, when using a MultiMode AFM, one can inject fluid into one of the syringe ports in the fluid cell.

Sometimes one wants the solution above the sample to be purely buffer solution, without the biomolecules or other biomaterials that are in the solution. In these cases, one can place the sample on the substrate in the AFM in a volume of 1–5 μL, place a buffer drop of 30–35 μL on the cantilever in the fluid cell, and then quickly turn over the fluid cell onto the substrate as in the example above. When using a very small sample volume, one will of course want to be speedy about getting the sample submerged in buffer on the cantilever before the sample on the substrate dries up. The sample can also be rinsed to remove loosely bound biomaterials before placing the cantilever with buffer solution over it.

If one wants to change fluids while the sample is in a MultiMode AFM, it is usually best to work with an O ring. Here, too, however, one can do limited fluid changing without using an O ring. One way to do this is to have two 1-mL syringes in the ports of the MultiMode fluid cell—one empty syringe and one filled with the new solution. The old solution can be sucked into the empty syringe, followed by a cautious injection of new solution from the filled syringe— ~50 μL will be sufficient. Repeating this a few times will give a fairly complete exchange of fluid—but one must be careful not to inject too much fluid, or it will flow over the edge of the sample and onto the scanner.

Similarly, one can change fluid in other open fluid cells by using syringes with needles for injecting and removing solutions in the space between the cantilever and the substrate.

A much finer system for pumping fluids into the fluid cell during imaging has been developed by our group, using computer-controlled fluid changes and microliter volume injections that can be carried out with little or no disruption of the image whose capture is in progress. We will discuss this option later.

B. Imaging with an O Ring

For more serious imaging with the MultiMode AFM under a series of fluids, the O ring is unavoidable. One can improve the O ring somewhat by slicing off the outer edge with a razor blade or scalpel to decrease the outer diameter of the O ring. Much better O rings with a new cross section have been designed by Johannes H. Kindt at UCSB. Hopefully these will soon become commercially available to the AFM community.

With an O ring to enclose the fluid cell, one will want to think about automating the flow of fluids into and out of the fluid cell. This can be done either with the computer-controlled system described later or with a gravity flow system using syringe barrels and valves as described by Thomson _et al._ (1996). The flow rate for this system is measured by collecting the effluent into a beaker on an electronic balance.

C. Removing Bubbles from the Cantilever

Air bubbles are a problem when they sit on the cantilever. If one is having imaging problems, it is wise to check for bubbles, which can cause imaging problems in fluid.

Sometimes air bubbles can be removed simply by lifting the fluid cell and lowering it down again over the sample. Or the fluid cell can be removed from the AFM and tapped gently to dislodge bubbles. Another way to dislodge bubbles is to flush fluid in and out of the fluid cell with a syringe attached to the port of the fluid cell.

One can also easily degas solutions before use, if there is a persistent problem with air bubbles. To degas a solution, pull it into a syringe, hold a piece of parafilm over the end of the syringe with your finger, and pull on the plunger of the syringe until air bubbles form. Tap the syringe against the edge of the lab bench while pulling on the plunger to dislodge bubbles from the walls of the syringe, and they will rise into the air space above the solution. Bubbles will, of course, be a problem whenever one injects cold fluids into the fluid cell, so temperature equilibration of fluids is important.

D. Imaging Modes

The two standard AFM imaging modes are tapping and contact. Our labs almost always uses the tapping mode, which reduces lateral forces. For some samples, the contact mode is preferred. It is often easier to see substructural detail in the contact mode when imaging flat samples where lateral forces are small. For example, two-dimensional protein arrays, including membrane protein arrays, are usually imaged in the contact mode (Czajkowsky and Shao, 1998; Engel et al., 1997; Müller et al., 1998; Yang et al., 1994), although the tapping mode AFM can give comparable resolution in the hands of an experienced user (Moller et al., 1999).

E. Imaging Parameters

In the late 1980s, a newcomer to the AFM field said he had expected to find that using the AFM was rather like using a toaster (Hermann Gaub, personal communication). Instead, he found it to be more like playing a violin. Although the AFM is becoming more toaster-like with its new improvements and more user experience, it is still somewhat violin-like. Therefore the user will find it useful to experiment with gains, setpoint, scan speed, drive frequency, and drive amplitude to find the best conditions for each new sample. The details presented in the following paragraphs about imaging parameters are to be used as a guide, not as a strict protocol.

When determining the frequency response for cantilevers in fluid, automatic tuning methods may not work. A plot of amplitude versus frequency shows multiple peaks in fluid. The highest peak does not necessarily produce optimal imaging. With experience, one usually finds that a particular location in these peaks gives good images. This location is often 5–10% below the peak frequency of the selected peak. A good technique for newcomers is to check the imaging quality at or slightly below the peak frequency of a few of the peaks until satisfactory results are obtained. If imaging quality starts to degrade while using the same cantilever, one can check the cantilever tuning again and readjust the drive frequency. Our lab typically uses 100-micron-long, narrow, V-shaped, silicon–nitride cantilevers in Plexiglas fluid cells. With these home-made Plexiglas fluid cells, the optimal peak frequency is typically close to 13 kHz. With glass fluid cells, the primary cantilever oscillations are at lower frequencies, near 9 kHz. These oscillations are all in the envelope of the thermal resonance frequency for the cantilever in fluid (Schaffer et al., 1996).

After the correct frequency is determined, the gains can be optimized. One can start scanning with the integral gain set at 1.2 and the proportional gain twice as large. With these values one can usually tell whether an image is obtainable or if one needs to change the frequency or the tip. Proportional gains are significantly less sensitive than integral gains, so first the integral gain must be adjusted only until the image is optimal. Then the proportional gain must be adjusted. This gain ends up being about two or three times the integral gain. We commonly use an integral gain between 1 and 3 in fluid, though we have used much higher gains on occasion.

The optimum imaging setpoint is selected by lifting off the surface completely while scanning, then slowly approaching until an image is formed. With dry AFM, pushing harder on the sample will often give a sharper image. In aqueous AFM, samples can be particularly soft, so minimal forces are often optimal. One may want to rescan the same area with a larger scan size to ensure one has not scraped the surface.

The imaging setpoint correlates with the drive voltage. In general smaller drive voltages are good for imaging relatively flat samples such as DNA and proteins on mica, while larger drive voltages are good for imaging relatively thick, sticky, or soft samples such as cells. With small-drive voltages, the setpoints for low-force imaging will be 0.5 V or less; with large-drive voltages, the setpoints for low-force imaging will be 1–2 V or higher.

In general, the scanning speed does not need to be changed when going from dry to aqueous samples. We usually use a scan speed of 2–4 Hz.

F. Cantilevers

As mentioned earlier, our lab typically used 100-micron-long, narrow, V-shaped, silicon–nitride cantilevers. EBD tips, oxide-sharpened Si–N tips and normal pyramidal Si–N tips have all been used successfully. Before fluid tapping was possible, we observed that EBD tips gave less sample damage (Hansma *et al.*, 1993).

The 200-micron-long, wide V-shaped cantilevers have a similar spring constant to the 100-micron narrow cantilevers. Their larger size makes them easier for beginners to use, and we have used them successfully for many samples. These 200-micron cantilevers are probably preferable to the 100-micron cantilevers for users who do not have a scanner with "vertical engage" such as the original MultiMode scanners from Digital Instruments. For older MultiMode scanners that need to be leveled manually, it is easier to get the longer cantilevers level enough with respect to the surface; with the shorter cantilevers, even a small sample tilt can cause the corner of the cantilever chip to hit the sample instead of the cantilever tip. Feeler gauges are useful for leveling these older scanners.

Other soft cantilevers for imaging in fluid are 400-micron-long rod-shaped silicon cantilevers and V-shaped silicon cantilevers from Park Scientific (now ThermoMicroscopes, Sunnyvale, CA).

G. Effects of Different Aqueous Solutions on AFM Imaging

Much of the challenge with biological AFM in fluid is in finding a good aqueous buffer solution that supports the biological activity of interest and also keeps the biomolecules well enough immobilized on the substrate for good imaging but not so tightly bound as to be inactive. In our group we explored and succeeded in imaging in liquid different biological macromolecules such as laminin, chaperonins, DNA, and DNA–protein complexes.

We investigated the three-dimensional arrangement and dynamic motion of laminin-1 (Ln-1) molecules (Chen *et al.*, 1998). Laminins are a family of extracellular matrix glycoproteins that play an active role in tissue development and maintenance. Four different buffers at pH 7.4 were used: high-salt MOPS buffer (20 mM MOPS, 5 mM MgCl$_2$, 150 mM NaCl), low-salt MOPS buffer (20 mM MOPS, 25 mM NaCl, 5 mM MgCl$_2$), PBS in 5 mM MgCl$_2$ (10 mM phosphate buffer, 2.7 mM KCl, 137 mM

NaCl), and Tris buffer (50 mM Tris, 150 mM NaCl, 5 mM MgCl$_2$). The two MOPS buffers (low-salt and high-salt) were the best for imaging substructures in individual Ln-1 molecules. The lower panels in Fig. 3 show the flexibility and mobility of Ln-1 arms in high-salt MOPS buffer (physiological conditions). Sometimes, imaging in a high-salt buffer was not as easy as imaging in a low-salt buffer. The images appeared less well defined, perhaps because the molecules were more weakly attached to the mica. Imaging Ln-1 in PBS or Tris buffer was very difficult, and the images were poor. In contrast to the successful imaging of Ln-1 in fluid, other basement-membrane macromolecules such as collagen IV and heparan sulfate proteoglycan could not be imaged in the previously cited buffers, though they gave good images in air (Chen and Hansma, 2000).

Another example of proteins imaged in solution without additional treatment such as fixation is the *Escherichia coli* chaperonin GroEL and its co-chaperonin GroES. These proteins play important roles in helping proteins reach their native states. We were able to scan the same sample region without excessively disturbing the array of either the GroEL or the GroES molecules. The central channel of the protein was resolved in many of the molecules. The best results were obtained when the protein arrays were imaged in 50 mM Hepes (pH 7.5), 50 mM KCl, and 10 mM MgCl$_2$. These preliminary results with a commercial AFM were followed by analyses of protein dynamics with a prototype small-cantilever AFM (Viani *et al.*, 2000).

To study biological processes such as DNA–protein interactions in fluid with the AFM, one has to compromise between strongly bound DNA, essential for good imaging conditions, and loosely bound DNA, required for reactions with other molecules such as enzymes. We have found, after exploring several buffers containing salts of divalent inorganic cations, that DNA molecules bind tightly enough to mica if the solution contains 1 mM concentrations of Ni(II), Co(II), Zn(II), or Mn(II) (Hansma and Laney, 1996). This finding was valuable for demonstrating the activity of *E. coli* RNA polymerase (RNAP) on mica (Kasas *et al.*, 1997). With varying Zn^{2+} concentration in the buffer solution, the DNA molecules bound loosely enough to be translocated by the RNAP and also with sufficient strength to be imaged with the AFM (e.g., Fig. 2). Although our labs have favored the use of divalent transition metal salts, the Bustamante lab has successfully imaged RNAP complexes in buffers without these salts (Bustamante *et al.*, 1999; Guthold *et al.*, 1999). This is another example of the "violin-playing" nature of AFM imaging at present.

We have observed other DNA–enzyme processes in the AFM, including reactions of DNA with other polymerases (Argaman *et al.*, 1997; Hansma *et al.*, 1999) and with DNaseI (Bezanilla *et al.*, 1994; Hansma, 2000; Hansma *et al.*, 2000). DNA degradation by DNaseI is a robust process in the AFM that makes it useful for testing new instrumentation such as the automated fluid-handling system of Fig. 4. After stable imaging in the presence of Ni(II), the buffer containing Mn(II) (the divalent cation required for the enzyme activity) and the DNaseI solution were injected with this system, yielding the results shown in Fig. 5.

Observing other DNA–enzyme interactions is an avenue of progress that provides many opportunities for new development in the instrumentation and new strategies in the imaging conditions.

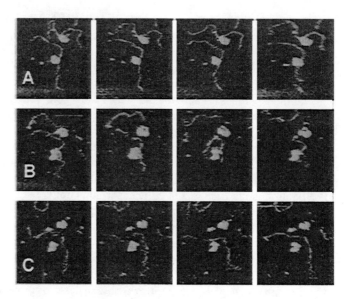

Fig. 2 Two active complexes of DNA with RNAP under fluid in an AFM. *E. coli* RNAP transcription complexes were prepared with a 1047-bp DNA template (Guthold *et al.,* 1999; Kasas *et al.,* 1997). (A), (B), and (C) each show a series of four consecutive images at 42-s intervals. (A) DNA strands move near the surface in Zn(II) buffer. (B) 3.5–6 min after the last image in (A). RNAP transcribes and/or detaches from DNA strands after NTPs are introduced. (C) 6–8 min after the last image in (B); Zn(II) buffer is reintroduced. Note that the image quality deteriorates in Zn(II)-free buffer and improves as Zn(II) buffer is reintroduced [see Hansma and Laney (1996)]. DNA images are 310 nm × 330 nm. (See Color Plate.)

3 dry laminin molecules:

3 sequential images of 1 wet laminin molecule, moving:

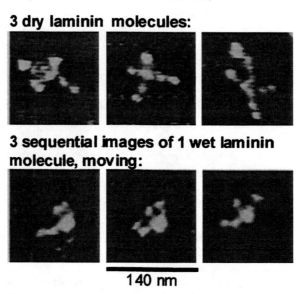

140 nm

Fig. 3 AFM imaging of laminin molecules in air shows submolecular structure in the laminin arms (top row). In the sequential images, a single laminin molecule in aqueous solution waves its arms (bottom row). (See Color Plate.)

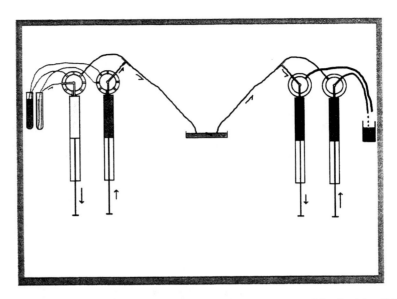

Fig. 4 The setup of the fluid-handling system. On the left are the pump-modules that inject fluid from different source solutions. On the right are the additional pump-modules sucking solution from an open fluid cell at the same rate. In the center is the fluid chamber around the sample with the cantilever above the sample. (See Color Plate.)

H. When To Image in Fluid

Fluid imaging is essential if one wants to see something happening, such as moving DNA molecules in the complexes with RNA polymerase in Fig. 2 (Hansma, 1999; Kasas *et al.*, 1997) or the motion of the laminin arms in Fig. 3. Another useful type of AFM in fluid is force mapping or force–volume (FV) imaging (Brown and Hoh, 1997; Radmacher *et al.*, 1994) (Fig. 6). This FV image of three synaptic vesicles with dark spots in their centers shows darker and lighter regions that correspond to harder and softer regions, respectively (Laney *et al.*, 1997). We were surprised to find that these vesicles were harder in their centers than at their edges, unlike most cells and other soft things (imagine, for example, a pillow). One can see that the vesicle centers are harder or stiffer than their peripheries because they are dark like the mica surface (though not nearly as hard as mica).

I. When Not To Image in Fluid

One does not want to get carried away with imaging in fluid, though, to the exclusion of imaging in air. It is of course usually easier to get stable images in air than in fluid, and air images also often have better resolution. For example, in Fig. 3, the images of laminin in fluid show the arms moving, but the images in air show the substructure in the laminin arms in much greater detail (Chen *et al.*, 1998).

Another example where imaging in air has proved to be more useful than imaging in fluid is the Ni(II)-mediated condensation of the DNA, poly (dG–dC)*(dC–dG)

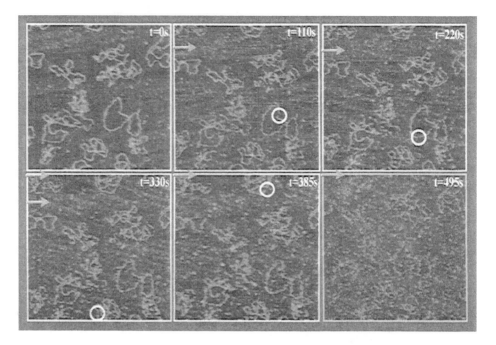

Fig. 5 Enzymatic degradation of single DNA molecules in the AFM. A field of DNA molecules (0.5 μg/mL of BlueScript plasmid DNA) in a buffer containing 20 mM Hepes, 5 mM MnCl$_2$, pH 7.6, continuously pumped at 5 μL/s. After the injection of DNaseI into the same buffer, the degradation of the molecules can be observed; arrows indicate frame and position in frame where the 10-μL injections occurred. The circles highlight new cuts in DNA molecules. The scan size is 1 μm $\times$ 1 μm; the z range is 7 nm. All imaging was done on a Nanoscope III Multimode-AFM (Digital Instruments). The microscope was operated in the fluid tapping mode using cantilever oscillation frequencies between 10 and 20 kHz. (See Color Plate.)

Height   **Force-Volume**

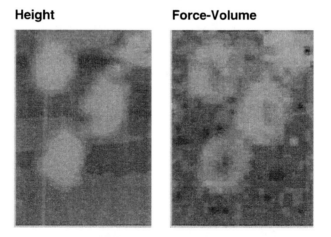

Fig. 6 Three cholinergic synaptic vesicles. Height image (left) and force–volume (FV) image (right) of three synaptic vesicles from the electric organ of *Torpedo*. The centers of the vesicles are harder or stiffer than the edges of the vesicles (see Laney *et al.*, 1997). (See Color Plate.)

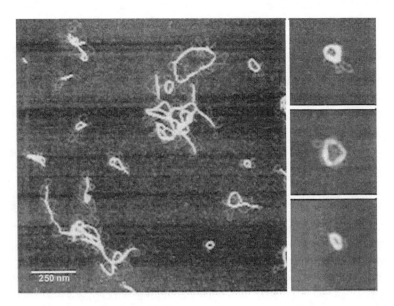

Fig. 7 These condensed DNA structures in air (left) and fluid (right) are similar. The side loops on these DNA condensates can be imaged more stably in air. Poly (dG–dC)*(dG–dC) condensed with 1 mM NiCl$_2$ to form loopy toroids. Left: A typical field of condensates was imaged with tapping mode AFM in air. Right: These three toroids were found in aqueous tapping mode AFM images. The scale bar applies to all images.

(GC-DNA)(Fig. 7, (Hansma 1999; Sitko, in preparation)). With this system, the observed structures were similar in air and in fluid. Because the DNA condensates bound strongly and irreversibly to the mica, they did not move or condense further during AFM in fluid. Therefore it was easier and no less useful to image these condensates in air instead of in fluid.

IV. Molecular Force Probing

A relatively new application for probe microscopy in fluid deserves special mention. One of the most dramatic examples of this new application is the unfolding of individual titin protein molecules (Fisher *et al.*, 1999; Rief, Gautel *et al.*, 1997). Other examples include tensile pulling of double-stranded DNA molecules (Lee *et al.*, 1994) and single polysaccharide molecules (Rief, Oesterhelt *et al.*, 1997), measuring the strength of single covalent bonds (Grandbois *et al.*, 1999) and ligand–receptor or ligand–ligand interactions (Dammer *et al.*, 1995, 1996; Florin *et al.*, 1994). This AFM application is sometimes called force spectroscopy (Rief, Oesterhelt *et al.*, 1997). Here it is referred to as molecular force probing (MFPing) to distinguish it from probe microscopic techniques that require scanning.

MFPing essentially involves measurements of force versus distance characteristics for single or multiple molecules stretched between the cantilever tip and the substrate in the AFM contact mode. The molecules being probed are attached through covalent or noncovalent interactions to the substrate and to the tip of the cantilever. A large array of techniques can be employed to achieve this goal. Usually, the tip of the AFM probe is first pressed against substrate, which has the material of interest deposited onto it. After a short incubation time (on the order of 1 s, to allow the molecules of interest bind to the tip) single or multiple pulls are performed. Both the tip and the substrate can be modified and/or functionalized to allow more specific attachment of material of interest to both working surfaces (Lee *et al.*, 1994; MacKerell and Lee, 1999; Rief, Oesterhelt *et al.*, 1997).

During each pull, the tip and cantilever move away from and toward the substrate. The primary form of the raw data from an MFP experiment is the deflection of the laser beam versus the distance of cantilever movement in the z direction. These data can be transformed into force versus distance plots using the Hooke's law formula f (Force) $= kx$ where x is the distance that the tip of cantilever was deflected and k is the spring constant for the cantilever. The spring constant for the cantilever can be estimated by a few different methods (Cleveland *et al.*, 1993; Sader *et al.*, 1995) if the instrument used for pulling does not have a built-in system for estimating the spring constant. One also calibrates the microscope to find the sensitivity of the cantilever, which is piezo-voltage per nanometer of cantilever deflection on a hard surface. This can be done by first manually approaching the surface with the tip and then pressing the tip against the surface until it is sharply deflected. Then one can perform a single pull in the away-then-back-to-the-surface direction and record the deflection of the laser beam versus the distance of the cantilever movement. This graph serves as a calibration curve for cantilever deflection values. When these calibrations are done, the MFP is ready for actual pulling experiments. After data are converted into force versus distance graphs as in Fig. 1, the best fitting model can be found for each separate pulling event (for example, worm-like chain model for titin domains unfolding or DNA stretching). From this model one can calculate corresponding contour lengths and persistence lengths for each observed event (Fisher *et al.*, 1999).

Numerous modifications of MFPing can be used to study intermolecular as well as intramolecular interactions. Elasticity of biopolymers, protein, and nucleic acid folding, interactions between biomolecules and receptor–ligand interactions, and forces of covalent and noncovalent bonds are a few examples of problems that can be studied with the MFP technique. Although almost any standard AFM can be used to perform some MFP experiments, the molecular force probe (MFP) from Asylum Research (Santa Barbara, CA) is dedicated specifically for nano-pulling experiments. The MFP has both hardware and software advantages over conventional AFM. The main hardware advantage is the improved control over the z position of the cantilever relative to the sample due to an absolute position sensor. This can be crucial in pulling experiments as repetitive pulling events often have to be performed without touching the substrate while, at the same time, approaching very close to the substrate. This is also a major software advantage of the MFP as compared to the AFMs we are acquainted with: that the software can perform repetitive molecular pulls without touching the substrate between pulls. The MFP's IGOR software (Wavemetrics, Lake Oswego, OR) can also be easily modified for specific experiments through macro commands written by the researcher or obtained from the growing MFP community.

In a typical experiment, a long, narrow, V-shaped silicon–nitride cantilever is mounted on the holder by applying a small speck of vacuum grease in the middle of mounting depression. Care is taken to prevent an excess of vacuum grease from contaminating the cantilever or holder's optical surface. When changing the cantilever, one should carefully remove traces of old vacuum grease from the holder by flushing it with ethanol and blotting it dry. Care should be exercised not to scratch the optical surface of cantilever holder (directly underneath and in front of the cantilevers), as this can impede the performance of the MFP.

The sample is prepared by depositing the material of interest on a transparent or translucent substrate such as a glass microscope slide or a gold-coated slide. Sometimes it works best to dry the sample onto the slide so that it is firmly attached and then add a drop of fluid before imaging. The tip often bonds sufficiently to the molecules of the sample by simply being pressed onto the sample surface. With pulling experiments on dextran, the initial approach was to specifically attach the dextran to the tip and substrate via biotin/streptavidin and gold/thiol linkages, but it turned out that simply drying the dextran on the surface was sufficient for strong binding (Rief, Oesterhelt et al., 1997). Pulling experiments can also be performed in air instead of in fluid, but the measured forces in air will be dominated by the meniscus forces from the thin water layer that covers surfaces in air (Drake et al., 1989).

For pulling in fluid, small drops of fluid are placed on both the sample and the cantilever prior final MFP assembly. After the spring constant and the sensitivity of the cantilever have been determined, single or multiple pulls are performed and recorded automatically. It is good practice to first repeat a pull like the one used to calibrate the cantilever sensitivity and to save it, as internal standard for the surface position and cantilever sensitivity, in your data file.

V. Advanced Fluid Handling

Finally, we present a new fluid-handling technique that we have developed to study the responses of biological systems to changing environmental conditions.

Since the first AFM studies of dynamic processes in fluid, controlled fluid exchange has been a challenge. Solutions to this challenge, as described earlier, have included not only direct injection of a new fluid by hand but also the gravity flow method (Thomson et al., 1996). The hand-injection method obscures the image at least in the moment of the exchange. The exact imaging area is often lost altogether in the disturbance caused by injecting fluid during imaging.

Gravity flow is a rather quiet, but at the same time, static method. Gravity flow is also tedious to optimize. The exact flow rate depends strongly on the physical setup of the system, such as fluid levels, the diameter of tubing, and the viscosity of the different fluids. This uncertainty makes exact timings difficult. Furthermore, changing conditions inside the AFM fluid cell, caused by the different flowrates of solutions, can cause thermal drift and cantilever bending especially in small cantilevers, thus obscuring the image. Another limitation of gravity flow is that it requires a closed fluid cell, which is not available for some systems and, where available, is often problematic (leaks, limited scan size, high load on scanner).

The ideal fluid-exchange system runs different solutions continuously and in small, controlled portions at a given rate and low noise. In addition, it can remove the fluid at exactly the same rate from an open fluid cell, therefore maintaining the fluid level in an open low-volume cavity. This versatility is ideal for use with a variety of probe microscopes, many of which have only open fluid cells.

We have constructed a system with the potential to combine all these properties, using four high-precision, computer-controlled syringe pumps (Fig. 4). Unlike other pumps, these pump modules have proven to be very quiet. These pumps show minimal or no disturbance in nanometer-scale images of DNA plasmids on a mica surface while continuously running a buffer solution at flow rates up to 15 μL/s. The long-term stability of the flowrate was shown to be better than 200 nL/min, resulting in volume changes inside the fluid cell of less than $\pm 6 \mu$L in 1/2 h.

A. Principle of Operation

Two syringe pumps inject fluid into the AFM fluid cell in an alternating manner, providing continuous in-flow. The modules are equipped with computer-controlled valves that can connect the syringe to one of multiple ports. While one module is running fluid into the fluid cell, the other module switches over to a different port and refills at a suction rate that is higher than that in the flow rate into the cell. Then, it switches back to the port connected to the fluid cell and waits for the module currently running to complete its stroke. When that happens, the now refilled module takes over and starts running. This changeover needs to be well synchronized to avoid glitches in the continuous flow and, hence, in the AFM image. The in-flow modules have eight ports to choose from so that the fluid to be injected into the AFM can be selected from various sources, and the source can be changed without interrupting the continuous flow (Fig. 4).

Using more complex protocols, it is possible to make single-microliter injections of one source while continuously running another. This is achieved using a layering technique: first, an empty syringe refills with the buffer that is running continuously. Then, a small volume of the fluid to be injected is slowly layered on top of the buffer solution inside the syringe (Fig. 4, left). Now, when this syringe starts running solution into the fluid cell again, the small volume layer is injected first. Experiments with stained solutions showed that mixing inside the syringe is minimal if the layering is performed slowly.

For continuous fluid removal from an open fluid cell, two more syringe pumps are necessary. These will aspirate fluid at the same rate as the injection modules dispense it. The flow rates must be well controlled, because errors accumulate over time and would eventually overflow or dry out the fluid cell. Microprocessor-controlled and stepper-motor-driven syringe pumps provide such accuracy. By changing the ratio between injection and aspiration slightly, the system can also compensate for evaporation.

Controlling the pump modules requires a microcomputer (PC) that generates the complex timings in real time and provides a structured, reliable user interface. We used a 200-MHz Pentium-II Laptop PC.

The software was written in the Turbo-Pascal dialect Delphi for Windows, and its use is straightforward. The user enters parameters such as flow rate and fluid source. The flow is controlled with Start, Stop, and Pause buttons. To switch fluids, one selects a different fluid source from a menu. For a low-volume injection, one selects the source and volume, and then hits the Inject button.

Because the software knows the volume of the tubing between the syringes and the fluid cell, it can exactly predict the time of change inside the AFM fluid cell. Whenever the user chooses to change the fluid running, a countdown is displayed that announces the exact (better than 1 s) time of fluid change inside the fluid cell. The validity of these timing predictions was tested by switching between DI water and Hepes buffer solution while continuously running force versus distance curves on a mica surface. This method (Hermann Gaub, personal communication) shows an abrupt change in the shape of the force curve when changing from DI water to buffer. The moment of force-curve change correlated well with the change time announced by the fluid-handling software.

The ability to exactly predict the time of injection or fluid change makes the system suitable for dynamic studies of biological single-molecule systems, such as environment-dependent force spectroscopy of single molecules, or the time-resolved activity of enzymes on DNA, as described in the following.

B. DNase Digesting DNA—A Fluid-Handling Example

Figure 5 shows an image series of Bluescript plasmid DNA on mica that is successively digested by DNaseI. A Hepes–Mn buffer is continuously pumped at 5 μL/s. Each arrow marks the time of injection of 10 μL volume of DNase solution. This method allows full control over the speed of the reaction and makes scanning for extended periods of time between sample exposures to DNase possible.

C. Outlook

The first results with this system are very encouraging. The system is, in principle, adaptable to any AFM or probe microscope on the market, including closed-cell systems such as the Digital Instruments Multimode AFM, as well as open-cell systems like either the DI Dimension series or the new molecular force probe by Asylum Research. It may well be suited for non-AFM applications that require quiet fluid handling. Further improvements are currently under way to make it more versatile. Two of these improvements are a temperature controller using a low-volume heat exchanger between the in-flow pumps and the fluid cell and an in-line pH probe. These improvements will make the system even more suitable for biological applications. A second in-flow pump pair would allow one to continuously mix between two different fluids. This would greatly simplify either the investigation of dependencies between an environmental variable and a sample property or the search for good imaging conditions for a new sample. Another possible application to be considered is cell culturing inside an AFM fluid cell. We hope that this new technique will soon become more broadly available.

VI. Conclusion

The examples pictured here are just a few of the many valuable research applications for AFM in aqueous fluids. The challenge of biological AFM in fluid comes with a great potential benefit: AFM looks at biomolecules on surfaces, and living organisms are filled with surfaces. Therefore, "surface biology" may well be the biological frontier of the millennium, replacing the "test tube biology" that has generated such a vast amount of valuable knowledge in the last century.

Acknowledgments

We thank Paul Hansma for his pivotal work on instrumentation for AFM in fluid. This work was supported by NSF Grants MCB 9604566 (LP), MCB 9982743 (HH, JS, EO), DMR9632716 (NB), and DMR 96-22169 (JK, MV).

References

Argaman, M., Golan, R., Thomson, N. H., and Hansma, H. G. (1997). Phase imaging of moving DNA molecules and DNA molecules replicated in the atomic force microscope. *Nucl. Acids Res.* **25**, 4379–4384.

Bezanilla, M., Drake, B., Nudler, E., Kashlev, M., Hansma, P. K., and Hansma, H. G. (1994). Motion and enzymatic degradation of DNA in the atomic force microscope. *Biophys. J.* **67**, 2454–2459.

Bezanilla, M., Manne, S., Laney, D. E., Lyubchenko, Y. L., and Hansma, H. G. (1995). Adsorption of DNA to mica, silylated mica and minerals: characterization by atomic force microscopy. *Langmuir* **11**, 655–659.

Brown, H. G., and Hoh, J. H. (1997). Entropic exclusion by neurofilament sidearms: a mechanism for maintaining interfilament spacing. *Biochemistry* **36**(49), 15,035–15,040.

Bustamante, C., Guthold, M., Zhu, X., and Yang, G. (1999). Facilitated target location on DNA by individual Escherichia coli RNA polymerase molecules observed with the scanning force microscope operating in liquid. *J. Biol. Chem.* **274**(24), 16,665–16,668.

Chen, C. H., Clegg, D. O., and Hansma, H. G. (1998). Structures and dynamic motion of laminin-1 as observed by atomic force microscopy. *Biochemistry* **37**, 8262–8267.

Chen, C. H., and Hansma, H. G. (2000). Basement Membrane Macromolecules: Insights from Atomic Force Microscopy. *J. Struct. Biol.* **131**, 44–55.

Cleveland, J. P., Manne, S., Bocek, D., and Hansma, P. K. (1993). A nondestructive method for determining the spring constant of cantilevers for scanning force microscopy. *Rev. Sci. Instrum.* **64**, 403–405.

Czajkowsky, D. M., and Shao, Z. (1998). Submolecular resolution of single macromolecules with atomic force microscopy. *Febs Lett.* **430**(1–2), 51–54.

Dammer, U., Hegner, M., Anselmetti, D., Wagner, P., Dreier, M., Huber, W., and Guentherodt, H.-J. (1996). Specific antigen/antibody interactions measured by force microscopy. *Biophys. J.* **70**, 2437–2441.

Dammer, U., Popescu, O., Wagner, P., Anselmetti, D., Guentherodt, H.-J., and Misevic, G. N. (1995). Binding strength between cell adhesion proetoglycans measured by atomic force microscopy. *Science* **267**, 1173–1175.

Drake, B., Prater, C. B., Weisenhorn, A. L., Gould, S. A., Albrecht, T. R., Quate, C. F., Cannell, D. S., Hansma, H. G., and Hansma, P. K. (1989). Imaging crystals, polymers, and processes in water with the atomic force microscope. *Science* **243**(4898), 1586–1589.

Engel, A., Schoenenberger, C. A., and Muller, D. J. (1997). High resolution imaging of native biological sample surfaces using scanning probe microscopy. *Curr. Opin. Struct. Biol.* **7**(2), 279–284.

Fisher, T. E., Oberhauser, A. F., Carrion-Vazquez, M., Marszalek, P. E., and Fernandez, J. M. (1999). The study of protein mechanics with the atomic force microscope. *Trends Biochem. Sci.* **24**(10), 379–384.

Florin, E.-L., Moy, V. T., and Gaub, H. E. (1994). Adhesion forces between individual ligand-receptor pairs. *Science* **264**, 415–417.

Grandbois, M., Beyer, M., Rief, M., Clausen-Schaumann, H., and Gaub, H. E. (1999). How strong is a covalent bond? *Science* **283**(5408), 1727–1730.

Guthold, M., Zhu, X., Rivetti, C., Yang, G., Thomson, N. H., Kasas, S., Hansma, H. G., Smith, B., Hansma, P. K., and Bustamante, C. (1999). Direct observation of one-dimensional diffusion and transcription by escherichia coli RNA polymerase. *Biophys. J.* **77**(4), 2284–2294.

Hansma, H. G. (1999). Varieties of imaging with scanning probe microscopes. *Proc. Natl. Acad. Sci. U.S.A.* **96**, 14,678–14,680.

Hansma, H. G. (2001). Surface biology of DNA by atomic force microscopy. *Ann. Rev. Physical Chemistry* **52**, 71–92.

Hansma, H. G., Bezanilla, M., Zenhausern, F., Adrian, M., and Sinsheimer, R. L. (1993). Atomic force microscopy of DNA in aqueous solutions. *Nucl. Acids Res.* **21**(3), 505–512.

Hansma, H. G., Golan, R., Hsieh, W., Daubendiek, S. L., and Kool, E. T. (1999). Polymerase activities and RNA structures in the atomic force microscope. *J. Struct. Biol.* **127**, 240–247.

Hansma, H. G., and Laney, D. E. (1996). DNA binding to mica correlates with cationic radius: Assay by atomic force microscopy. *Biophys. J.* **70**, 1933–1939.

Hansma, H. G., Pietrasanta, L. I., Golan, R., Sitko, J. C., Viani, M., Paloczi, G., Smith, B. L., Thrower, D., and Hansma, P. K. (2000). Recent highlights from atomic force microscopy of DNA. *Biological Struct. Dynamics Convers.* **11**, 271–276.

Hansma, H. G., Revenko, I., Kim, K., and Laney, D. E. (1996). Atomic force microscopy of long and short double-stranded, single-stranded and triple-stranded nucleic acids. *Nucl. Acids Res.* **24**, 713–720.

Kasas, S., Thomson, N. H., Smith, B. L., Hansma, H. G., Zhu, X., Guthold, M., Bustamante, C., Kool, E. T., Kashlev, M., and Hansma, P. K. (1997). *E. coli* RNA polymerase activity observed using atomic force microscopy. *Biochemistry* **36**, 461–468.

Laney, D. E., Garcia, R. A., Parsons, S. M., and Hansma, H. G. (1997). Changes in the elastic properties of cholinergic synaptic vesicles as measured by atomic force microscopy. *Biophys. J.* **72**, 806–813.

Lee, G. U., Chrisey, L. A., and Coulton, R. J. (1994). Direct measurement of the forces between complementary strands of DNA. *Science* **266**, 771–773.

Lyubchenko, Y. L., Jacobs, B. L., and Lindsay, S. M. (1992). Atomic force microscopy of reovirus dsRNA: a routine technique for length measurements. *Nucl. Acids Res.* **20**(15), 3983–3986.

MacKerell, A. D. Jr., and Lee, G. U. (1999). Structure, force, and energy of a double-stranded DNA oligonucleotide under tensile loads. *Eur. Biophys. J.* **28**(5), 415–426.

Moller, C., Allen, M., Elings, V., Engel, A., and Muller, D. J. (1999). Tapping-mode atomic force microscopy produces faithful high-resolution images of protein surfaces. *Biophys. J.* **77**(2), 1150–1158.

Müller, D. J., Fotiadis, D., and Engel, A. (1998). Mapping flexible protein domains at subnanometer resolution with the atomic force microscope. *Febs Lett.* **430**(1–2), 105–111.

Radmacher, M., Cleveland, J. P., Fritz, M., Hansma, H. G., and Hansma, P. K. (1994). Mapping interaction forces with the atomic force microscope. *Biophys. J.* **66**, 2159–2165.

Rief, M., Gautel, M., Oesterhelt, F., Fernandez, J. M., and Gaub, H. E. (1997). Reversible unfolding of individual titin immunoglobulin domains by AFM. *Science* **276**(5315), 1109–1112.

Rief, M., Oesterhelt, F., Heymann, B., and Gaub, H. E. (1997). Single molecule force spectroscopy on polysaccharides by atomic force microscopy. *Science* **275**(5304), 1295–1297.

Sader, J. E., Larson, I., Mulvaney, P., and White, L. R. (1995). Method for the calibration of atomic force microscope cantilevers. *Rev. Scientific Instrum.* **66**(7), 3789–3798.

Schaffer, T. E., Cleveland, J. P., Ohnesorge, F., Walters, D. A., and Hansma, P. K. (1996). Studies of vibrating atomic force microscope cantilevers in liquid. *J. Appl. Phys.* **80**(7), 3622–3627.

Thomson, N. H., Kasas, S., Smith, B., Hansma, H. G., and Hansma, P. K. (1996). Reversible binding of DNA to mica for AFM imaging. *Langmuir* **12**, 5905–5908.

Viani, M. B., Pietrasanta, L. I., Thompson, J. B., Chand, A., Gebeshuber, I. C., Kindt, J. H., Richter, M., Hansma, H. G., and Hansma, P. K. (2000). Probing protein-protein interactions in real time. *Nat. Struct. Biol.* **7**(8), 644–647.

Yang, J., Mou, J., and Shao, Z. (1994). Structure and stability of pertussis toxin studied by in situ atomic force microscopy. *FEBS Lett.* **338**, 89–92.

CHAPTER 11

Supported Lipid Bilayers as Effective Substrates for Atomic Force Microscopy

Daniel M. Czajkowsky and Zhifeng Shao

Department of Molecular Physiology and Biological Physics
University of Virginia School of Medicine
Charlottesville, Virginia 22908

I. Introduction

It seems commonly known that obtaining images with atomic force microscopy (AFM) is relatively easy: The technology has been developed to such a level that the microscope is, to a large extent, no more complicated to operate than a light microscope. Nonetheless, obtaining high-resolution AFM images of biological samples under solution can be, in fact, quite challenging. This difficulty is frequently not because of any intrinsic characteristic of the biological specimen but rather because of the problem of preparing the sample in a manner suitable for AFM imaging.

In particular, there are two qualities that samples must have to obtain the highest resolution. First, they must be attached to a flat substrate with sufficient strength to withstand the lateral forces imparted by the tip. Although these lateral tip interactions can be minimized by imaging in the tapping mode, higher resolution images are, in general, always produced with more tightly bound samples, whatever the imaging mode. Second, the sample must be clean. This often overlooked requirement is necessary not only to prevent contamination (and thus enlargement) of the tip during imaging but also, importantly, to enable an accurate interpretation of the image once obtained, since the molecules of interest may not have a topography much different from that of aggregates of denatured proteins. There is also a tendency for closely packed samples to yield images with higher resolution (Czajkowsky and Shao, 1998; Shao et al., 1996), but at the moment, the only necessary conditions of AFM samples seem to be that they are clean and tightly attached to a substrate.

However, because of these requirements, there is a strict limitation on the choice of substrate to those that have a high affinity for the molecular complexes under study but a low affinity for all others, particularly aggregates of denatured proteins. By far, the most popular substrate in AFM is muscovite mica, a layered aluminosilicate that can be cleaved with tape to produce a fresh atomically flat surface (Bailey, 1984). Yet with so many complexes capable of interacting with mica, the requirement for clean samples shifts to a necessity for clean stock solutions, which may be challenging. Moreover, if the interaction with the sample is too weak, there are only a few possible modifications of the mica surface which might improve adhesion (Lyubchenko and Shlyakhtenko, 1997; Shlyakhtenko et al., 1999)

This chapter will describe methods to prepare an alternative set of substrates, supported lipid membranes, that are approximately as flat as mica but that can be better tailored to the properties of a particular sample by simply using lipids with appropriate headgroups. With this better control over sample adhesion, the attachment of unwanted molecules can be reduced, and so an additional advantage of such substrates is often cleaner samples.

II. Preparation of the Supported Bilayer Substrates

Supported lipid bilayers, as their name implies, are simply single bilayers bound to a hard surface such as mica. These bilayers were initially developed as model cell membrane systems, and indeed much of their current use, including in AFM investigations, is in studies of the structure and functioning of integral membrane proteins. However, as these surfaces are flat, robust, relatively straightforward to prepare, and possess an easily modified reactive surface, they have been increasingly used as direct substrates for water-soluble biological samples as well.

In what follows, we describe three methods to prepare these substrates. For all experiments, the lipids were purchased from Avanti Polar Lipids (Alabaster, AL).

A. Fusing Small Vesicles

In this method (Fig. 1), small unilamellar vesicles (SUVs) are added to a mica substrate, where they spontaneously bind, rupture, and then fuse to form the supported bilayer.

When phospholipids are first hydrated, they rapidly self-assemble into multilamellar vesicles (each resembling an onion). Since the diameters of these vesicles are comparable

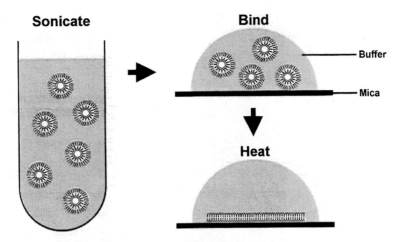

Fig. 1 Preparation of the supported bilayer by fusing vesicles with mica. Sonication of hydrated lipids produces small unilamellar vesicles, which can bind and, if the lipids are in the fluid phase, fuse onto mica to form the supported bilayer. Heating above the main transition temperature may be needed if the lipids are in the gel phase at room temperature. In addition, if the lipids are anionic, 1 mM Ca^{2+} may be required to promote adhesion of the vesicles to the negatively charged mica surface.

to the wavelength of light, the solution is cloudy. The SUVs, with diameters typically smaller than 100 nm, are formed by sonication of this suspension until the solution becomes transparent.

In a typical preparation, the lipids, initially dissolved in chloroform, are added to the bottom of a round-bottom disposable culture tube, and the organic solvent is evaporated under nitrogen, leaving a thin film of lipids. A low-salt solution (for example, 20 mM NaCl) is next added so that the final lipid concentration is 0.5 to 1 mg/ml. The low ionic strength ensures minimal competition of cations with the mica surface. The tube is then sealed under nitrogen (to prevent oxidation of the lipids) and placed in a bath sonicator, usually for 1 h for unsaturated lipids and 5 h for saturated lipids, until the solution is clear. A droplet ($\sim$40 μl for a mica size of $\sim$3 $\times$ 4 mm^2) of this vesicle solution is next added to a freshly cleaved fragment of mica at room temperature. After incubating for 1 h, the sample is washed with buffer to remove excess vesicles. If the lipids are in the fluid phase at room temperature, the supported bilayer will have formed at this point. If the lipids are in the gel phase at room temperature, it is necessary to incubate the sample above the main transition temperature for approximately 30 min to promote the rupture and fusion of the otherwise intact vesicles.

Figure 2 shows a supported lipid bilayer of the saturated zwitterionic lipid dipalmitoylphosphatidylcholine (DPPC) formed by vesicle fusion. The dark region in this image is a defect in the bilayer produced by scanning the AFM tip at a high rate and with a large applied force, prior to acquiring this image. For fluid-phase bilayers, which cannot be permanently displaced by a scanning tip, the presence of the membrane can be verified by incubating with bovine serum albumin (BSA) (adding 1 μl of 1 mg/ml stock solution

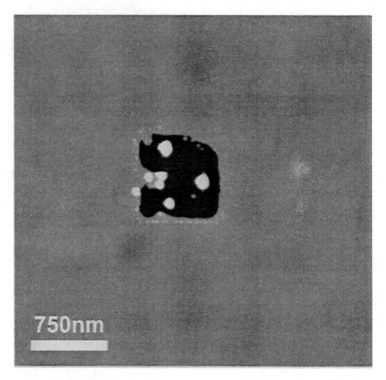

Fig. 2 AFM image of a supported DPPC bilayer at room temperature. The dark region is a defect in the bilayer created by scanning at a high force and a high scan rate prior to acquiring this image. Such tip-induced defects are, however, not stable in fluid phase bilayers. With these, the presence of the bilayer can be demonstrated by adding bovine serum albumin, as described in the text.

for 30 min followed by washing with buffer). By mechanisms not fully understood, this protein either creates or stabilizes defects in these bilayers so that the presence of the bilayer is revealed as planar thicker regions separating local clusters of BSA bound directly to mica.

The previously cited procedure should work for any lipids that form SUVs, except when higher concentrations of anionic lipid (greater than 20%) are used. In this case, 1 mM Ca^{2+} should be included in the buffer to enable adsorption of the vesicles to the negatively charged mica surface. After the supported bilayer has formed, the calcium can be washed away without damage to the supported bilayer, probably as a result of the low diffusion of ions through the bilayer.

B. Using a Langmuir Trough

When phospholipids are added to an air/water interface, they remain at the interface, oriented with their hydrophilic headgroups facing the solution and the hydrophobic tails facing the air. A Langmuir trough is an instrument that enables the controlled deposition

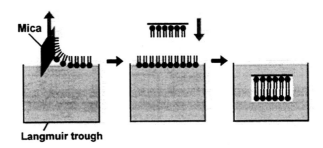

Fig. 3 Preparation of the supported bilayer using a Langmuir trough. The bilayer is prepared by depositing two separate lipid monolayers onto the mica surface as mica passes through the air/water interface of the trough twice.

of such monolayers at a chosen density (determined from a measurement of the surface pressure) onto a substrate such as mica.

To prepare the supported bilayer (Fig. 3), the mica fragment is first submerged within the aqueous solution in the trough, and the air/water interface is cleaned with suction. The lipids are then dissolved in hexane/ethanol (9/1 vol/vol), chloroform/methanol (2/1 vol/vol), or chloroform to 1 mg/ml and applied to the air/water interface. After waiting for several minutes for the organic solvent to evaporate (15 min with hexane/ethanol and 1 h with chloroform-containing solvent), the monolayer is slowly compressed (initially at 160 cm^2/min but then at less than 15 cm^2/min for smaller areas) to a final surface pressure of either $\sim$32 mN/m for unsaturated lipids or $\sim$45 mN/m for saturated lipids. Bilayers have fewer defects if prepared with a higher surface pressure, but monolayers collapse when the pressure is too large. Feedback is engaged at this point to maintain a constant surface pressure during the rest of the procedure. The mica fragment is slowly raised through this interface (0.4 cm/min), depositing the lipid monolayer onto the substrate (the headgroups are facing mica). This slow rate of deposition has been found to be crucial, as the quality of the sample degrades with faster rates.

After replacing the subphase, a similar procedure is used to prepare the second monolayer. The monolayer-coated mica fragment is horizontally lowered through the interface (at the same 0.4 cm/min) forming the supported bilayer. Comparable results are obtained with vertical deposition, but this typically produces more defects in the supported membrane. In addition, the samples are more homogeneous when the mica fragment is small and when only a few samples are prepared at once.

C. Using a Langmuir Trough and Small Teflon Wells

In this method (Fig. 4), a monolayer is first deposited onto a mica surface using a Langmuir trough, and then small wells in blocks of Teflon are used to prepare the second leaflet and to incubate with the sample. While the organic solvent in the first monolayer is evaporating, the second monolayer is formed at the air/water interface in the Teflon well using a Hamilton syringe (usually applying a $\sim$1-μl droplet of the 1 mg/ml lipid solution

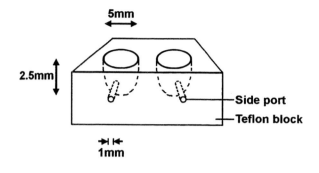

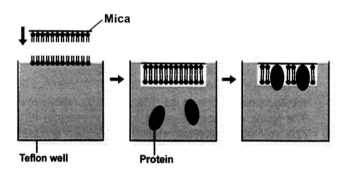

Fig. 4 Preparation of the supported bilayer using a Langmuir trough and small Teflon wells. The first monolayer is deposited onto mica using the Langmuir trough as described in Section II,B. This monolayer-coated mica is then lowered onto the second monolayer formed at the air/water interface in the Teflon well. The sample, here a protein that inserts into the bilayer, is then injected through the side port. The upper image is a schematic diagram of the wells typically used to prepare these samples.

in chloroform/methanol (2/1 vol/vol)). After depositing the first monolayer onto mica (as described in the preceding section), the monolayer-coated mica fragment is then immediately lowered horizontally onto the monolayer in the Teflon well to form the supported bilayer. The sample is next injected into the well through the side port, and the whole Teflon block is placed in a humid chamber during incubation. To remove the samples, the Teflon block is submerged within a large container of buffer, and the samples are transferred into Petri dishes.

III. Examples of Applications

The strength of the bilayer–sample interaction is determined by the nature of the lipid headgroups. In general, there are two different types of interaction that have been used to mediate sample adhesion: specific interactions with ligand-linked lipids and electrostatic interactions. Examples of both are described in the following.

A. Bilayers Containing Ligand-Linked Lipids

One of the most readily identifiable biological samples in AFM is the B-subunit of cholera toxin (CTB) (Czajkowsky and Shao, 1998; Shao *et al.*, 1996). Pathologically, CTB mediates the association of the complete cholera toxin with cell membranes through interactions with its lipid ganglioside receptor, G_{m1} (Spangler, 1992). Figure 5A shows an unprocessed image of CTB tightly adsorbed onto a bilayer composed of DPPC/G_{m1}

Fig. 5 Unprocessed AFM images of water-soluble proteins attached to bilayers containing ligand-linked lipids. (A) Image of the cholera toxin B subunit on DPPC/G_{m1} (90/10 mol%), showing the pentameric structure commonly revealed in AFM topographs. (B) Image of the two-dimensional crystal of streptavidin bound to a fluid bilayer with 10 mol% biotinylated lipid. This lipid was derivatized by covalently attaching the biotin moiety to the tertiary amine on the phosphatidylethanolamine headgroup, following similar methods to attach labels onto arginine and lysine residues.

(90/10 mol%), revealing both the characteristic pentameric architecture and the 1-nm pore of the CTB oligomers in the AFM topographs. To prepare this sample, the bilayer was formed by vesicle fusion and ~0.5 μg of the toxin (List Biochemicals, Campbell, CA) was incubated with the substrate for ~1 h, followed by a wash with buffer (a simple 20 mM NaCl solution).

Images of this type of sample were among the first to illustrate the usefulness of supported lipid bilayers as substrates in AFM (Yang *et al.*, 1993; Mou *et al.*, 1995). However, there are, naturally, only a limited number of samples whose endogenous ligand is a lipid. Nonetheless, for those ligands that can be attached to tertiary amines (following methods used to incorporate labels onto lysine and arginine residues), a variety of ligand-linked lipids can be synthesized, attaching the particular ligand to the terminal amino moiety of the phosphatidylethanolamine headgroup.

Biotinylated lipids, derivatized in this way, are commercially available and supported bilayers containing 10 mol% *N*-biotinyl-dipalmitoylphosphatidylethanolamine and 45/45 mol% dioleoylphosphatidylcholine/dioleoylphosphatidylethanolamine were prepared using the Langmuir trough and small Teflon wells. Injection of 1 μl of 1 mg/ml streptavidin into the well, followed by an incubation period of 1 h produced two-dimensional crystals of streptavidin (Sigma Chemicals, St. Louis, MO) on these bilayers (Fig. 5B), similar to what has been observed in other studies (Darst *et al.*, 1991; Yatcilla *et al.*, 1998; Scheuring *et al.*, 1999).

Recently, nickel-chelating lipids have been synthesized and used to prepare two-dimensional crystals of histidine-tagged proteins on lipid monolayers in several electron microscopy studies (Schmitt *et al.*, 1994; Dietrich *et al.*, 1995). These specimens should likewise be useful in AFM investigations.

An alternative approach, not yet widely explored, is to covalently attach F_{ab} fragments to the lipid headgroup and take advantage of a strong antibody–antigen interaction to mediate sample adhesion. In addition, lipids with either glutaraldehyde-functionalized headgroups or thiol headgroups are also commercially available (Avanti Polar Lipids, Alabaster, AL), and it should be possible to use these lipids to covalently attach samples to the bilayer. Although there would likely be a loss of specificity in the attachment, such a surface may be necessary when no other effective substrate can be found.

B. Bilayers Containing Charged Lipids

DNA and DNA-binding proteins are some of the most intensively studied samples with AFM (Bezanilla *et al.*, 1994; Bustamante and Rivetti, 1996; Kasas *et al.*, 1997; Lyubchenko and Shlyakhtenko, 1997; Schulz *et al.*, 1998; Ellis *et al.*, 1999). To date, the best resolution of DNA by AFM has been obtained with samples adsorbed to a cationic lipid bilayer (Mou *et al.*, 1995). Images of such specimens, as Figs. 6A and 6B illustrate, show, at large scan sizes, a densely packed single layer of tightly adsorbed DNA, and, at smaller scan sizes, the 3.4-nm periodic striations of the major groove along individual DNA strands.

The bilayer was prepared using a Langmuir trough, depositing DPPC in the first leaflet (directly facing mica) and dipalmitoyl-trimethyl-ammonium-propane, a cationic lipid in

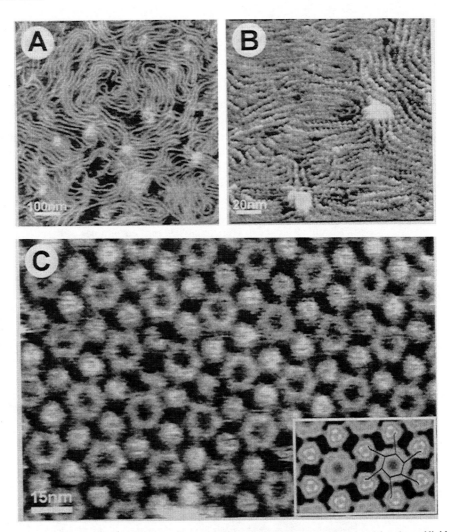

Fig. 6 High-resolution AFM images of biological molecules adsorbed to bilayers containing charged lipids. (A) Large scan size image of DNA adsorbed to a cationic bilayer showing a high surface coverage of DNA. (B) At smaller scan sizes, the 3.4-nm periodicity of the major groove can be directly seen in individual DNA strands with such samples. (C) Unprocessed image of *Helicobacter pylori* VacA adsorbed to anionic lipid membranes. Both the fluidity and the negative charge of these bilayers were necessary elements of this substrate to achieve samples that yielded images with a high resolution.

the gel phase at room temperature, in the second leaflet (facing the bulk solution). Next, 10 μl of 20 μg/ml lambda DNA (Sigma Chemicals, St. Louis, MO) was added to the 10 mM Tris, 1 mM EDTA, pH 8.0, buffer covering the bilayer. Finally, these samples were incubated for 1 h at 50°C (above the transition temperature of the lipids) to increase adsorption and to permit close-packing of bound DNA.

The ability to use fluid membranes to facilitate the preparation of highly condensed samples is one of the additional advantages of these substrates. This is clearly illustrated in samples of the *Helicobacter pylori* cytotoxin, VacA, which can be adsorbed to bilayers prepared with the Langmuir trough and Teflon wells (Czajkowsky *et al.*, 1999). This toxin was found to bind only to bilayers containing negatively charged lipids at low pH. When the pH was raised to ~7, the membrane-associated VacA hexamers were able to readily diffuse on the bilayer to form large patches of two-dimensional crystal, which could be imaged to high resolution (Fig. 6C). These are the clearest images of VacA, produced by any technique to date, and have led to a better understanding of how this cytotoxin damages cells (Czajkowsky *et al.*, 1999; Iwamoto *et al.*, 1999). This approach has also been successfully applied to a number of other pore-forming toxins, such as Staphylococcal α-hemolysin (Czajkowsky *et al.*, 1998), aerolysin (unpublished observations), and perfringolysin O (unpublished observations). In addition, this method has been employed to prepare closely packed samples of F-actin on cationic membranes (unpublished observations).

IV. Summary

Supported lipid bilayers offer a diverse set of substrates for AFM investigations of both water-soluble samples and integral membrane proteins. Although their amphipathic nature is necessary for the latter specimens, it is their robustness and easily changed surface characteristics that make these surfaces particularly attractive for the former ones.

When starting any AFM investigation of water-soluble biological complexes, it is probably best to try mica first, owing to its remarkable effectiveness with a wide range of samples. Yet, if mica should prove inadequate, supported lipid bilayers are, as demonstrated here, a sensible second choice.

Acknowledgments

We wish to thank Dr. Jianxun Mou for the image of CTB and Dr. Tim Cover for generously providing VacA. This work was supported by grants from the National Institutes of Health, National Science Foundation, and the American Heart Association.

References

Bailey, S. W. (1984). *Rev. Mineral.* **13,** Micas.

Bezanilla, M., Drake, B., Nudler, E., Kashlev, M., Hansma, P. K., and Hansma, H. G. (1994). Motion and enzymatic degradation of DNA in the atomic force microscope. *Biophys. J.* **67,** 2454–2459.

Bustamante, C., and Rivetti, C. (1996). Visualizing protein-nucleic acid interactions on a large scale with the scanning force microscope. *Annu. Rev. Biophys. Struct.* **25,** 395–429.

Czajkowsky, D. M., Iwamoto, H., Cover, T. L., and Shao, Z. (1999). The vacuolating toxin from *Helicobacter pylori* forms hexameric pores in lipid bilayers at low pH. *Proc. Natl. Acad. Sci. U.S.A.* **96,** 2001–2006.

Czajkowsky, D. M., and Shao, Z. (1998). Submolecular resolution of single macromolecules with atomic force microscopy. *FEBS Lett.* **430,** 51–54.

Czajkowsky, D. M., Sheng, S. T., and Shao, Z. (1998). Staphylococcal alpha-hemolysin can form hexamers in phospholipid bilayers. *J. Mol. Biol.* **276,** 325–330.

Darst, S., Ahlers, M., Meller, P., Kubalek, E., Blankenburg, R., Ribi, H., Ringsdorf, H., and Kornberg, R. (1991). Two-dimensional crystals of streptavidin on biotinylated macromolecules. *Biophys. J.* **59,** 387–396.

Dietrich, C., Schmitt, L., and Tampe, R. (1995). Molecular-organization of histidine-tagged biomolecules at self-assembled lipid interfaces using a novel class of chelator lipids. *Proc. Natl. Acad. Sci. U.S.A.* **92,** 9014–9018.

Ellis, D. J., Dryden, D. T. F., Berge, T., Edwardson, J. M., and Henderson, R. M. (1999). Direct observation of DNA translocation and cleavage by the EcoKI endonuclease using atomic force microscopy. *Nature Struct. Biol.* **6,** 15–17.

Iwamoto, H., Czajkowsky, D. M., Cover, T. L., Szabo, G., and Shao, Z. (1999). VacA from *Helicobacter pylori*: a hexameric chloride channel. *FEBS Lett.* **450,** 101–104.

Kasas, S., Thomson, N. H., Smith, B. L., Hansma, H. G., Zhu, X., Guthold, M., Bustamante, C., Kool, E. T., Kashlev, M., and Hansma, P. K. (1997). *Escherichia coli* RNA polymerase activity observed using atomic force microscopy. *Biochemistry* **36,** 461–468.

Lyubchenko, Y. L., and Shlyakhtenko, L. S. (1997). Visualization of supercoiled DNA with atomic force microscopy in situ. *Proc. Natl. Acad. Sci. U.S.A.* **94,** 496–501.

Mou, J., Czajkowsky, D. M., Zhang, Y. Y., and Shao, Z. (1995). High-resolution atomic-force microscopy of DNA—The pitch of the double helix. *FEBS Lett.* **371,** 279–282.

Mou, J., Yang, J., and Shao, Z. (1995). Atomic force microscopy of cholera toxin B-oligomers bound to bilayers of biologically relevant lipids. *J. Mol. Biol.* **248,** 507–512.

Scheuring, S., Müller, D. J., Ringler, P., Heymann, J. B., and Engel, A. (1999). Imaging streptavidin 2D crystals on biotinylated lipid monolayers at high resolution with the atomic force microscopy. *J. Microsc.* **193,** 28–35.

Schmitt, L., Dietrich, C., and Tampe, R. (1994). Synthesis and characterization of chelator-lipids for reversible immobilization of engineered proteins at self-assembled lipid interfaces. *J. Am. Chem. Soc.* **116,** 8485–8491.

Schulz, A., Mücke, N., Langowski, J., and Rippe, K. (1998). Scanning force microscopy of *Escherichia coli* RNA polymerase σ^{54} holoenzyme complexes with DNA in buffer and in air. *J. Mol. Biol.* **283,** 821–836.

Shao, Z., Mou, J., Czajkowsky, D. M., Yang, J., and Yuan, J. Y. (1996). Biological atomic force microscopy: what is achieved and what is needed. *Adv. Phys.* **45,** 1–86.

Shlyakhtenko, L. S., Gall, A. A., Weimer, J. J., Hawn, D. D., and Lyubchenko, Y. L. (1999). Atomic force microscopy imaging of DNA covalently immobilized on a functionalized mica substrate. *Biophys. J.* **77,** 568–576.

Spangler, B. D. (1992). Structure and function of cholera-toxin and the related escherichia-coli heat-labile enterotoxin. *Microbiol. Rev.* **56,** 622–647.

Yang, J., Tamm, L. K., Tillack, T. W., and Shao, Z. (1993). New approach for atomic force microscopy of membrane proteins, the imaging of cholera toxin. *J. Mol. Biol.* **229,** 286–290.

Yatcilla, M. T., Robertson, C. R., and Gast, A. P. (1998). Influence of pH on two-dimensional streptavidin crystals. *Langmuir* **14,** 497–503.

CHAPTER 12

Cryo–Atomic Force Microscopy

Sitong Sheng and Zhifeng Shao

Department of Molecular Physiology and Biological Physics
University of Virginia School of Medicine
Charlottesville, Virginia 22908

I. Introduction

Despite the success of atomic force microscopy (AFM) in obtaining high-resolution images of various biological specimens in solution (Shao *et al.,* 1995, 1996; Engel *et al.,* 1997; Muller *et al.,* 1998), the condition to have a sample that either is closely packed (Yang and Shao, 1993; Mou *et al.,* 1995; Mou, Czajkowsky *et al.,* 1996) or is arranged in two-dimensional arrays is almost always required (Czajkowsky *et al.,* 1998; Muller *et al.,* 1997). When isolated macromolecules are imaged, the achievable resolution is often much lower, due to the intrinsic flexibility of these structures (A-Hassan *et al.,* 1998; Erie *et al.,* 1994; Ill *et al.,* 1993; Roberts *et al.,* 1995). Even with the tapping mode AFM where the lateral force generated by the scanning tip is greatly reduced (Hansma *et al.,* 1994), the improvement in lateral resolution remains moderate (Fritz *et al.,* 1995). For much larger structures, such as intact cells or organelles, the documented resolution is even lower, often in the range of tens to hundreds of nanometers (A-Hassan *et al.,*

1998; Schaus and Henderson, 1997; Schneider *et al.*, 1997; Smythe and Warren, 1991). To overcome these difficulties and to expand the range of high-resolution AFM to include supramolecular complexes, imaging at cryogenic temperatures (cryo-AFM) was thought to be an effective approach (Prater *et al.*, 1991). In combination with quick freezing and deep etching, the integrity of biological structures should be well preserved and dynamic states could also be captured for AFM imaging. With improvements in the sharpness of the AFM tip (Bustamante *et al.*, 1994; Dai *et al.*, 1996; Hafner *et al.*, 1999; Keller and Chi-chung, 1992; Sugawara *et al.*, 1999) and the use of novel, more sensitive force-detection techniques, this approach should allow reliable nanometer resolution imaging on almost any type of biological structure to be achieved. Therefore, the method of cryo-AFM should be particularly suited for applications in cell biology where close packing and specimen uniformity are extremely difficult to achieve, and molecules in their native environment are preferred for understanding their functions.

In this chapter, we describe the necessary instrumentation for achieving AFM imaging at cryogenic temperatures. We will discuss considerations that have led to the design of a cryo-AFM under ambient pressure and present examples of application of this method to several typical specimens. In conclusion, we will discuss necessary future improvements that may significantly enhance the performance of the cryo-AFM in resolving molecular details at the single-molecule level.

II. Designs and Instrumentation

A. General Considerations

To design a practical cryo-AFM, the operating temperature of the system is the first consideration. Unlike applications in materials science, liquid nitrogen temperature appears to be perfectly suited for biological specimens. This is because, based on measurements of several enzymes at low temperature, there is a glass-like transition at around 200 K, below which the molecule should be in a more rigid state (Dorrington, 1979; Iben *et al.*, 1989; Perutz, 1992). This simplifies the system design considerably, because a conventional double-walled, intermediate vacuum insulated dewar should be adequate for housing such a system without extensive heat leaking into the system, in contrast to the more complicated helium temperature systems (White, 1979).

For most cryogenic instruments, high vacuum, which was also explored for AFM early on, is the most popular approach (Prater *et al.*, 1991). However, in such a system, the specimen is often at the lowest temperature and tends to accumulate surface contaminants over time. Since the AFM is essentially a surface probe, such contamination has created serious problems for reliable imaging in the past (Prater *et al.*, 1991). Because the adhesion force is often increased by surface contamination, both the tip and the specimen are damaged during scanning, leading to a much lower resolution (Han *et al.*, 1995; Yang and Shao, 1993). A simpler alternative is to house the AFM in a liquid nitrogen dewar under ambient pressure (Mou *et al.*, 1993). In such a design, the pool of liquid nitrogen retained at the bottom of the dewar serves as an enormous cold trap, and the continuous venting of the nitrogen gas cleans the system over time. As a result of this continuous purge and adsorption by liquid nitrogen, the environment in the dewar can be extremely clean and free from "airborne" contaminants. Since a large volume of liquid nitrogen can be stored in the system, the heat capacity is relatively large to ensure that temperature drift remains very small (Han *et al.,* 1995; Sheng and Shgao, 1998).

B. Instrumentation

The previously cited concept has been successfully implemented in a practical system that allows both contact mode and tapping mode AFM imaging. A schematic illustration is shown in Fig. 1. The AFM, including a piezo-tube scanner, is enclosed in a sealed

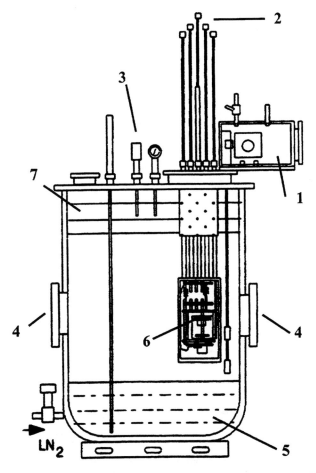

Fig. 1 Schematic illustration of the cryo-AFM operated under ambient pressure in dewar. The major components are indicated as the following: (1) specimen preparation chamber, (2) adjustment shafts, (3) safety valve, (4) quartz window, (5) liquid nitrogen, (6) AFM, and (7) copper plates.

chamber which is suspended through thin-wall stainless-steel tubings to the top flange. The remotely operated control shafts used to align the laser and the photo-diode detector to the cantilever, as well as tip engagement, are in these tubings. In the chamber, the AFM is further suspended by several steel springs tuned to ~1 Hz with magnetic damping to isolate the AFM from mechanical vibrations of the environment. To minimize any electronic noise coupling to the signal and to improve the laser stability, the laser driver, based on field-effect transistors, and the preamplifier are built right into the optical head. For simplicity in design, the laser diode is also mounted on the AFM head. Such simple designs apparently provide a stable platform sufficient for high-resolution imaging. It should be mentioned that the scanning range is reduced to about one-third of the range at room temperature for the same size piezo-tube, due to its reduced sensitivity (Mou *et al.*, 1993). Furthermore, the time response in the laser driver circuit should be optimized, because the electrostatic build up in the dewar can often lead to the failure of laser diodes. For the optical assembly, glass or quartz lenses should be used to minimize any misalignment due to differential shrinkage. We do not find it necessary to use Invar as the primary material for the AFM frame, and both brass and aluminum have yielded acceptable performance, based on the quality of the images. This has reduced the cost considerably.

The cantilever is preloaded by a spring clip to the tip holder which is mounted magnetically on the base plate of the AFM. For tapping mode operations, a small piezo-element is embedded below the magnet, which is driven directly by a NanoScope IIIa controller. The electrical leads to the piezo-element are separated from other signal leads to minimize "cross-talk" between weak signals and high-voltage signals. The optical head can be lifted manually from the scanner base to allow exchanging and self-centering of either the cantilever holder or the specimen, which is also magnetically mounted. Both the specimen and the cantilever can be replaced, using the motor-driven track connected to the specimen preparation chamber at the top. Additional positions are also installed in the dewar to allow temporary storage of cantilevers and specimens at the cryogenic temperature. The specimen chamber is made of transparent Plexiglas which allows a direct visualization of all operations conducted through the latex gloves mounted on two sides of the chamber. AFM adjustments and exchange of other components are monitored through a CCD camera placed outside the quartz windows on the wall of the dewar.

C. Initial Characterizations

Experimental results show that this system performed extremely well. The temperature stability is better than 4 mK/min, which allows a slow frame rate without detectable distortion (Han *et al.*, 1995). Figure 2 shows an image of NaCl crystals deposited on mica by precipitation, obtained at a scan rate of 1–2 Hz with commercial Si_3N_4 cantilevers ($k = 0.03$ N/m). At room temperature, extensive distortion is often found at this slow scan rate. However, to obtain such high resolution, bubbling of liquid nitrogen must be eliminated. The simplest approach is to elevate the pressure inside the dewar to a few pounds per square inch (Mou *et al.*, 1993). At this higher pressure, the boiling temperature of liquid nitrogen is also slightly higher (Hayat, 1989). However, since the heat leak into

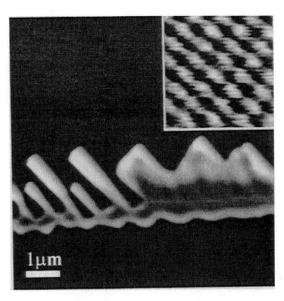

Fig. 2 NaCl crystals imaged by cryo-AFM at ~80 K. Inset: high-resolution image of the crystal (Lattice: ~0.4 nm).

the dewar is small and the heat capacity of the system is enormous, it requires many hours of heat accumulation before this new boiling temperature is reached. Before this point, bubbling is absent. After this point, the pressure can be relieved and the dewar can be repressurized to initiate another "quiet" period. To ensure safety, an automatic relief valve should be installed and set at 5 psi or lower.

Although the tip holder is designed to be compatible with standard cantilever chips, not all cantilevers are suitable for applications at cryogenic temperatures. In particular, for small k constant cantilevers, those without metal coating should be used, because the coated cantilevers have exhibited extensive bending due to differential shrinkage. For stiffer cantilevers, this is not a serious concern since the bending is quite small. For tapping mode imaging, despite the reduced efficiency, the piezo-element below the tip holder is sufficient to drive the cantilever into oscillation (see Fig. 3 for a tuning curve).

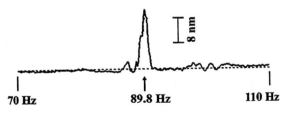

Fig. 3 A representative tuning curve obtained in the cryo-AFM for the cantilever of $k = 3.2$ N/m.

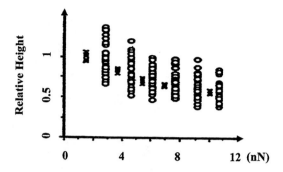

Fig. 4 Relative height measured over individual molecules in cryo-AFM at ~80 K. It is noted that data for IgG (o) are more scattered than DNA (x) because of their random orientation on mica. In solution, a probe force greater than a few nanonewtons is sufficient to cause extensive damage to the specimen. Here, even at 12 nN, stable images are still obtained, although at a lower resolution.

It is not difficult to generate amplitudes up to 30 nm for cantilevers with a k constant of 3 N/m. The change in the k value is moderate due to the temperature change (Han *et al.,* 1995).

An important validation of this technique is whether biological molecules would have a greater stiffness at low temperatures (Shao and Zhang, 1996). Therefore, we measured the height of individual molecules versus applied force for both DNA and IgG. As shown in Fig. 4, both types of specimens remained compressible at about 80 K. However, the extent of compression is much reduced when compared to that at room temperature. Although the exact Young's modulus is difficult to estimate, its value falls in the range of India rubber (Sheng and Shao, 1998). Therefore, biological structures at cryogenic temperatures should be more stable and less prone to damage at the molecular level.

III. Applications in Structural Biology

To date, the cryo-AFM has been applied to a number of biological specimens, obtaining reliable and reproducible images at nanometer resolution without any signal averaging (Han *et al.,* 1995; Shao and Zhang, 1996; Zhang *et al.,* 1996, 1997; Sheng and Shao, 1998; Shao and Sheng, 1999; Shao *et al.,* 2000; Chen *et al.,* 2000; Zelphati *et al.,* 2000). These initial results clearly demonstrated the validity of the cryo-AFM and its power in single-molecule imaging. Several examples are discussed here to illustrate the usefulness of this technique.

A. Imaging Individual Molecules

Cryo-AFM images in Fig. 5 demonstrate the range of the specimens that cryo-AFM has been successfully applied to. In preliminary studies of proteins involved in the activation of complements (Janeway and Travers, 1994), two molecules were imaged

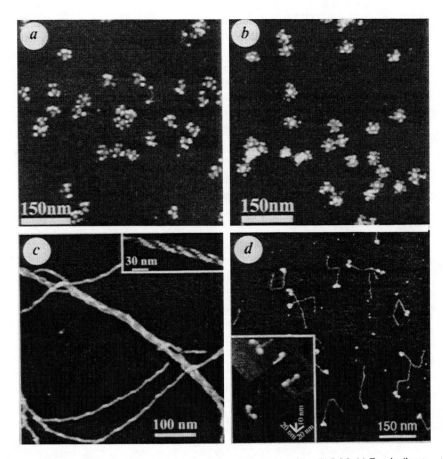

Fig. 5 Representative cryo-AFM images of single molecules: (a) C1q, (b) IgM, (c) F-actin (inset: actin bundles at high resolution), (d) sooth muscle myosin (inset: myosin molecules at high resolution).

with cryo-AFM. The large, flexible hexameric protein, C1q, is shown in Fig. 5a. The six IgM binding domains, which are connected to the stem by a triple helix, are clearly resolved in this case (Shao and Zhang, 1996; Shao and Sheng, 1999). This molecule apparently has a high intrinsic flexibility, because the molecular conformation is seen to vary from molecule to molecule. The orientation of the molecule on mica also appears to be random, indicating that mica does not select a particular part of the molecule for adsorption, unlike the case of GroEL (Mou, Sheng *et al.,* 1996). Interestingly, another related large molecule, IgM, seems to take a preferred orientation on mica (Fig. 5b). It is noted that the center of this pentameric molecule protrudes out from the plane of the Fab domains. This conformation is not the same as that of the current model for IgM (Janeway and Travers, 1994), which may have significant implications for its function. Even though this result must be further corroborated with other approaches, the advantage of a direct three-dimensional profile of a large molecule is clearly demonstrated by these images. With some improvements in resolution, data obtained by cryo-AFM may even allow for a reasonable attempt at constructing an atomic model based on homologies to other immunoglobulins (Sondermann *et al.*, 2000). The uniqueness of cryo-AFM was also demonstrated in studying key players involved in smooth muscle contraction. As shown in Fig. 5c, filamentous actin is imaged with cryo-AFM. The intrinsic high contrast (signal-to-noise ratio) of cryo-AFM allows not only the clear visualization of the isolated F-actin but also the detailed arrangement of individual filaments in actin bundles. Furthermore, high-resolution imaging also resolves individual monomers (Fig. 5c, inset) as well as the helical handiness of F-actin (Shao *et al.*, 2000). Not in one case is the proposed left-handed F-actin observed (Bustamante *et al.*, 1994). Another protein of the smooth muscle studied with cryo-AFM is the smooth muscle myosin (Fig. 5d) (Zhang *et al.*, 1997). In the image shown here, not only are the two heads and the long coiled-coil tail well resolved but also the regulatory domains within the myosin heads (Fig. 5d, inset). This level of resolution should already be sufficient for elucidating the anticipated interactions between the two heads within the myosin molecule (Trybus, 1994). This study further shows that the stability of the coiled-coil tail is also sensitive to the ion concentration in the solution, suggesting that electrostatic repulsion must be shielded when these molecules are assembled into the thick filament. It is noted that in all these examples no image averaging is applied because the heterogeneous nature of the specimen precludes the application of this technique. Yet, the high contrast still allows the resolution of submolecular details reproducibly. Therefore, cryo-AFM should be preferred for single-molecule imaging.

In preparing these specimens, the molecules are allowed to adsorb to a mica surface at a concentration in the range of 10–20 μg/ml and are then rinsed with a desired buffer to remove molecules in free suspension. After this step, the specimen is quickly rinsed with deionized water, and the excess solution is quickly removed by a stream of clean nitrogen in the Plexiglas chamber prior to its transportation into the cryo-AFM. Since water is extremely difficult to remove from the mica surface (Uchihashi *et al.*, 2000), these specimens should retain a very thin layer of water on their surface. With high-force scanning on clean mica, this adsorbed water layer has been shown to be no more than 1 nm; however, this water layer can still limit the achievable resolution. In addition, during the removal of the excess solution, the surface tension can also lead to a reduced height for most molecules. Therefore, it is highly preferable to prepare these specimens using the deep-etch method discussed in the next section.

B. Resolving Surface Details of Large Assemblies

The cryo-AFM is also effective when it is applied to larger structures. An image of influenza virus is shown in Fig. 6a. It is seen that the virus is ruptured due to the last rinse with water which caused the virus envelope to burst under osmotic pressure. The viral genome is also seen to spill out from the rupture hole. To reveal surface features, the details of the image are extracted from the data, with a method similar to high-pass filtering (Shao and Zhang, 1996), which is shown in Fig. 6b. With cryo-AFM, interesting surface details are also resolved with red blood cells. To prevent cell from lysis due to water rinse, a slight fixation (0.05% glutaraldehyde for 2–3 s) was used.

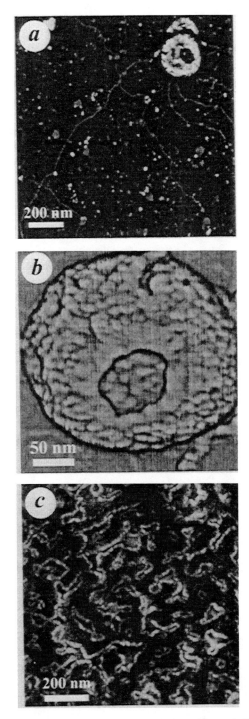

Fig. 6 Cryo-AFM images of large structures. (a) Influenza virus at a large scale. Notice the released genome due to the osmotic shock during sample preparation. (b) Some surface features are resolved at higher resolution. (c) Surface of red blood cells from rabbit.

A few cells can be found intact and the concave shape is resolved (Zhang *et al.*, 1996; Sheng and Shao, 1998). At higher resolution, a corrugated surface structure is resolved (Fig. 6c) which formed enclosed boundaries. Although the nature of these structures has not yet been identified, it is worth mentioning that similar results were also found by electron microscopy (Glaeser *et al.*, 1966) but largely ignored in the literature. These results indicate that a higher resolution is normally attainable in the cryo-AFM for these extremely soft specimens. Similar studies at room temperature rarely exceeded 100 nm in resolution (Schneider *et al.*, 1997; A-Hassan *et al.*, 1998).

IV. Deep Etching as the Preferred Sample Preparation Method

It is well documented that dehydration, whether complete or nearly complete, can alter biological structures. Therefore, quick freezing combined with deep etching was developed as the preferred method of specimen preparation for electron microscopy (Willison and Rowe, 1980). Obviously, these techniques can also be modified and applied to cryo-AFM. Among the various approaches, the so-called sandwich technique is among the simplest and easiest (Losser and Armstrong, 1990). With this method, a small droplet of solution, preferably in low-salt buffers, is applied to a piece of freshly cleaved mica. The amount of solution should be controlled to have a thin layer of solution, no more than 20 μm thick. Then, a clean piece of cover glass is placed on top of the mica and the "sandwich" is immersed in a liquid cryogen, such as liquid nitrogen or liquid ethane. If the solution layer is thin enough, the cooling rate should be sufficient to preserve the structures of interest (Van Venrooij *et al.*, 1975). The frozen specimen is then mounted on a spring-loaded specimen holder (Fig. 7) and transported into the dewar.

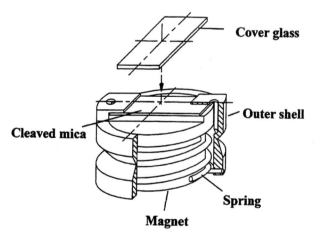

Fig. 7 An illustration of the special specimen holder that can accept "sandwich"-type frozen specimens. The diameter of the holder is 15 mm.

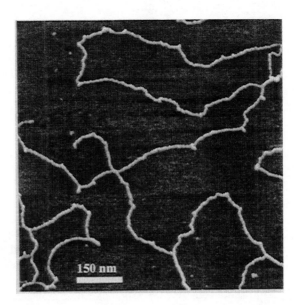

Fig. 8 Cryo-AFM image of deep-etched lambda-phage DNA. DNA is adsorbed to spermidine-treated mica.

Even though complete etching can be achieved under ambient pressure in liquid nitrogen vapor (Sheng and Shao, unpublished results), we found it much easier to use a small vacuum chamber built into the dewar. After the cover glass is removed, the specimen can be fully etched at about −80°C in 10–20 min. The use of the cover glass is absolutely necessary, because any direct contact with cryogen would cause significant contamination of the specimen. Sample cleanliness is always required for high resolution. The entire procedure is not much different from that for preparing specimens for electron microscopy.

Our preliminary results show that an immediate improvement is the height of the molecules. Figure 8 shows an image of deep-etched lambda-phage DNA. The average height measured from this image is 1.8 nm, which is 50% higher than those measured from pre-dehydrated specimens (Han *et al.,* 1995). The effect on proteins and other structures should be more significant but remains to be characterized.

V. New Directions

Based on the successful initial applications of this new technique, an immediate development is to introduce the method of freeze fracture in the cryo-AFM. As shown earlier (Heuser, 1983), an oblique fracture angle should allow the fracture plane to intersect with the specimen at various heights, thus allowing a direct probing of internal structures of a large complex. Since no metal shadows are required, a higher resolution should be possible on the native surface. Freeze fracture should also facilitate the study of integral membrane proteins in the cryo-AFM. An intriguing possibility is to repeatedly image the exposed surface after the removal of the exposed structure followed by limited etching. This is similar to sequential sectioning (Willison and Rowe, 1980), but the "section" thickness can be much smaller, and it is the remaining block that is imaged.

It is also possible to improve the spatial resolution into the sub-nanometer range with individual molecules or complexes, since the surface structure is known to be well preserved with quick freezing (Henderson *et al.*, 1990; Booy *et al.*, 1991; Yeager *et al.*, 1994; Avila-Sakar and Chiu, 1996). However, to achieve this, sharper tips must be used which should not be much greater than a few atoms. Although even single-atom tips can be reliably fabricated by *in situ* build up with some crystalline metals (Binh and Marien, 1988), the real difficulty lies in the fact that such tips cannot sustain the impact of contact and are fractured before any image can be obtained. In fact, even with the current Si_3N_4 tips which have an apex of a few nanometers (Sheng and Shao, 1998), tip fracture is often observed in the cryo-AFM upon initial engagement. An effective solution to circumvent this problem is to use the non-contact-imaging mode (Albrecht *et al.*, 1991), which was successfully implemented for materials science with an extraordinary resolving power (Franz *et al.*, 2000; Lantz *et al.*, 2000). Such an approach can also be applied to biological imaging. Therefore, we expect that the combination of "single-atom" tips and noncontact AFM should push the resolution on single molecules into the sub-nanometer range with the same reproducibility and robustness as other structural techniques in the near future.

Acknowledgment

This work is supported by grants from NIH, NSF, and the American Heart Association. We also thank Mr. Gang Huang for Fig. 7; Dr. L. K. Tamm for influenza virus; and Dr. D. M. Czajkowsky for helpful discussions.

References

A-Hassan, E., Heinz, W. F., Antonik, M. D., D'Costa, N. P., Nageswaran, S., Schoenenberger, C. A., and Hoh, J. H. (1998). Relative microelastic mapping of living cells by atomic force microscopy. *Biophys. J.* **74,** 1564–1578.

Albrecht, T., Grutter, P., Horne, D., and Rugar, D. (1991). Frequency modulation detection using high Q cantilevers for enhanced force microscopy sensitivity. *J. Appl. Phys.* **69,** 668–673.

Avila-Sakar, A. J., and Chiu, W. (1996). Visualization of beta sheets and side chain clusters in 2-dimensional periodic arrays of streptavidin on phospholipid monolayers by electron crystallography. *Biophys. J.* **70,** 57–68.

Binh, V. T., and Marien, J. (1988). Characterization of microtips for scanning Tunneling Microscopy. *Surf. Sci.* **202,** L539–549.

Booy, F. P., Newcomb, W. W., Trus, B. L., Brown, J. C., Baker, T. S., and Steven, A. C. (1991). Liquid-crystalline, phage-like packing of encapsidated DNA in Herpes Simplex virus. *Cell* **64,** 1007–1015.

Bustamante, C., Erie, D., and Keller, D. (1994). Biochemical and structural applications of scanning probe microscopy. *Curr. Opin. Struct. Biol.* **4,** 750–760.

Chen, C., Sheng, S., Shao, Z., and Guo, P. (2000). Dimer as building block to assemble pRNA hexamer that gears bacterial virus phi29 DNA translocating machinery. *J. Biol. Chem.* **275,** 17,510–17,516.

Czajkowsky, D. M., Sheng, S., and Shao, Z. (1998). Staphylococcal α-hemolysin can form hexamers in phospholipid bilayers. *J. Mol. Biol.* **276,** 325–330.

Dai, H. J., Hafner, J. H., Rinzler, A. G., Colbert, D. T., and Smalley, R. E. (1996). Nanotubes as nanoprobes in scanning probe microscopy. *Nature* **384,** 147–150.

Dorrington, K. (1979). "The Mechanical Properties of Biological Materials." Cambridge University Press, Cambridge.

Engel, A., Schoenenberger, C. A., and Muller, D. J. (1997). High resolution imaging of native biological sample surfaces using scanning probe microscopy. *Curr. Opin. Struct. Biol.* **7**, 279–284.

Erie, D. A., Yang, G., Schultz, H. C., and Bustamante, C. (1994). DNA bending by Cro protein in specific and nonspecific complexes: implications for protein site recognition and specificity. *Science* **266**, 1562–1566.

Franz, J., Giessibl, S., Hembacher, S., Bielefeldt, H., and Mannhart, J. (2000). Subatomic features on the silicon (111)-(7 × 7) surface observed by atomic force microscopy. *Science* **289**, 422–425.

Fritz, M., Radmacher, M., Cleveland, J. P., Allersma, M. W., Stewart, R. J., Gieselmann, R., Janmey, P., Schmidt, C. F., and Hansma, P. K. (1995). Imaging globular and filamentous proteins in physiological buffer solution with tapping mode atomic force microscopy. *Langmuir* **11**, 3529–3535.

Glaeser, R. M., Hayes, T., Mel, H., and Tobias, C. (1966). Membrane structure of O_sO_4 fixed erythrocytes viewed 'face on' by electron microscopy techniques. *Exp. Cell Res.* **42**, 467–477.

Hafner, J. H., Cheung, C. L., and Lieber, C. M. (1999). Direct growth of single walled carbon nanotube scanning probe microscopy tips. *J. Am. Chem. Soc.* **121**, 9750–9751.

Han, W., Mou, J., Sheng, J., Yang, J., and Shao, Z. (1995). Cryo atomic force microscopy: A new approach for biological imaging at high resolution. *Biochemistry* **34**, 8215–8220.

Hansma, P. K., Cleveland, J. P., Radmancher, M., Walters, D. A., Hillner, P. E., Bezanilla, M., Fritz, M., Vie, D., Hansma, H. G., Prater, C. B., Massie, J., Fukunaka, L., Gurley, J., and Elings, V. (1994). Tapping mode atomic force microscopy in liquids. *Appl. Phys. Lett.* **64**, 1738–1740.

Hayat, M. A. (1989). "Principles and Techniques of Electron Microscopy: Biological Applications," 3rd Edition. CRC Press, Boca Raton, FL.

Henderson, R., Baldwin, J. M., Ceska, T. A., Zemlin, F., Beckmann, E., and Dowining, K. H. (1990). Model for the structure of bacteriorhodopsin based on high-resolution electron cryo-microscopy. *J. Mol. Biol.* **213**, 899–929.

Heuser, J. (1983). Procedure for freeze drying molecules adsorbed to mica flakes. *J. Mol. Biol.* **169**, 155–195.

Iben, J. E. T., Braustein, D., Doster, W., Frausenfelder, H., Hong, M. K., Johnson, J. B., Luch, S., Ormos, P., Schulte, A., Steinback, P. J., Xie, A., and Young, R. D. (1989). Glassy behavior of a protein. *Phys. Rev. Lett.* **62**, 1916–1919.

Ill, C., Keivens, V., Hale, J., Nakamura, K., Jue, R., Cheng, S., Melcher, E., Drake, B., and Smith, M. (1993). A COOH-terminated peptide confers regiospecific orientation and facilitates atomic force microscopy of an IgG1. *Biophys. J.* **64**, 919–924.

Janeway, C., and Travers, P. (1994). "Immunobiology." Garland Publishing, London.

Keller, D., and Chi-chung, C. (1992). Imaging steep, high structures by scanning force microscopy with electron beam deposited tips. *Surf. Sci.* **268**, 333–339.

Lantz, M. A., Hug, H. J., Van Schendel, P. J. A., Hoffmann, R., Martin, S., Baratoff, A., Abdurixit, A., Güntherodt, H. J., and Gerber, C. H. (2000). Low temperature scanning force microscopy of the Si(111)-(7 × 7) surface. *Phys. Rev. Lett.* **84**, 2642–2645.

Losser, K. E., and Armstrong, C. F. (1990). A simple method for freeze-drying of macromolecular complexes. *J. Struct. Biol.* **103**, 48–56.

Mou, J., Czajkowsky, D. M., Sheng, S., Ho, R., and Shao, Z. (1996). High resolution surface structure of E. coli GroES oligomer by atomic force microscopy. *FEBS Lett.* **381**, 161–164.

Mou, J., Sheng, S., Ho, R., and Shao, Z. (1996). Chaperonins GroEL and GroES: views form atomic force microscopy. *Biophys. J.* **71**, 2213–2221.

Mou, J., Yang, J., and Shao, Z. (1993). An optical detection low temperature atomic force microscope at ambient pressure for biological research. *Rev. Sci. Instrum.* **64**, 1483–1488.

Mou, J., Yang, J., and Shao, Z. (1995). Atomic force microscopy of cholera toxin B-oligomers bound to bilayers of biologically relevant lipids. *J. Mol. Biol.* **248**, 507–512.

Muller, D. J., Engel, A., Carrascosa, J. L., and Velez, M. (1997). The bacteriophage phi29 head-tail connector imaged at high resolution with the atomic force microscope in buffer solution. *EMBO J.* **16**(10), 2547–2553.

Muller, D. J., Fotiadis, D., and Engel, A. (1998). Mapping flexible protein domains at subnanometer resolution with the atomic force microscope. *FEBS Lett.* **430**, 105–111.

Perutz, M. F. (1992). Protein function below 220K. *Nature* **358**, 548.

Prater, C. B., Wilson, M. R., Garnaes, J., Masie, J., Elings, V. B., and Hansma, P. K. (1991). Atomic force microscopy of bilogical samples at low temperature. *J. Vac. Sci. Technol.* **B9**, 989–991.

Roberts, C., Williams, P., Davies, J., Dawkes, J., Stefon, J., Edwards, J., Haymes, C., Bestwick, C., Davies, M., and Tendler, S. (1995). Real space differentiation of IgG and IgM antibodies deposited on microtiter wells by scanning force microscopy. *Langmuir* **11**, 1822–1826.

Rogers, W., and Glaser, M. (1993). Distributions of proteins and lipids in erythrocyte membrane. *Biochemistry* **32**, 12591–12598.

Schaus, S. S., and Henderson, E. R. (1997). Cell viability and probe-cell membrane interactions of XR1 glial cells imaged by atomic force microscopy. *Biophys. J.* **73**, 1205–1214.

Schneider, S., Sritharan, K., Geibel, J., Oberleithner, H., and Jena, B. (1997). *Proc. Natl. Acad. Sci. U.S.A.* **94**, 316–321.

Shao, Z., Mou, J., Czajkowsky, D. M., Yang, J., and Yuan, J. (1996). Biological atomic force microscopy: what is achieved and what is needed. *Adv. Phys.* **45**, 1–86.

Shao, Z., and Sheng, S. (1999). Resolving spatial conformations of immuno-proteins with cryo-atomic force microscopy. *Microsc. Microanal.* 5(Suppl. 1), 1008–1009.

Shao, Z., Shi, D., and Somlyo, A. V. (2000). Atomic force microscopy of filamentous actin. *Biophys. J.* **78**, 950–958.

Shao, Z., Yang, J., and Somlyo, A. P. (1995). Biological atomic force microscopy: from microns to nanometers and beyond. *Annu. Rev. Cell Dev. Biol.* **11**, 241–265.

Shao, Z., and Zhang, Y. (1996). Biological cryo atomic force microscopy: A brief review. *Ultramicroscopy* **66**, 141–152.

Sheng, S., Czajkowsky, D., and Shao, Z. (1999). AFM tips: How sharp are they? *J. Microsc.* **195**, 1–5.

Sheng, S., and Shao, Z. (1998). Biological cryo atomic force microscopy: instrumentation and applications. *Jap. J. Appl. Phys.* **37**, 3828–3833.

Smythe, E., and Warren, G. (1991). The mechanisim of receptor-mediated endocytosis. *Eur. J. Biochem.* **202**, 689–699.

Sondermann, P., Huber, R., Oosthuizen, V., and Jacob, U. (2000). The 3.2-Å crystal structure of the human IgG1 Fc fragment-FcγIII complex. *Nature* **406**, 267–273.

Sugawara, Y., Minobe, T., Orisaka, S., Uchihashi, T., Tsukamoto, T., and Morita, S. (1999). Non-contact AFM images measured on Si(111) root 3 × root 3-Ag and Ag(111) surfaces. *Surf. Int. Anal.* **27**, 456–461.

Trybus, K. M. (1994). Regulation of expressed truncated smooth muscle myosins. Role of the essential light chain and tail length. *J. Biol. Chem.* **269**, 20819–20822.

Uchihashi, T., Tanigawa, M., Ashino, M., Sugawara, Y., Yokoyama, K., Morita, S., and Ishikawa, M. (2000). Identification of B-form DNA in an ultrahigh vacuum by noncontact-mode atomic force microscopy. *Langmuir* **16**, 1349–1353.

Van Venrooij, G. E., Aertsen, A. M., Hax, W. M., Ververgaert, P. H., Verhoeven, J. J., and Van der Vorst, H. A. (1975). Freeze-etching: Freezing velocity and crystal size at different locations in samples. *Cryobiology* **12**, 46–61.

White, G. K. (1979). "Experimental techniques in low temperature physics." Clarendon Press, Oxford.

Willison, M., and Rowe, A. (1980). Replica, shadowing and freeze-etching techniques. *In* "Practical Methods in Electron Microscopy" (A. Glauert, Ed.), Vol. 8. Elsevier/North-Holland Biomedical Press, Amsterdam.

Yang, J., and Shao, Z. (1993). The effect of probe force on resolution in atomic force microscopy of DNA. *Ultramicroscopy* **50**, 157–170.

Yeager, M., Berriman, J. A., Baker, T. S., and Bellamy, A. R. (1994). Three-dimensional structure of the rotavirus haemagglutinin VP4 by cryo-electron microscopy and difference map analysis. *EMBO J.* **13**, 1011–1018.

Zelphati, O., Liang, X., Nguyen, C., Barlow, S., Sheng, S., Shao, Z., and Felgner, P. (2000). PNA dependent gene chemistry: stable coupling of peptides and oligonucleotides to plasmid DNA. *BioTech.* **28**, 304–316.

Zhang, Y., Shao, Z., Somlyo, A. P., and Somlyo, A. V. (1997). Cryo atomic force microscopy of smooth muscle myosin. *Biophys. J.* **72**, 1308–1318.

Zhang, Y., Sheng, S., and Shao, Z. (1996). Imaging biological structures with the cryo atomic force microscope. *Biophys. J.* **71**, 2168–2176.

CHAPTER 13

Conformations, Flexibility, and Interactions Observed on Individual Membrane Proteins by Atomic Force Microscopy

Daniel J. Müller[*,†] and Andreas Engel[*]

[*]M. E. Müller Institute, Biocenter
University of Basel
CH–4056 Basel, Switzerland

[†]Max-Planck-Institute of Molecular Cell Biology and Genetics
D–01307 Dresden, Germany

I. Introduction

The applicability of atomic force microscopy (AFM) (Binnig *et al.*, 1986) for imaging biological objects in their aqueous environment was already demonstrated shortly after the invention of this technique (Drake *et al.*, 1989). Although similar and higher resolution can be obtained by electron microscopy and X-ray crystallography, the excellent signal-to-noise ratio of AFM topographs allows the direct imaging of native proteins (Czajkowsky *et al.*, 1998; Fotiadis *et al.*, 1998; Karrasch *et al.*, 1994; Mou *et al.*, 1995, 1996; Müller, Engel *et al.*, 1997; Müller, Schabert *et al.*, 1995; Schabert *et al.*, 1995; Scheuring, Ringler *et al.*, 1999). Progress to enhance the resolution of native proteins was achieved in several laboratories by optimizing instrumentation (Cheung *et al.*, 2000; Han *et al.*, 1997; Hansma *et al.*, 1994; Viani *et al.*, 1999), sample preparation (Czajkowsky *et al.*, 1998; Karrasch *et al.*, 1993; Mou *et al.*, 1995; Müller, Amrein *et al.*, 1997; Scheuring, Müller *et al.*, 1999; Yang *et al.*, 1993), and image acquisition (Möller *et al.*, 1999; Müller, Fotiadis *et al.*, 1999) methods.

The forces interacting between the AFM stylus and the biological sample are mainly covered by electrostatic and van der Waals interactions (Butt, 1991, 1992). Force–distance curves, which are acquired by approaching the sample to the stylus while measuring its deflection, reveal the nature of these interactions (Butt *et al.*, 1995) generated when imaging is done in buffer solution (Müller and Engel, 1997; Rotsch and Radmacher, 1997). According to the Derjaguin–Landau–Verwey–Overbeek (DLVO) theory (Israelachvili, 1991), these forces can be minimized by adjusting the electrolyte concentration and pH of the buffer. After such adjustment (Müller, Fotiadis *et al.*, 1999), topographs of protein surfaces recorded by contact mode AFM routinely allow their structural details to be determined with a lateral resolution better than 1 nm and a vertical resolution of about 0.1 nm (Fotiadis *et al.*, 2000; Heymann *et al.*, 2000; Müller and Engel, 1999; Müller, Sass *et al.*, 1999; Scheuring, Müller *et al.*, 1999; Scheuring, Ringler *et al.*, 1999; Seelert *et al.*, 2000).

Images showing the surface topography of biomolecules in solution not only reveal the object in its most native state but also exhibit an outstanding signal-to-noise (*S/N*) ratio. Striking images have been recorded that show submolecular features of single biomolecules (Fotiadis *et al.*, 2000; Seelert *et al.*, 2000). Minute structural changes at their molecular surface can be detected with sufficient time resolution to monitor conformational changes involved in biological processes. Imaging of a statistically significant number of single proteins by AFM allows their structural variability to be assessed and multivariate statistical classification to be applied to unravel the principal modes of the protein motion (Müller *et al.*, 1998). Such structural changes can be induced in a controlled manner to identify flexible protein structures (Müller, Buldt *et al.*, 1995; Müller, Engel *et al.*, 1997; Scheuring, Ringler *et al.*, 1999). However, most interestingly, AFM enables conformational changes of single proteins and of their assemblies to be observed directly (Müller, Baumeister *et al.*, 1996; Müller and Engel, 1999; Müller, Sass *et al.*, 1999; Müller, Schoenenberger *et al.*, 1997).

With the AFM tip a "nanotool" is available that allows molecules to be dissected (Fotiadis *et al.*, 1998; Hansma *et al.*, 1992; Hoh *et al.*, 1993) and the forces between single molecules to be measured (Fisher *et al.*, 1999). In single-molecule force-spectroscopy experiments, protein complexes are tethered to both support and AFM tip and the two moved apart. This technique has been employed to measure forces between pairs of interacting biological molecules (Dammer *et al.*, 1995, 1996; Florin *et al.*, 1994; Lee, Chrisey *et al.*, 1994; Lee, Kidwell *et al.*, 1994b; Moy *et al.*, 1994), the elasticity of polysaccharides (Marszalek *et al.*, 1998; Rief, Oesterhelt *et al.*, 1997) and DNA (Lee, Chrisey *et al.*, 1994; Rief *et al.*, 1999; Smith *et al.*, 1996), and forces required for the unfolding of titin domains (Carrion-Vazquez *et al.*, 1999; Oberhauser *et al.*, 1998; Rief, Gautel *et al.*, 1997).

Recently, protein complexes were imaged before and after the removal of individual subunits using the AFM tip as a nanotweezer (Fotiadis *et al.*, 1998). Encouraged by these results, the single-molecule imaging and single-molecule force-spectroscopy capabilities of the AFM have been combined to provide novel insights into the inter- and intramolecular interactions of proteins (Müller, Baumeister *et al.*, 1999; Oesterhelt *et al.*, 2000). In this combined technique the secondary structural elements facing the aqueous solution are first imaged. After this, the protein is selected and the AFM tip is tethered to its C- or N-terminal region. Pulling on the polypeptide end allowed the forces required to unfold the tertiary and secondary structure of the protein to be measured (Oesterhelt *et al.*, 2000). Subsequent imaging at subnanometer resolution allowed the unambiguous correlation of the force spectroscopy curve to the individual protein regions vacant in the second scan.

This review focuses on a selection of the various applications and methodological developments used to investigate membrane proteins with the AFM. We will discuss examples where the surface and secondary structure elements of biological macromolecules have been revealed at subnanometer resolution. Such topographs allow mapping the flexibility and unraveling the conformational variability of proteins. AFM topographs are compared to structural data yielded by other methods. Most importantly, the significance of the observed conformational changes to protein function and to interactions of the protein with its environment will be discussed. Finally, we will discuss recent approaches combining single-molecule AFM imaging and single-molecule force spectroscopy.

II. High-Resolution AFM Imaging

Biological specimens are frequently imaged in their dehydrated state by AFM. Besides the fact that only buffer solution presents a physiologically relevant environment to maintain the structure and function of most biological macromolecular systems, it has been repeatedly shown that air dying can cause lasting structural deformations (Baumeister *et al.*, 1986; Dubochet *et al.*, 1988; Kellenberger *et al.*, 1982). Additionally, the thin water film present on the sample surface at ambient conditions (humidity >50%) forms a capillary meniscus at the AFM stylus on imaging. Such capillary forces (10–100 nN!) exceed all other attractive forces by orders of magnitude, leading to additional sample deformation, thereby limiting the resolution. Although experimentally slightly more demanding than scanning in air, samples should be imaged in aqueous buffer solutions, simply because this warrants the best preservation of the biological structure and allows the tip–sample interactions to be precisely controlled. Progress in AFM imaging has been achieved recently by minimizing the tip–sample interaction forces which cause sample deformation (Müller, Sass *et al.*, 1999). In the following, the use of contact mode (Müller, Fotiadis *et al.*, 1999) and tapping mode (Möller *et al.*, 1999) AFM imaging to minimize sample deformation and reproducibly obtain high-resolution AFM topographs is described.

A. Contact Mode Imaging

In a simplified model, electrostatic and van der Waals forces govern the tip sample interactions in aqueous solution. Hydrophilic surfaces of biological specimen are charged in water, leading to long-range electrostatic double-layer (EDL) interactions, which can be attractive or repulsive, depending on the surface charges which, in turn, depend on the pH. These interactions can conveniently be monitored by approaching the stylus to the sample and withdrawing it periodically. Attractive or repulsive forces are revealed by the deflection of the cantilever (Butt *et al.*, 1995; Heinz and Hoh, 1999). The interaction length of these forces can be controlled by screening the surface charges with electrolytes. Since the AFM tip (silicon nitride, Si_3N_4) is negatively charged at neutral pH and protein layers are often negatively charged as well, the electrostatic forces are frequently repulsive. In biological systems, van der Waals interactions do not depend on the ionic strength, decay rapidly, and are always attractive. The DLVO theory (Israelachvili, 1991) quantitatively describes these forces and allows the interactions between a spherical stylus and a planar sample to be modeled, providing critical information for the optimization of imaging conditions (Müller, Fotiadis *et al.*, 1999).

Although suppliers specify AFM tip radii of 10–50 nm, topographs of flat biological surfaces that exhibit a resolution of better than 1 nm have been acquired routinely (Fotiadis *et al.*, 2000; Mou *et al.*, 1996; Müller and Engel, 1999; Müller, Sass *et al.*, 1999; Müller, Schabert *et al.*, 1995; Schabert *et al.*, 1995; Scheuring, Müller *et al.*, 1999; Scheuring, Ringler *et al.*, 1999; Seelert *et al.*, 2000; Walz *et al.*, 1996). Therefore, the AFM tips employed most likely had a single nanometer-sized asperity that protruded sufficiently to contour the finest surface structures. However, such a small asperity would be expected to exert a prohibitively high pressure on the underlying structure inducing its deformation and reducing the resolution (Fig. 1A). Only contact of the sample with the large area of the whole tip would reduce the pressure on the macromolecules to a reasonable level (Shao et al., 1996). Developing this model further, electrolytes can be used to adjust the tip–sample interactions, provided the electrostatic forces are repulsive. Because of their long-range interaction, the electrostatic double-layer force does not contribute to submolecular structures observed by the high-resolution AFM topograph (Fig. 2). Ideally, the scanning AFM tip then surfs on a cushion of the long-range electrostatic repulsion while the small asperity is in contact contouring fragile details of the biological surface (Fig. 1B) (Müller, Fotiadis et al., 1999).

B. Tapping Mode Imaging

Contact mode AFM is not suitable for imaging weakly immobilized structures such as single macromolecules since these are often pushed away by the AFM stylus while raster scanning the surface (Karrasch et al., 1993). Tapping mode AFM (TMAFM) has been developed (Hansma et al., 1994; Putman et al., 1994; Zhong et al., 1993) to overcome this disadvantage. In tapping mode, the AFM stylus oscillates only touching the sample at the end of its downward movement, which reduces the contact time and the friction forces compared to contact mode AFM. This has allowed a variety of macromolecules to be observed (Bezanilla et al., 1994; Dunlap et al., 1997; Fritz et al., 1995; Guthold et al., 1999; Kasas et al., 1997; Lyubchenko and Shlyakhtenko, 1997; Martin et al., 1994; Radmacher et al., 1995; Shlyakhtenko et al., 1998) which could not be imaged before. To date, unfortunately the TMAFM images lack the spatial resolution exclusively acquired using the AFM in the contact mode.

Nevertheless, improvements have recently been reported; optimization of the TMAFM imaging parameters allowed faithful high-resolution topographs of native proteins to be obtained (Möller et al., 1999). For high-resolution imaging the TMAFM was operated at drive frequencies close to the resonance frequency of the cantilever when immersed in buffer solution. This enabled controlling precisely the cantilever oscillation. Generally, the TMAFM drive frequency used to activate the cantilever oscillation is limited by the fluid drive frequency determined by the liquid cell holding the cantilever immersed in buffer solution (Schäffer et al., 1996). However, recent developments, such as the magnetically activated cantilever (MAC) mode AFM (Han et al., 1997), allow the cantilever to be activated directly over an extended frequency range allowing the resonance frequency of various cantilevers to be adjusted and their oscillation to be precisely controlled. To achieve efficient coupling between the drive voltage and the cantilever amplitude response, the cantilever was brought within a distance of 8 μm to the sample. High-resolution imaging was achieved using drive amplitudes between 1 and 8 nm and a free amplitude damping of $\approx$1 nm.

Topographs of the hexagonally packed intermediate (HPI) layer from Deinococcus radiodurans show (Fig. 3) that the signal-to-noise ratio and the sensitivity of TMAFM are clearly sufficient to resolve submolecular details of proteins (Möller et al., 1999). The averages calculated from the TMAFM topographs of the HPI layer and of the

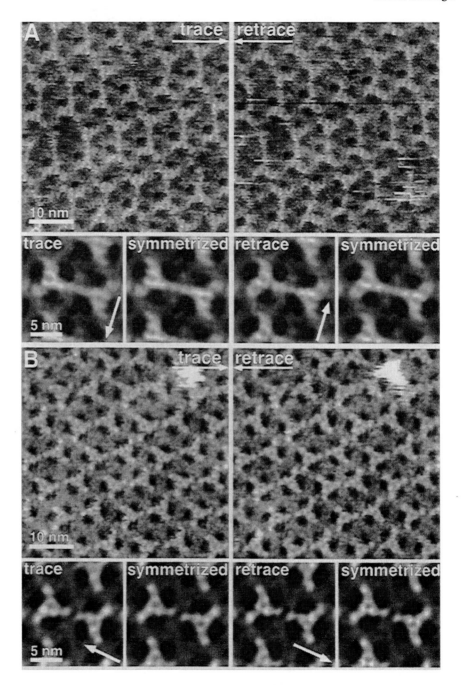

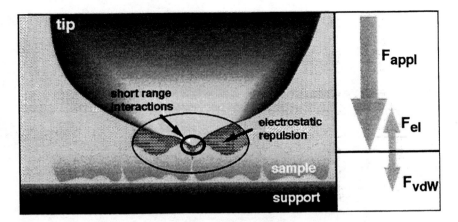

Fig. 2 Forces interacting between AFM tip and biological sample in buffer solution. The electrostatic double-layer (EDL) force interacts via long-range forces with a relatively large area of the macromolecular assembly. In contrast, the short range van der Waals attraction and Pauli repulsion interact between individual microscopic protrusions of the AFM tip and the biological sample. The force effectively interacting at the AFM tip apex is a composite of all interacting forces. If the EDL force is negligible ($F_{el} \approx 0$) or eliminated, the effective force is equal to the sum of the applied force and of the attractive van der Waals force $|F_{eff}| = |F_{appl} + F_{vdW}| < |F_{appl}|$. Being the opposite of sign, a sufficiently high EDL force will partially compensate the applied force. Thus, under these conditions, the effective force is smaller than the applied force $|F_{eff}| = |F_{appl} + F_{el} + F_{vdW}| < |F_{appl}|$. At minimized interacting forces, possible structural deformation of the flexible biological sample will be minimized (Müller, Fotiadis et al., 1999). Additionally, minimized forces reduce the contact area between AFM tip and sample which will enhance the lateral and vertical resolution (Weihs et al., 1991).

purple membrane exhibited a lateral resolution lying between 1.1 and 1.5 nm which is close to the range achievable using the contact mode (Müller, Fotiadis et al., 1999). However, comparison of standard deviation (SD) values calculated from both TMAFM and contact mode AFM topographs showed that the noise of TMAFM topographs was higher. Interestingly, the flexibility of protein substructures as indicated by the maximum of the SD map appears to be independent of the imaging mode used.

Fig. 1 Contact mode AFM imaging of native membrane proteins at subnanometer resolution. (A) Periplasmic surface of native porin OmpF simultaneously recorded in trace and in retrace direction. Imaged in 50 mM MgCl$_2$, 50 mM KCl (10 mM Tris–HCl, pH 7.6) the electrostatic double-layer (EDL) interaction was not eliminated. Correlation averages (insets; $n = 157$) of each direction show structural differences. (B) Periplasmic surface recorded after adjusting the EDL damping. When imaging in 300 mM KCl (10 mM Tris–HCl, pH 7.6) the EDL interaction between AFM stylus and periplasmic surface exhibited a long range ($\approx$5–10 nm) repulsive force of close to 0.1 nN. The scanning AFM stylus was thus damped by the long-range EDL forces allowing the soft protein structure to be reproducibly contoured at subnanometer resolution (Fig. 2). Structural differences between correlation averages (insets; $n = 104$) of both scanning directions are minimized. While surface structures recorded in (A) do not correlate well with the atomic structure of OmpF porin (see Fig. 12) topographs observed in (B) show excellent agreement. Both topographs were imaged using contact mode at applied forces of 0.1 nN and at scan frequencies of 15.4 Hz. Vertical brightness range of topographs corresponds to 1.2 nm (raw data) and 1 nm (insets). Topographs (A) and (B) are displayed as reliefs tilted by 5°.

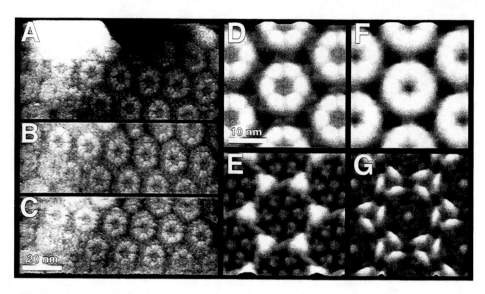

Fig. 3 Reproducible high-resolution imaging of native proteins using tapping mode AFM. (A) Topograph of the outer surface of the HPI layer obtained in 200 mM KCl, pH 7.8, 20 mM Tris-HCl. (B) and (C) Same surface area imaged after 8 and 16 min, respectively. (D) sixfold symmetrized correlation average from 97 unit cells (10.8% rms deviation from sixfold symmetry). (E) Sixfold symmetrized standard deviation map. (F) Averaged topograph of the outer HPI-layer surface recorded using contact mode AFM. The correlation average was sixfold symmetrized [top; $n = 79$; 2.8% rms deviation from sixfold symmetry (Müller *et al.,* 1998)]. (G) Sixfold symmetrized standard deviation map of (F). Full gray level ranges: 4 nm (A, B, and C), and 3 nm (D and F). Minima and maxima of the standard deviation were 0.62 and 0.95 nm for (E) and 0.11 and 0.34 nm for (G). Scale bars: 20 nm (A, B, and C) and 10 nm (E and G). TMAFM imaging conditions: HPI layer adsorbed onto highly ordered pyrolytic graphite: original scan size 185 nm, scan frequency 3.8 Hz, drive frequency 9.3 kHz, drive amplitude 244 mV, setpoint 311 mV. The contact mode topograph was recorded in buffer solution at applied forces of ≈100 pN. All images are displayed as reliefs tilted by 5°.

III. Identification of Membrane Proteins

A. Identification by Antibody Labeling

In most AFM studies, it appears difficult to identify the biological macromolecule and its surfaces. This information, however, is a fundamental pre-requisite for structural and functional interpretation of biological structures. Antibodies exhibit an extremely high affinity to their antigen, and thus can be used to label proteins, viruses, tissues, cell surfaces, and other macromolecular structures.

In our AFM experiments, native purple membrane was allowed to adsorb from buffer solution onto a freshly cleaved mica surface (Müller, Amrein *et al.,* 1997). After being rinsed with the same buffer solution, the purple membrane patches were routinely observed in the fluid cell of the AFM (Fig. 4A). At low magnification, no topographic difference between individual membranes was detected. To distinguish the two surfaces of purple membrane, antibodies directed against the bacteriorhodopsin C-terminus

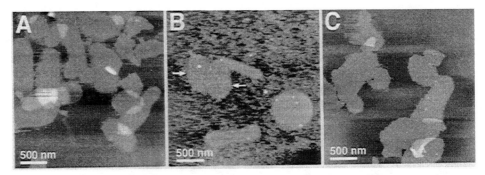

Fig. 4 Immunoatomic force microscopy of purple membrane. (A) Native purple membrane adsorbed flatly onto freshly cleaved mica. (B) Antibodies were added to the buffer solution after adsorption of the native membranes yielding some densely labeled membranes (arrows), while others remained unlabeled. The antibodies were directed against the C-terminus of bacteriorhodopsin located on the cytoplasmic purple membrane surface. Thus, the labeled membranes (arrows) exposed their cytoplasmic surface toward the aqueous solution. (C) AFM topograph of papain-digested purple membrane leading to the removal of the C-terminus. After incubation with antibodies, digested purple membrane remained untextured even after extending the reaction time for antibody binding from 1 to 24 h at room temperature. Imaging buffer: 150 mM KCl, 10 mM Tris, pH 8. Forces applied to the AFM tip were between 0.1 and 0.2 nN; scan frequency was 3.5 Hz. Vertical brightness range of contact mode topographs corresponds to 15 nm.

located on the cytoplasmic bacteriorhodopsin surface were injected into the buffer solution (Müller, Schoenenberger *et al.*, 1996). After addition of antibodies, some of the membranes were labeled exhibiting a rough surface (Fig. 4B), whereas others remained as smooth as the membranes shown in Fig. 4A. To further establish the specificity of antibody labeling, the C-terminus was removed by papain digestion of purple membrane. Papain-digested purple membrane exhibited the same dimensions and smooth surface texture as native purple membrane. Consistent with the absence of antibody binding in the dot immunobinding assays no decoration of the digested membranes was observed (Fig. 4C), even after extending the incubation time with the antibodies from 1 to 24 h (Müller, Schoenenberger *et al.*, 1996). These results indicated that purple membranes have been specifically labeled by the antibodies directed to the C-terminus when the extracellular surface was in contact with the mica. Thus, those smooth patches of purple membrane not labeled by the antibody were oriented with their extracellular surface toward the AFM tip.

B. Identification by Replacement of Polypeptide Loops

The replacement of individual polypeptide loops results in structural changes of the membrane protein surface which can be observed by AFM.

Eight polypeptide residues of the EF loop of bacteriorhodopsin (Fig. 5A) were replaced with 24 residues from the third cytoplasmic loop of bovine rhodopsin (Rho EF; Fig. 5B) which is important in signal transduction from the retinal in rhodopsin to its G-protein, transducin (Helmreich and Hofmann, 1996). The third cytoplasmic loop of

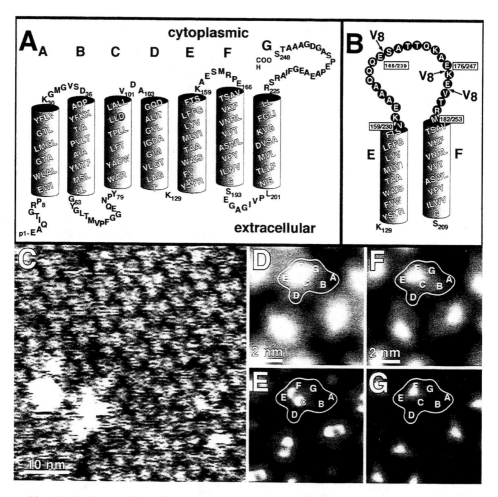

Fig. 5 Imaging a polypeptide loop grafted onto bacteriorhodopsin. (A) Secondary structural model of bacteriorhodopsin from *Halobacterium salinarum*. (B) The polypeptide loop connecting transmembrane α-helices E and F (loop EF) of the bacteriorhodopsin molecule was replaced by loop EF from bovine rhodopsin to produce the mutant IIIN (indicated by the filled circles, the numbering in the boxes give the residue in IIIN first and the residue in the rhodopsin loop second). V8 protease cleaves this loop after the glutamates indicated by the arrows (see Fig. 6). (C) Topograph of the mutant IIIN containing the rhodopsin EF loop. (D) Threefold symmetrized correlation average of mutant IIIN trimer. (E) Standard deviation map of the average. (F) Threefold symmetrized correlation average of the bacteriorhodopsin trimer imaged elsewhere (Müller, Sass *et al.*, 1999). (G) Standard deviation map of the bacteriorhodopsin trimer. The outlined BR monomer represents a section close to the cytoplasmic surface of the lipid membrane, and the positions of the transmembrane α-helices A to F were obtained after merging six atomic models of BR (Heymann *et al.*, 1999). Vertical brightness range of contact mode topographs corresponds to 1.8 nm. Minima and maxima of the SD maps were 0.32 and 0.43 nm for (E) and 0.07 and 0.19 nm for (G) respectively.

rhodopsin (Rho EF) also interacts with rhodopsin kinase, which phosphorylates light-activated rhodopsin, and with arrestin, which displaces transducin from light-activated phosphorylated rhodopsin. To directly observe the rhodopsin loop, purple membrane containing the mutant bacteriorhodopsin (called IIIN) was imaged by AFM under physiological conditions to a resolution of 0.7 nm (Fig. 5C). It was found that the modification of loop EF changed neither the crystallographic lattice nor the extracellular surface (Heymann *et al.*, 2000). This was not unexpected, because fragments of bacteriorhodopsin separated in the EF loop can be reconstituted with the bacteriorhodopsin chromophore (Kataoka *et al.*, 1992; Liao *et al.*, 1984; Sigrist *et al.*, 1988). Thus, the bacteriorhodopsin framework was not affected by the loop replacement which provided a stable foundation for studying the Rho EF loop. The major difference in the topographs between the cytoplasmic surfaces of the mutant and bacteriorhodopsin purple membrane is the much larger EF loop projecting toward the C-terminus (Figs. 5D and 5F).

C. Identification by Removal of a Polypeptide Loop

From Fig. 5 it is clear that structural changes of membrane proteins induced by the replacement of individual polypeptide loops can be directly observed by AFM under physiological conditions. Alternatively, individual loops can be removed to identify structural features of membrane protein surfaces.

Digestion of the rhodopsin loop EF from mutated IIIN bacteriorhodopsin (Fig. 5B) with V8-protease did not affect the purple membrane crystallinity (Fig. 6A). The AFM topograph showed a significant reduction in the major protrusion compared to the undigested surface (Fig. 6B) and the ends of helices E and F became clearly visible. This structural change was consistent with mass spectrometry indicating that a 10-residue fragment of loop Rho EF had been removed (Heymann *et al.*, 2000). Interestingly, AFM topographs of purple membrane did not show any indication of the largest polypeptide residue located at the cytoplasmic surface, the C-terminus (24 aa), and the simplest interpretation is that this is too unstructured to allow imaging.

D. Identification by Removal of Polypeptide Ends

This alternative method to identify the surfaces of membrane proteins is illustrated by the selective cleavage of terminal polypeptide sequences of aquaporin Z (AqpZ) from *E. coli* (Scheuring, Ringler *et al.*, 1999) and of major intrinsic protein (MIP) from sheep lenses (Fotiadis *et al.*, 2000).

Figure 7A shows the AFM topograph of recombinant AqpZ tetramers reconstituted into a bilayer and assembled into a two-dimensional (2D) crystal. The AqpZ had an N-terminal fragment of 26 amino acids located on the cytoplasmic surface. After overnight treatment with trypsin, the N-terminal fragment had been removed from the protein at the trypsin cleavage site Arg26. While the uncleaved sample allowed only one AqpZ surface to be imaged, the digested sample clearly showed substructures of the cytoplasmic (circle) and extracellular (square) AqpZ surface (Fig. 7B).

Topology prediction and antibody labeling of MIP places the approximately 5-kDa C-terminal region on the cytoplasmic surface of the lens fiber cell membrane. The native

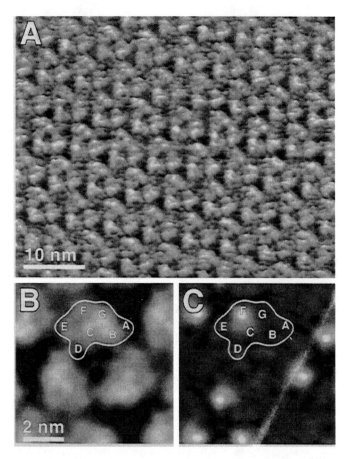

Fig. 6 Imaging bacteriorhodopsin after removal of the EF loop. (A) Topograph of the bacteriorhodopsin surface after cleavage of the rhodopsin EF loop with V8 protease. For cleavage sites of the V8 protease, see Fig. 5B. The topograph is displayed as a relief tilted by 5°. (B) Threefold symmetrized average of the bacteriorhodopsin trimer imaged in (A). (C) Standard deviation map of the average. Vertical brightness range of contact mode topographs corresponds to 1 nm. Minima and maxima of the SD map was 0.1 and 0.17 nm.

cytoplasmic surface of MIP tetramers, reconstituted into a bilayer and assembled into a 2D crystal, exhibited maximum globular protrusions of 0.8 ± 0.1 nm (Fig. 7C). After removal of the C-terminal tail with carboxypeptidase Y the cytoplasmic surface changed its appearance (Fig. 7D). The cytoplasmic surface appeared coarser, and the averaged structure revealed the partial loss of four prominent protrusions leaving a central cavity within the MIP tetramer. This structural change is emphasized by the difference map (Fig. 7F) calculated between the unit cell of the digested (Fig. 7D) and of the native (Fig. 7E) cytoplasmic MIP surface. It is important to note that neither the extracellular surface of AqpZ nor that of MIP appeared to be structurally affected by the enzymatic digestion of the cytoplasmic surface.

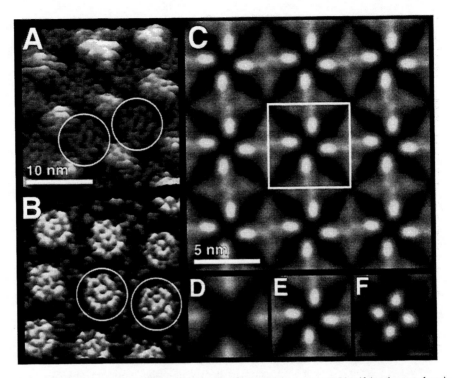

Fig. 7 Identification of protein structures by removal of polypeptide ends. (A) Identifying the cytoplasmic surface of aquaporinZ (AqpZ). Topograph of AqpZ tetramers assembled into a 2D crystal. In this crystal form, each tetramer is neighbored by four tetramers oriented in the opposite direction with respect to the membrane plane. B, The same crystal imaged after trypsin digestion. Since trypsin removes most of the 26-amino-acid long C-terminal region, the surface structure of the cytoplasmic surface changed drastically. The extracellular surface of the AqpZ tetramers was unchanged by this treatment (circles). Vertical brightness range of contact mode topographs corresponds to 3 nm. Topographs are displayed as reliefs tilted by 5° (images courtesy of Simon Scheuring, University of Basel). (C) Imaging the removal of the C-terminal region of the major intrinsic protein (MIP). Averaged topograph of the cytoplasmic MIP surface imaged in buffer solution. The MIP tetramers from sheep eye lenses were reconstituted into a lipid bilayer where they assembled into a 2D crystal (Fotiadis *et al.,* 2000). The same unit cell (white square) contained one MIP tetramer and had a side length of 6.4 nm. (D) Unit cell of the cytoplasmic MIP surface after removal of the C-terminal region. (E) Unit cell of the native cytoplasmic MIP surface. (F) Difference map calculated between topographs of digested (D) and of native MIP revealing the location of the C-terminal regions of the cytoplasmic surface; major protrusion. Vertical brightness range of contact mode topographs corresponds to 1 nm (images courtesy of Dimitrios Fotiadis, University of Basel).

From the previous AFM measurements it follows that the conformation of the AqpZ N-terminal fragment (26 amino acids) was too flexible to be imaged with subnanometer resolution (blurred protrusion) but was structurally sufficiently stable to distort the scanning stylus, thereby preventing the visualization of other substructures on the cytoplasmic surface. The C-terminal fragment of MIP existed in a structurally more stable conformation and was imaged by the AFM stylus. In contrast, the C-terminal region of bacteriorhodopsin, consisting of 25 amino acids, is not observed by AFM and does not influence the visualization of surrounding substructures by AFM (compare to Fig. 6). The results illustrate that the polypeptide ends of distinct proteins exist in conformations of different stability; the conformation of the C-terminal region of MIP is stable enough to be reproducibly imaged at subnanometer resolution; the N-terminal region of AqpZ is structurally less stable than the C-terminal end of MIP but more stable than the disordered C-terminal domain of bacteriorhodopsin (Belrhali *et al.*, 1999; Essen *et al.*, 1998; Grigorieff *et al.*, 1996; Heymann *et al.*, 1999; Luecke *et al.*, 1999b; Mitsuoka *et al.*, 1999) which is not detected by AFM (Müller, Sass *et al.*, 1999). Accordingly, a structural change caused by the cleavage of a polypeptide can only be observed by the AFM if it could be reproducibly detected before of its removal.

IV. Observing the Oligomerization of Membrane Proteins

α-Hemolysin is a water-soluble protein that undergoes several conformational changes from the time it is released from *Staphylococcus* until it interacts with a plasma membrane. Initially hemolysin is a monomer, which undergoes oligomerization into a homo-oligomeric ring finally inserting into the lipid bilayer forming a pore. Interestingly, this pore which facilitates water permeation across the membrane can be genetically engineered to sense a range of different organic molecules (Gu *et al.*, 1999). For some years it has been discussed whether the hemolysin oligomer exists in a hexameric or in a heptameric stoichiometry since different techniques have shown different oligomeric states of the complex (Gouaux *et al.*, 1994; Song *et al.*, 1996). Avoiding problems which may arise determining the oligomeric stoichiometry of proteins imaged by diffracting techniques, the high signal-to-noise ratio of the AFM allows subunits of protein complexes to be imaged directly and their oligomeric stoichiometry to be determined.

Staphylococcal α-hemolysin inserted into phospholipid bilayers is shown in Fig. 8. The AFM topograph shows unambiguously the hexameric state of the oligomeric complex which assembled into a two-dimensional array (Czajkowsky *et al.*, 1998). Interestingly, additional data have been recently published on α-hemolysin mutants locked in their open state (Malghani *et al.*, 1999). In contrast to the previously described data, the mutants appear to exist in a heptameric stoichiometry. As already pointed out (Czajkowsky *et al.*, 1998), it may be possible that α-hemolysin may form stable oligomers which differ in their stoichiometry. However, it remains a challenge to determine those factors that influence the oligomeric state of biochemically indistinguishable α-hemolysins.

ATP synthases are large protein complexes that convert the energy of a transmembrane proton (or Na^+) gradient into the biological energy source ATP. Its integral membrane complex F_o ($\sim$170 kDa) couples the transmembrane flow of protons to the rotation of a molecular stalk (Kato-Yamada *et al.*, 1998; Noji *et al.*, 1997; Sabbert *et al.*, 1996). The rotational force expels the spontaneously formed ATP from the three catalytic sites of the water-soluble F_1 complex ($\sim$400 kDa). While the catalytic subcomplex $\alpha_3\beta_3\gamma$ as well the isolated subunits δ and ε of F_1 have been solved to atomic resolution (Abrahams *et al.*,

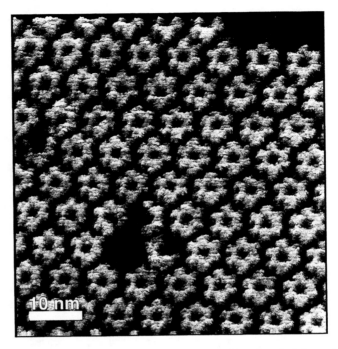

Fig. 8 Contact mode AFM topograph of α-hemolysin oligomers. The α-hemolysin inserted into the phospholipid bilayer in 10 mM sodium phosphate (pH 7.2) at room temperature and assembled into a 2D lattice. Raw image displayed as relief tilted by 5°. The scale bar represents 7.5 nm. Topograph displayed as relief tilted by 5° (image courtesy of Daniel Czajkowsky, University of Virginia).

1994; Bianchet *et al.,* 1998; Wilkens *et al.,* 1995; Wilkens *et al.,* 1997), the structure of the F_o complex still awaits elucidation. To gain insight into the mechanochemical coupling synthesizing ATP, the arrangement of the transmembrane F_o complex assembled from subunits I_1, II_1, III_x, and IV_1 in chloroplast ATP synthase (CF_oF_1) or from subunits a_1, b_2, c_x in bacterial and mitochondrial ATP synthase (EF_oF_1) is a matter of investigation. Several subunits, III_x and (c_x), form the "proton turbine" of the ATP synthase. The mechanism determining the exact number of subunits, III_x, however, is a topic of debate and remains to be answered.

Atomic force microscopy of the III_x oligomer of the most abundant ATP synthase from chloroplast revealed the surface at a sufficient resolution to allow the number of III subunits to be counted. Thus, topographs of the reconstituted cylindrical complex assembled from subunit III of the chloroplast F_oF_1-ATP synthase provided compelling evidence that this proton-driven turbine comprises 14 subunits (Fig. 9) (Seelert *et al.,* 2000). This finding is in contrast to the stoichiometry of the *E. coli* F_o complex which is postulated to be a dodecamer of subunit c, mainly based on crosslinking experiments (Jones *et al.,* 1998), genetic engineering (Jones and Fillingame, 1998), and model building (Dmitriev *et al.,* 1999; Groth and Walker, 1997; Rastogi and Girvin, 1999). Interestingly, X-ray analyses of yeast F_oF_1-ATP synthase crystals yielded a decameric complex (Stock *et al.,* 1999),

Fig. 9 Proton-driven rotor of the chloroplast ATP synthase reconstituted into a membrane bilayer. As shown by this unprocessed topograph, this cylindrical oligomer comprises 14 subunits (Seelert *et al.*, 2000). The dense packing of oligomers required an alternating orientation vertical to the membrane plane. Thus, the distinct wide and narrow rings represent the two surfaces of the cylindrical complex. Imaging buffer: 25 mM MgCl$_2$, 10 mM Tris–HCl, pH 7.8. Vertical brightness range of contact mode topograph corresponds to 2 nm. The raw image is displayed as a relief tilted by 5°.

indicating that polymorphic stoichiometries of F$_o$ complexes may have a biological origin which is not yet understood.

V. Unraveling the Conformational Variability of Membrane Proteins

A. Force-Induced Conformational Changes

The cytoplasmic bacteriorhodopsin surface, imaged with a force of 100 pN applied to the AFM stylus, revealed trimeric structures arranged in a trigonal lattice of 6.2 ± 0.2 nm side length (Fig. 10A, top; (Müller, Sass *et al.*, 1999). Each subunit in the trimer features a particularly pronounced protrusion extending 0.83 ± 0.19 nm above the lipid surface. This protrusion is associated with the loop connecting α-helices E and F (Fig. 10B; (Müller, Buldt *et al.*, 1995)). Increasing the applied force to about 200 pN during imaging changed the AFM topographs significantly. The prominent EF loops were bent away and the shorter loops of the bacteriorhodopsin monomers were visualized (Figs. 10A, bottom; and 10D). This conformational change was fully reversible (Müller, Buldt *et al.*, 1995),

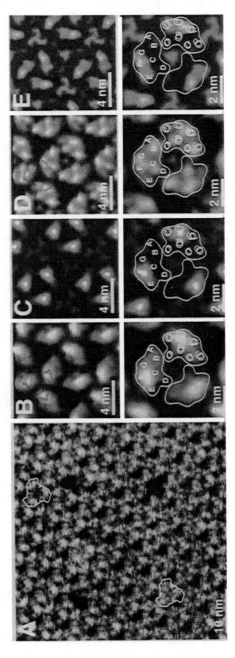

Fig. 10 Force-induced conformational change of the cytoplasmic purple membrane surface. (A) At the top of the topograph the force applied to the AFM stylus was 100 pN. While scanning the surface line by line, the force was increased until it reached 150 pN at the bottom of the image. This force-induced conformational change of bacteriorhodopsin was fully reversible (Müller, Büldt et al., 1995). Correlation averages of the cytoplasmic surface recorded at 100 pN (B) and at 200 pN (D). The correlation averages are displayed in perspective view (top, shaded in yellow brown) and in top view (bottom, in blue) with a vertical brightness range of 1 nm and exhibited 9.2% (B) and 14.1% (D) RMS deviations from threefold symmetry. Structural flexibilities were accessed by SD maps (C and E corresponding to B and D, respectively) which had a range from 0.08 (lipid) to 0.19 nm (EF loop region). Surface regions exhibiting a SD above 0.12 nm are superimposed in red-to-white shades in top of figure (B and D). The contact mode topograph was recorded in buffer solution (100 m*M* KCl, 10 m*M* Tris–HCl, pH 7.8). The outlined bacteriorhodopsin trimer representing sections close to the cytoplasmic surface of the lipid membrane was obtained after merging six atomic models of bacteriorhodopsin (Heymann *et al.*, 1999). Topographs (A), (B), and (E) are displayed as relief tilted by 5°.

suggesting that loop EF is a rather flexible element of the bacteriorhodopsin molecule. At this force of 200 pN, the maximum height difference between the protein and the lipid membrane was 0.64 ± 0.12 nm. Four distinct protrusions were recognized in almost every monomer, and a further distinct protrusion was present at the center of the trimers. The calculated diffraction pattern of this topograph documents an isotropic resolution out to 0.45 nm (not shown).

While the standard deviation of the height measurements was around 0.1 nm for most morphological features of the topography, the EF loop exhibited an enhanced SD of 0.19 nm (Fig. 10C), consistent with the high-temperature factor observed by electron microscopy (Grigorieff et al., 1996) and the structural variation among the atomic bacteriorhodopsin models (Heymann et al., 1999). When the major protrusion representing loop EF had been pushed away by applying a force of 200 pN to the stylus, the cytoplasmic surface of the bacteriorhodopsin molecule appeared different and exhibited details of the shorter loops connecting helices AB and CD (Fig. 10D).

The protrusion between helices F and G together with the minor elevation between helices E and F likely represents what remained structured from loop EF and the protruding parts of helices E and F that are compressed by the AFM stylus (Fig. 10D). However, it cannot be excluded that the protrusion between helices F and G included a small part of the C-terminal domain. This uncertainty arises because the AFM height signal in this area exhibited a significant standard deviation (Fig. 10E; red shaded in Fig. 10D). The other protrusions in the AFM topograph may be assigned by comparison with the atomic models derived from the bacteriorhodopsin trimer (see following section). In these models, helix B protrudes out of the bilayer, and helix A ends below the bilayer surface. Therefore, the protrusion close to helix B is likely to represent the short loop connecting helices A and B (Fig. 10D). In addition, the discrete protrusion between helices C and D corresponds to their connecting loop. A further protrusion of 0.2 nm height was present at the threefold axis of the bacteriorhodopsin trimer and probably arises from structured lipid molecules (Grigorieff et al., 1996).

To further analyze the conformations of the cytoplasmic surface, the unit cells of topographs recorded at applied forces of 100 and 200 pN were extracted, aligned with respect to a reference, and classified by principal component analysis (Frank et al., 1987; van Heel, 1984). The threefold symmetrized averages of the major classes shown in Figs. 11A to 11E reveal the movement of the flexible structures. The classes A, B, and C, D were closely related to the force gradient. Increasing the force to 120 pN resulted in a slight deformation of the EF loop and enhanced the details of the surrounding protein structure (Fig. 11A; compare to Fig. 10B). Increasing the force to approximately 150 pN further pushed the EF loop away (Fig. 11B), whereas at about 180 pN the conformational change of the loop was complete (Figs. 11C to 11E). A central protrusion was apparent in some bacteriorhodopsin trimers when imaged at 180 pN (Figs. 11C and 11E). Most probably, this protrusion represented lipid headgroups which were absent or disordered in some bacteriorhodopsin trimers. Increasing the applied force to 300 pN resulted in deformation of the peripheral protrusions of the trimer. The structural information of these areas was lost (Müller et al., 1998), and when imaged at applied forces above 300 pN the bacteriorhodopsin trimers were irreversibly deformed (data not shown).

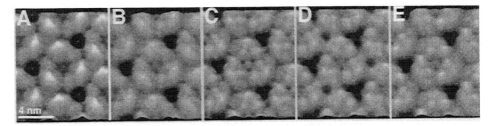

Fig. 11 Structural variability of the cytoplasmic bacteriorhodopsin surface. The threefold symmetrized averages were calculated from unit cells classified by multivariate statistical analysis using the algorithm kindly provided by J.-P. Bretaudiere (Bretaudiere and Frank, 1986). (A) PM imaged at slightly enhanced forces of 120 pN (compare to Fig. 11B). (B) Same membrane imaged at an applied force of approximately 150 pN. In (C), (D), and (E) three conformations of the membrane are imaged at approximately 180 pN. The last three averages differ in their central protrusion and in that of the EF loop (compare to Fig. 10). The correlation averages are displayed in perspective view with a vertical brightness range of 1 nm. Topographs are displayed as relief tilted by 5°.

VI. Comparing AFM Topographs to Atomic Models

A. Comparing Topographs of OmpF Porin to the Atomic Model

OmpF porin is present as stable trimeric structures in the outer membrane of *E. coli*. Each 340-amino acid (aa) OmpF monomer is folded into 16 antiparallel β-strands to form a large hollow β-barrel structure which perforates the membrane. The transmembrane pore facilitates the passages of hydrophilic solutes up to an exclusion size of ≈ 600 kDa (Nikaido and Saier, 1992). It is suggested that the charges of the porin channel primarily modulate the pore selectivity (Klebba and Newton, 1998; Schirmer, 1998), and recent calculations have shown that the OmpF pore may establish an electrical potential which increases with decreasing electrolyte concentration (Schirmer and Phale, 1999). As observed in the AFM topograph of the periplasmic surface (Fig. 12A), each trimer (outlined circle) is compromised of tripartite protrusions and three transmembrane channels that are separated by 1.2-nm-thick walls. The transmembrane channel has a characteristic elliptical cross section of $a = 3.4$ nm and $b = 2.0$ nm. The arrows point out individual polypeptide loops of a few aa size each connecting two antiparallel ß-strands lining the transmembrane pore. Most features recorded in this AFM topograph were correlated directly to the atomic model of the OmpF (Cowan *et al.*, 1992) surface rendered at 0.3 nm resolution (Fig. 12B). Correlation averaging of the porin trimer enhanced common structural details among individual trimers (Fig. 12C) but blurred variable areas of the subdomains (compare to porin trimers shown in raw data, Fig. 12A). However, the characteristic shape of the transmembrane channel was more pronounced showing an elliptical cross section of $a = 3.4$ nm and $b = 2.0$ nm. Structural areas of the periplasmic OmpF trimer surface exhibiting an enhanced variability are recovered by the standard deviation map of the average (Fig. 12D). Enhanced values of the SD map can directly correlate to surface structures which are expected to have enhanced flexibility (Fig. 12B).

Fig. 12 AFM topograph and atomic model of OmpF porin. (A) The unprocessed topograph exhibits features that are recognized in the atomic model of the periplasmic OmpF porin surface, rendered at 0.3 nm resolution (B). The circle outlines one porin trimer. Short β-turns comprising only a few amino acids are sometimes distinct (arrows). (C) Correlation average of the rectangular unit cell of porin. (D) SD map of the correlation average. All data are displayed in perspective view with a vertical brightness range of 1 nm (A, B, and C). Vertical brightness range of contact mode topographs corresponds to 1 nm. Structural flexibilities accessed by the SD map ranged from 0.06 to 0.12 nm. Images are displayed as reliefs tilted by 5°.

B. Structure and Flexibility of the Bacteriorhodopsin Surface

Information about the surfaces of bacteriorhodopsin have been derived from electron crystallography (Grigorieff *et al.*, 1996; Mitsuoka *et al.*, 1999) and AFM (Figs. 10 and 11) and X-ray diffraction (Belrhali *et al.*, 1999; Essen *et al.*, 1998; Luecke *et al.*, 1998; Sato *et al.*, 1999) at high resolution. This provided an excellent opportunity to assess the quality of the AFM topographs recorded from purple membrane and to understand the implications of combining AFM data with the other structure determination methods. Six bacteriorhodopsin atomic models were combined and compared to the AFM

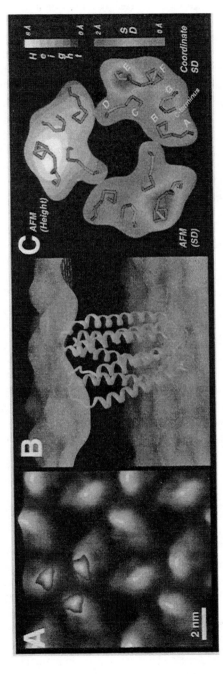

Fig. 13 Quantitative analyses of the native cytoplasmic bacteriorhodopsin surface. (A) Correlation average of the AFM topograph recorded at an applied force of 100 pN (Müller, Sass *et al.*, 1999). Regions with enhanced flexibility are derived from SD maps and superimposed in red to white shades. The vertical brightness range of topograph corresponds to 1 nm. The raw image is displayed as a relief tilted by 5°. (B) Mapping the structural variance of bacteriorhodopsin on the atomic model and the AFM envelope. The atomic model is an average of six models derived from electron and X-ray crystallography, with the coordinate variance mapped from blue (low variance) to red (high variance). The surfaces are derived from the AFM height images, with the SD mapped onto each surface from blue (low SD) to red (high SD). The minimum separation between the surfaces is ~4 nm. Calculations are as given in Heymann *et al.*, 1999. (C) Cytoplasmic surface with each bacteriorhodopsin monomer displaying a different surface property. The surface loops are shown as backbone tracings colored according to the backbone coordinate root-mean-square deviation (SD) calculated after merging five different atomic models of bacteriorhodopsin (Heymann *et al.*, 1999). The gray scale image shows the height map determined by AFM (Fig. 2); the prominent protrusion is the EF loop. The colored monomers represent the coordinate variation (SD) between the atomic models and the SD of the height measured by AFM, respectively. Height and SD maps determined by AFM correlate amazingly well to corresponding data from X-ray and electron crystallography. (See Color Plate.)

data to determine the value and reliability of each source of information (Heymann et al., 1999).

Figure 13B shows one atomic model suspended within an envelope of the purple membrane reconstructed from the AFM data (Fig. 13A). The ribbon diagram is color coded according to the coordinate variance between the different atomic models, while the surfaces are mapped with the standard deviation of the AFM topographs. There is an excellent correspondence between the surface loops of the bacteriorhodopsin model and the AFM envelope. SD maps of the height measured by AFM corresponds well with both the relative distribution of B-factors of the atomic models and the coordinate variance between the models (Fig. 13C). This agrees with the notion that the major difference between the various structural studies lies in the surfaces.

In both the electron crystallography and the AFM experiments, a 2D crystal of bacteriorhodopsin close to its native state was imaged, allowing surface loops the maximum possible conformational freedom. In X-ray crystallography, the surface loops were resolved, but they were often involved in 3D crystal contacts and consequently their positions may not represent their true conformational state and variation in vivo. Specifically, the EF loop appears to adopt different conformations in the 2D assembly of bacteriorhodopsin molecules of purple membrane, while it has less conformational freedom in the 3D crystals. This agrees with electron paramagnetic resonance (EPR) spectroscopy of spin-labeled cysteine mutants which show a high mobility of the amino acid residues in the EF loop (Pfeiffer et al., 1999). In contrast to electron and X-ray crystallographic methods, the AFM was used to image surface structures of bacteriorhodopsin in buffer solution and at room temperature, i.e., under conditions resembling its physiological environment. The SD maps of bacteriorhodopsin had a higher peak for the longer and less structured loop EF than for the BC loop which forms a short β-sheet.

The N- and C-termini of bacteriorhodopsin were not resolved in any of the structural studies. This suggests high flexibility, and the atomic models and AFM data indicate that the N-terminus and most of the C-terminus are completely unstructured and averaged out (see Chapter 3).

VII. Conformational Changes of Native Membrane Proteins

A. Surface Structures Can Change upon Interaction with Adjacent Molecules

Recrystallization of bacteriorhodopsin in the presence of n-dodecyl trimethylammonium chloride (DTAC) yielded well-ordered 2D crystals (Michel et al., 1980). Topographs of these orthorhombic crystals showed bacteriorhodopsin dimers assembled into a rectangular lattice with a $p22_12_1$ symmetry and unit cell dimensions of $a = 5.8$ nm and $b = 7.4$ nm (Fig. 14A) (Michel et al., 1980). Accordingly, the bacteriorhodopsin dimers alternately had their cytoplasmic surface or their extracellular surface facing the stylus. The maximum height difference between the protrusions and the bilayer was 0.81 ± 0.09 nm. Surprisingly, it was not possible to induce conformational changes of the EF loops in this bacteriorhodopsin crystal form. Increasing the applied force of the

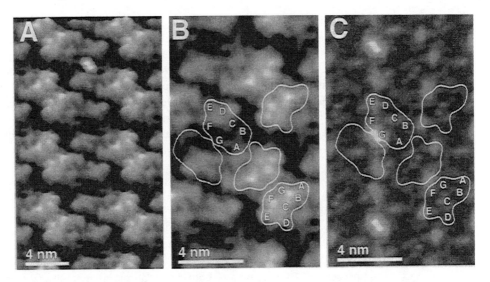

Fig. 14 Native bacteriorhodopsin assembled into an orthorhombic lattice. (A) In this crystal form ($p22_12_1$) the rows of bacteriorhodopsin dimers alternate to expose either their cytoplasmic or their extracellular surfaces to the aqueous solution. The correlation averages are displayed (B) in perspective view (left, shaded in yellow brown) and (C) in top view (right, in blue) with a vertical brightness range of 1 nm. (D) SD map of (C) having a vertical brightness range from 0.06 to 0.17 nm. Surface regions exhibiting a SD above 0.12 nm are superimposed in red to white shades in (B). The outlined regions are presented for comparison with the extracellular (yellow outline) and cytoplasmic (white outline) slides of bacteriorhodopsin as shown in Figs. 2 and 3. The contact mode topograph was recorded in buffer solution (100 mM KCl, 10 mM Tris–HCl, pH 7.8) at a loading force of 100 pN. Topographs A and B are displayed as reliefs tilted by 5°.

stylus resulted in the deformation of the whole protein surface rather than in the bending of a single loop, and it reduced the lateral resolution.

The arrangement of the protrusions on the cytoplasmic face of bacteriorhodopsin was very distinct when AFM topographs of the orthorhombic *in vitro* assemblies were analyzed (Figs. 14B and 14C). The protrusion of the AB loop was shifted by 0.3 nm compared to the trigonal unit cell (Fig. 10), now being located between the position of helices A and B (Fig. 14C). The short loop connecting helices C and D was observed as a discrete protrusion in the orthorhombic lattice, close to its position in the trigonal lattice. Remarkably, the EF loop was observed as a bean-shaped structure independent of the applied force (Fig. 14C). The triangular protrusion located between helices B and G may result from the C-terminus. None of these structures exhibited significant variability, indicating structural stabilization by the different packing arrangement in the orthorhombic compared to the trigonal lattice. An additional protrusion (Fig. 14C) was observed at the periphery of each bacteriorhodopsin monomer packed in the orthorhombic lattice, probably representing bound lipid molecules (Grigorieff *et al.*, 1996).

The observed structural changes suggest that the interactions of the cytoplasmic polypeptide loops depend on how the bacteriorhodopsin molecules associate. In the bacteriorhodopsin trimer, there is a crevice between helices A and B and helices E and

D of neighboring monomers (Fig. 14C; outlined). Lipid molecules in this crevice are stable (Grigorieff *et al.,* 1996) and stabilize the bacteriorhodopsin trimer by specific interactions with their lipid and headgroup moieties (Essen *et al.,* 1998). This crevice is not present in the orthorhombic bacteriorhodopsin assembly, and hence the different molecular interactions probably allow the displacement of the loop connecting helices A and B (Fig. 14C; white contours). Differences in helix E have also been observed in X-ray structures from different crystal forms. While the end of helix E was not resolved in the crystals grown in the cubic lipid phase (Edman *et al.,* 1999; Luecke, *et al.,* 1998; Luecke *et al.,* 1999a; Pebay-Peyroula *et al.,* 1997), helix E was stable and fully resolved in the structure by Essen *et al.* (1998), where crystal contacts along helices F and G occurred. Accordingly, it was concluded that in the orthorhombic bacteriorhodopsin crystals the interactions between helices F and G of two adjacent bacteriorhodopsin molecules affected both the structural appearance and the rigidity of both the EF loop and the C-terminal region. In addition, the protrusion of loop AB was shifted toward helix A, away from the intermolecular space, in the orthorhombic lattice.

Similarly, a change of the protrusions of the extracellular surface was observed on comparing AFM topographs of both native purple membrane and orthorhombic bacteriorhodopsin crystal. A detailed discussion of these structural changes is published elsewhere (Heymann *et al.,* 1999; Müller *et al.,* 2000; Müller, Sass *et al.,* 1999).

B. Functional Related Conformational Changes of the HPI layer

The multiple functions of the hexagonally packed intermediate (HPI) layer from *Deinoccocus radiodurans* are still a matter of debate (Fig. 15). As a typical member of the outermost surface (S) layers of bacteria (Sleytr, 1997; Sleytr *et al.,* 1993) evidence exists that the HPI layer stabilizes the cell shape and functions as a protective barrier against hostile factors from the environment, which nutrients, solutes, and waste products have to cross (Baumeister *et al.,* 1988). The HPI layer, like many other S-layer proteins, appears to be exceptionally resistant to protease treatment in a nondenaturated state. Furthermore, it is discussed whether the HPI layers of two neighboring bacteria form connexons to enable cell–cell communication (Baumeister and Hegerl, 1986). As revealed from electron microscopy and crystallography studies (Baumeister *et al.,* 1986; Engel *et al.,* 1982; Rachel *et al.,* 1986) the HPI layer consists of a single type of protomer, six of which form a pore. The stable framework of the HPI layer is created by protomer–protomer contacts between neighboring pores.

AFM topographs of the HPI layer reveals the structural difference between the outer (Fig. 16) and the inner surface (Fig. 17). Both surface structures show six protomers forming hexameric rings surrounding a central pore and arranged in a hexagonal lattice ($a = b = 18$ nm). Each protomer of a pore is connected to a protomer from the adjacent pore. These intermolecular links are thought to be the basis for the unusual stability of the HPI layer. While these distinct arms emanating from the hexamers exhibit a clockwise rotation at the outer surface, they exhibit an anticlockwise rotation at the inner surface. The outer surface of the HPI layer consists of donut-shaped hexamers,

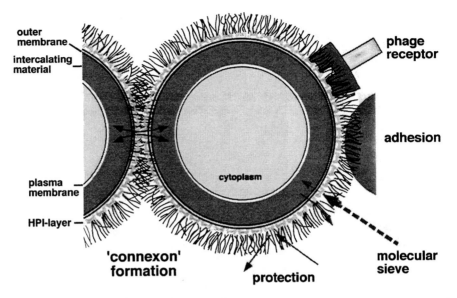

Fig. 15 Suggested functions of the hexagonally packed intermediate (HPI) layer from *D. radiodurans*. Although most of the functions of surface (S) layers, a family to which the HPI layer belongs, remain enigmatic they must be regarded as multifunctional. As well as providing the cells with a molecular sieve controlling the passage of molecules into and out of the cell by filtrating large molecules, S layers may mediate cell–cell contacts and communication, interact with other macromolecules such as phage receptors, or mediate adhesion to other surfaces. Schematic drawing adapted according to Baumeister and Hegerl (Baumeister and Hegerl, 1986).

featuring six V-shaped protrusions (Fig. 16), seen more clearly in the correlation average (Fig. 3F).

Most interestingly, it appears that individual pores of the inner HPI-layer surface can exist in either an unplugged (circles) or a plugged (squares) conformation (Fig. 17; top left). As demonstrated by time-lapse AFM topographs (Fig. 17; top right), the pores can reversibly change their conformation (Müller, Baumeister *et al.,* 1996). After translational and angular alignment of single pores form the HPI layer, a multivariate analysis of 330 pores from 10 different topographs was performed. The averages of the two major classes, both showing six subunits of the core and their emanating arms, exhibit either an "open" or a "closed" pore. Figure 17(bottom) displays a montage of the calculated and sixfold symmetrized topographies of both conformations. Protrusions located at the core were arranged on an equilateral hexagon of side length 4.0 ± 0.2 nm. The height difference to the emanating protrusions was 2.5 ± 0.2 nm, while the maximum height of the protrusions was 2.9 ± 0.3 nm. The depression in the open conformation of the core was 1.8 ± 0.5 nm, and the depression over the protrusion of the closed conformation was 1.0 ± 0.5 nm (Müller, Baumeister *et al.,* 1996). This suggests that the HPI layer serves as a molecular sieve with an open and a closed state (Fig. 17; bottom).

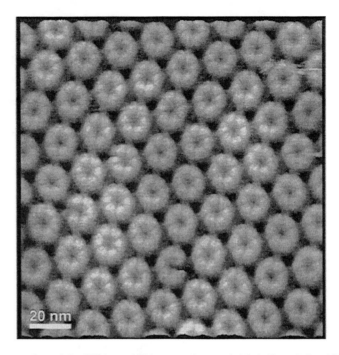

Fig. 16 Outer surface of the HPI layer. AFM topographs recorded in buffer solution (pH 7.8, 10 mM Tris–HCl). See Fig. 3 for the correlation average. Vertical brightness range of the contact mode topograph corresponds to 4 nm. Raw data are displayed as a relief tilted by 5°.

The lack of functional and structural data, however, prevents this hypothesis from being confirmed.

C. Functional Related Conformational Changes of OmpF Porin

The outer membranes of Gram-negative bacteria protect the cells from hostile factors such as proteolytic enzymes, bile salts, antibiotics, toxins, and low pH. Uptake of small nutrients and release of metabolites are facilitated by passive pores, the porins. *E. coli* outer membranes contain approximately 10^5 porins per cell that allow passage of small solutes, <600 Da (Nikaido and Vaara, 1985). In addition to nutrients, antibiotics that need to cross the outer membrane for binding to their targets diffuse through these aqueous pores (Nikaido, 1989). Conductance measurements have shown that the porin OmpF trimer exists in either open or closed states, depending on the transmembrane potential (Lakey, 1987; Schindler and Rosenbusch, 1978). The critical voltages ($V_c \approx 90$ mV for OmpF porin), above which channels close, is affected by the pH (Todt *et al.,* 1992), membrane-derived oligosaccharides (Delcour *et al.,* 1992), polysaccharides (Schindler and Rosenbusch, 1981), polycations (dela Vega and Delcour, 1995), and pressure (Le Dain *et al.,* 1996). Although the structures of several porins have been solved (Cowan

Fig. 17 Conformational change of individual pores of the HPI-layer. Top left: inner surface of the HPI layer as imaged by contact mode AFM in buffer solution. Protomers of the hexameric pores are clearly visible. Individual pores can exist in an open (circles) and a plugged (square) conformation. Top right: imaging of the same surface after 5 min demonstrates that some of the pores have changed their conformation. This conformational change is fully reversible and can be observed over hours. Bottom: surface relief of both the averaged open and closed conformations. The correlation averages of both the open and the closed conformations were sixfold symmetrized and then assembled in the montage. The distinct arms emanating from the cores of the hexamers exhibit an anticlockwise rotation. The distance between adjacent pores is 18 nm. The full grey level range of all contact mode topographs corresponds to 3 nm vertical distance. Average is displayed as a relief tilted by 5°.

et al., 1992; Schirmer, 1998; Weiss et al., 1991), the nature of channel closure is poorly understood. The transmembrane pore of OmpF is formed by 16 antiparallel β-strands that are connected by short loops on the periplasmic surface and long extracellular loops. One of the latter folds into the barrel, thus constricting the eyelet of the channel. This loop was proposed to move upon membrane potential changes (Schulz, 1993), thereby closing the channel. However, recent results exclude this movement of the constriction loop (Phale et al., 1997). Therefore, other effects must explain the closure of porin channels (Klebba and Newton, 1998).

The structurally and functionally well-characterized bacterial porin OmpF is an excellent structure to assess in depth the ability of the AFM to monitor conformational

Fig. 18 Conformational changes of porin OmpF. (A) Extracellular surface of OmpF. The comparison between the atomic model rendered at 0.3 nm resolution (left). An AFM topograph (right) illustrates that the loops which protrude 1.3 nm from the bilayer are flexible. The asterisks mark the twofold symmetry axis of the rectangular unit cells housing two porin trimers. (B) pH-dependent conformational change of the extracellular surface. At pH ≤ 3 the flexible loops reversibly collapse toward the center of the trimer, thereby reducing their height by 0.6 nm. (C) Conformational change of porin induced by an electrolyte gradient. The monovalent electrolyte gradient across the membrane was >300 mM. Similar to the pH-dependent conformational change, the extracellular domains reversibly collapsed onto the porin surface. Contact mode topographs exhibit a vertical range of 1.5 nm (A) and 1.2 nm (C and D) and are displayed as reliefs tilted by 5°.

changes. The short polypeptide loops at the periplasmic surface (Fig. 12) and the long loops at the extracellular surface (Fig. 18A) are observed on individual porin trimers in the AFM topographs (Fig. 18A; right). The variable extracellular loops protrude by 1.3 nm above the bilayer at neutral pH. Three conditions have been demonstrated to induce the collapse of this flexible domain toward the trimer center to form a structure with a height of only 0.6 nm (Müller and Engel, 1999): (i) application of an electric potential >200 mV across the membrane, (ii) generation of a K^+ gradient >0.3 M (Fig. 18B), and (iii) acidic pH (≤ 3) (Fig. 18C). The last condition suggests a protective function: *E. coli* cells passing through the acidic milieu of a stomach may survive longer by closing the outer membrane pores. The first condition, however, is compatible with results from black lipid membrane and patch-clamping experiments which demonstrated that porin acts as a voltage-gated channel (Delcour, 1997; Klebba and Newton, 1998).

VIII. Observing the Assembly of Membrane Proteins

The Schiff base of bacteriorhodopsin reacts with reagents such as hydroxylamine on illumination with light (Oesterhelt *et al.*, 1974). This chemical reaction results in the breakage of the Schiff base bond between the bacteriorhodopsin and the retinal yielding the apoprotein bacterioopsin and retinaloxime. Consequently, the absorption maximum of purple membrane at 568 nm diminishes, and the absorption maximum of retinaloxime at about 366 nm is observed (Oesterhelt *et al.*, 1974). These spectral changes depend upon the illumination time and reflect the photobleaching process of purple membrane (Fig. 19A). The loss of the Schiff base bond leads to structural changes in the apoprotein (Bauer *et al.*, 1976; Becher and Cassim, 1977). As observed using AFM, the process of photobleaching was associated with the disassembly of the purple membrane crystal into smaller crystals (Fig. 19B). The bleached purple membrane had entirely lost most of its crystalline nature (Fig. 19C). High-resolution topographs showed the progressive separation of bacteriorhodopsin trimers, first along distinct lattice lines and later all over the membrane. Furthermore, the topographs showed that the bacterioopsin molecules remained in their trimeric assembly during the entire photobleaching process. Regeneration of the photobleached membranes into fully active purple membrane resulted in the renewed association of the bacteriorhodopsin trimers into a trigonal crystal (Fig. 19D). The regenerated membranes exhibited similar diameter, thickness, and crystallinity to native purple membrane (Möller *et al.*, 2000).

From these results, it can be concluded that the transformation of bacteriorhodopsin into bacterioopsin changes the interactions between the trimers. Such interactions might result from changes in the tertiary structure of the protein. Since the bacteriorhodopsin trimer remains stable during the entire course of photobleaching, it might be concluded that major structural changes occur at the rim of the trimer were it interacts with adjacent lipids. These interfaces are lined by the transmembrane α-helices E and F and by helix G of bacteriorhodopsin to which the retinal is bound. Cleavage of the Schiff base is followed by both a reversible change of these interfaces and a disassembly of the two-dimensional purple membrane crystal.

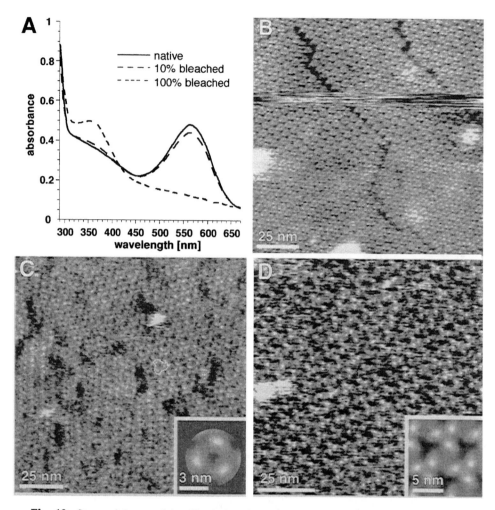

Fig. 19 Structural changes of photobleached purple membrane. (A) Absorption spectra of wild-type purple membrane (solid line) and of purple membrane photobleached in hydroxylamine to 10 and to 100% (dashed lines). (B) Topograph of the cytoplasmic surface of purple membrane photobleached to 10%. (C) Cytoplasmic surface of purple membrane photobleached to 100%. A single bacteriorhodopsin trimer is outlined in the raw data. The inset represents the correlation average of the bacteriorhodopsin trimer. (D) Cytoplasmic surface of purple membrane fully regenerated after the addition of retinal chromophore. The bacteriorhodopsin trimers reassembled into the 2D crystal of purple membrane. The inset represents the correlation average of the cytoplasmic surface with the bacteriorhodopsin trimer located at the center of the image. All images were recorded in buffer solution (150 mM KCl, 10 mM Tris–HCl, pH 7.8) at a loading force of 100 pN. Vertical full gray level range, 1.2 nm (images courtesy of Clemens Möller, University of Basel).

IX. Detecting Intra- and Intermolecular Forces of Proteins

A. Unzipping Single Protomers from a Bacterial Pore

Structural, chemical, and morphological studies have shown that S layers are one of the most primitive membrane structures developed during evolution. They cover the cell surface completely, are usually composed of a single protein, and are endowed with the ability to assemble into monomolecular arrays by an entropy-driven process (Sleytr *et al.*, 1993). Experimental results indicate that the integrity of the S-layer lattice is maintained by different combinations of weak bonds which are stronger than those binding to the underlying cell envelope component (Sleytr, 1997; Sleytr *et al.*, 1993).

The inner surface of the S layer of *Deinoccocus radiodurans*, the inner surface of the HPI layer, is directly attached to the underlying outer cell membrane (Fig. 15) via strong hydrophobic interactions (Thompson *et al.*, 1982), which are formed by fatty acid residues covalently linked to the hydrophobic N-terminal domain of the HPI layer and serve as a membrane anchor (Peters *et al.*, 1987). The C-terminal region located at the outer surface of the HPI layer is covalently linked to a $\approx$40-nm-thick carbohydrate coat. Naturally, these carboxyl tails help to attach the bacteria to favorable areas. Thus, forces applied to the carboxyl tail will directly pull on the individual protomers of the HPI layer. This in turn will react as a stabilizing framework distributing fractions of the pulling forces across the outer cell membrane. Thus, the force anchoring a single protomer in its hexameric assembly is of interest.

To gain insight into the interaction forces, individual protomers of the inner surface of the HPI layer were first imaged (Fig. 20A). The resolution of the topograph was sufficient to resolve the individual subunits of the hexameric pores and of their emanating arms. After imaging the HPI layer, the AFM stylus was attached to an individual protomer by an enforced stylus–sample contact to allow single-molecule force spectroscopy experiments (Fig. 20B). As visible from the force spectroscopy curve, a strong adhesion of the inner HPI layer surface to the silicon nitride stylus was observed on retracting the stylus approximately 15 nm from the contact point. This indicates that a molecular structure bridged the gap between the AFM stylus and the HPI layer. It is likely that this molecular bridge is mediated by the hydrophobic sequence segments near the N-terminus which itself also carries an alkyl moiety, a region thought to interact with the outer membrane (Peters *et al.*, 1987). Imaging of the HPI layer after recording force–extension curves allowed adhesion forces to be correlated to structural alterations (Fig. 20C). A single protomer was missing from one HPI hexamer. Using this approach, individual protomers of the HPI layer were found to be removed at pulling forces of $\approx$310 pN.

B. Unzipping an Entire Bacterial Pore

While weak adhesive forces were seen more often than those that lead to the extraction of an individual protomer, an even less frequent but amazing event is displayed in Fig. 21 (Müller, Baumeister *et al.*, 1999). Here the force–extension curve exhibits six major (200–300 pN) equally spaced peaks (Fig. 21B), while the corresponding control

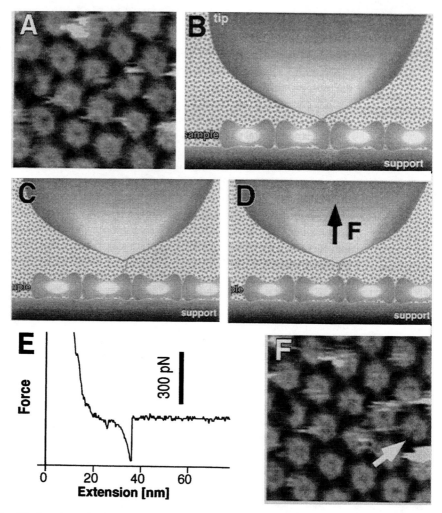

Fig. 20 Imaging and removing individual protomers of a bacterial pore. (A) Inner surface of the HPI layer showing hexameric pores. Individual pores exist in either plugged or unplugged conformations (compare to Fig. 18). (B) After recording the topograph the AFM stylus was pushed onto a single protomer for about 1 s and then retracted (C). (D) While separating HPI layer and AFM stylus the cantilever deflection was recorded. (E) Occasionally, the force–extension curve recorded showed a single-adhesion peak of ≈310 pN at an average distance of 18 nm from the surface. (F) The same surface area imaged after recording the force–extension curve shows a single protomer has been removed (arrow). Contact mode topographs exhibit a vertical range of 3 nm.

topograph shows that during retraction of the stylus an entire HPI hexamer has been zipped out of the S layer (Fig. 21C; compare to Fig. 20A). The equally spaced force peaks (Fig. 21B) indicate a strong interaction between protomers through a flexible link that has a length of 7.3 ± 1.6 nm, close to the thickness of the HPI layer. It is interesting to compare our results to the forces required to unfold immunoglobulin (Ig) segments

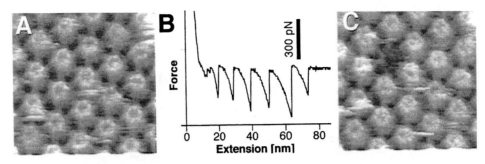

Fig. 21 Unzipping an entire pore of the HPI layer. (A) Inner surface of the HPI layer. (B) After recording the topograph, the AFM stylus was pushed onto a single pore of the HPI surface for about 1 s and retracted. Occasionally, the force–extension curve recorded showed a sawtooth-like pattern with up to six force peaks of about 300 pN. (C) The same surface area imaged after recording the force–extension curve shows an entire hexameric pore has been removed. The emanating arms of the adjacent pores to which the unzipped hexamer was connected are clearly visible. Contact mode topographs exhibit a vertical range of 3 nm and are displayed as relief tilted by 5°.

of native and recombinant titin (Rief, Gautel *et al.*, 1997). These force–extension curves exhibit peaks that are quite similar in shape and magnitude to those reported for the HPI layer. On fitting force–extension curves from the HPI layer with the wormlike chain (WLC) equation, the best fit was found for a persistence length of 0.4 nm, identical to the persistence length of the titin chain (Rief, Gautel *et al.*, 1997). This supports the idea of the flexible link being a polypeptide chain comprised of 26 residues with an extended length of 7.3 nm connecting HPI monomers.

In the experiments with recombinant titin, the maximum number of peaks seen in one force curve corresponded to the number of Ig segments (either four or eight). In analogy to this, a maximum of six peaks was observed for the hexameric HPI protein complex. However, the titin peaks were separated by 28–29 nm, compatible with the unfolding of a polypeptide chain comprising 89 residues. With the HPI layer, the retracting stylus appears to extract the first protomer, without unfolding it, by simply stretching the intermolecular link until the neighboring protomer is pulled out, eventually leading to the unzipping of a complete HPI hexamer. The fact that several protomers can be pulled out sequentially implies that the interaction forces within hexamers are stronger than the forces between them. Following this model, the extraction of each protomer involves breakage of the spoke that connects hexamers within the HPI layer.

X. Conclusions and Perspectives

A. Improvement of the AFM Technique, Image Acquisition, and Sample Preparation

In the last few years there has been amazing progress in the biological application of AFM. Imaging in aqueous solutions initiated by the Hansma group has been perfected, leading to images that exhibit a striking signal-to-noise ratio and that compete with the best results from electron microscopy. This development would not have been possible without the impact of various improvements of the AFM technique, AFM cantilevers and probes, image acquisition methods, and sample preparation.

Future experiments will be carried out with AFM probes, such as those currently being developed, that can acquire multiple signals (Cheung *et al.*, 2000; Lieberman *et al.*, 1994; Schurmann *et al.*, 2000). These probes will be important in the direct assessment of the relationship between structure and function of biomolecules. Photolysis induced by a local optical probe will release caged ATP locally to initiate biological processes, and function-related conformational changes of the biomolecules involved will be directly observed by using the same probe, such as the stylus of an AFM. Ligands will be deposited locally with a scanning micropipette, and electric stimuli will be applied to a single-voltage-gated channel, while the multifunctional stylus will allow the activated biomolecule to be observed during the work process.

B. Biochemical Identification of Macromolecular Structures

As in every microscopic technique, it is important to identify the supramolecular structures imaged. In this respect, antibodies can be used as specific labels, for example, to identify certain protein constituents and the surfaces of cell membranes. Antibodies once bound can be directly visualized by the AFM (Müller, Schoenenberger *et al.*, 1996). In the future, the smaller Fab fragments of antibodies may be employed to stoichiometrically label cell membranes, thus allowing the localization of polypeptides. In addition, various Fab fragments may be used to study the assembly of polypeptides.

When interpreting a protein topography recorded at submolecular resolution, it is necessary to identify both its surfaces and surface structures. AFM topographs revealed from native, untreated membrane proteins have clearly shown structural differences compared to topographs recorded after the replacement or the enzymatic removal of a polypeptide domain. These structural changes have allowed the surface, the C- or N-terminal region, and polypeptide loops of membrane proteins to be unambiguously identified. Some secondary structural elements exhibit a high intrinsic flexibility. The structural modification of these domains, however, may not be detected reproducibly by AFM.

C. Structural Information of AFM Topographs

The AFM topographs of native protein surfaces recorded in buffer solution at room temperature clearly show the conformations of the polypeptide loops. The signal-to-noise ratio of the topographs allows the observation of details on single proteins, and their major conformations can be classified. Standard deviation maps can be calculated to assess the structural variability of the surface structures, revealing the elasticity of single loops. These important improvements of the AFM application and data analysis provide evidence that the AFM not only fulfills the prerequisites to directly monitor function-related conformational changes of biological macromolecules (Drake *et al.*, 1989; Engel *et al.*, 1999; Müller and Engel, 1999; Müller, Schoenenberger *et al.*, 1997) but also characterizes dynamic aspects of protein structures such as their flexibility and variability. In the case of bacteriorhodopsin, protrusions representing single-polypeptide loops exposed to the aqueous solution have been shown to change their structure, variability and flexibility upon interactions occurring within the membrane composed of proteins and lipids (Müller, Sass et al., 1999).

D. Complementary Structural Information

Good correspondence between protein surface structures revealed from both AFM topographs and different structure determination techniques has been observed. In addition, each technique provides similar but also complementary information. X-Ray crystallography is the premier technique for atomic resolution, while electron crystallography examines the specimen in a more native environment at near-atomic resolution. AFM offers an even better assay of both surface structure and variation at submolecular resolution under physiological conditions. Thus, the combination of these techniques represents a complete structural analysis of the specimen.

E. Flexibility, Variability, and Conformation of Individual Proteins

The flexibility observed for protein structures is important for their function. For example, the observed helix F movement during the photocycle of bacteriorhodopsin (Dencher et al., 1989; Koch et al., 1991; Subramaniam et al., 1999) requires a conformational change in loop EF, which was found to exhibit enhanced flexibility in AFM experiments (Müller, Sass et al., 1999). Similarly, the long flexible extracellular loops of porin OmpF undergo conformational changes under conditions associated with the channel closure of the transmembrane pore. In the case of the inner HPI-layer surface, it was shown that the pore exhibits a region of enhanced flexibility which directly correlates to a central plug switching between two conformations (Müller et al., 1998). Of similar importance, however, is the possibility of unraveling the principle modes of motion of such flexible structures and thereby determining conformations associated with the working cycle of a protein.

F. Assembly of Membrane Proteins

Lipids, other biological molecules, and membrane proteins take part in dynamic clustering and form rafts that move within the fluid bilayer (Pralle et al., 2000) which are thought to function as platforms for the attachment of proteins which function in sorting and trafficking through the secretory and endocytic pathways (Brown and London, 1998; Simons and Ikonen, 1997). Furthermore, the assembly of membrane proteins into rafts appears to be of significant importance during signal transduction. It will be a great challenge within the forthcoming decade to understand the mechanisms driving this assembly. In this context, the observation of the dis- and reassembly of bacteriorhodopsin into fully active purple membrane can be seen as a step toward studying the formation of functional assemblies by membrane proteins (Möller et al., 2000). In the future, more complex biological systems will be investigated and may deliver insights into the biogenesis of membrane proteins, into interactions between similar or different membrane proteins, and into the formation of supramolecular complexes of membrane proteins.

G. Single-Molecule Imaging and Force Spectroscopy

Force spectroscopy is a new field that has already provided exciting data on protein–protein interactions as well as on protein folding. These rapid developments demonstrate the power of these novel types of molecular mechanics measurements and also suggest that AFM techniques are still open to new ideas and developments. As indicated by a pioneering paper (Hinterdorfer *et al.,* 1996), the combination of force measurements and imaging will allow receptors to be detected and localized on cell surfaces. The further development of this technique has already allowed imaging and manipulating the substructures of individual proteins co-using the AFM stylus as both an imaging tool and as a "nonotweezer" (Fotiadis *et al.,* 1998). Recently, this co-usage of the AFM stylus has been improved to image single proteins and to tether selected substructures to the AFM tip, allowing the molecular mechanics between their supramolecular assemblies to be measured. To unambiguously correlate the force spectra recorded, the resulting vacancy of the protein removed was imaged (Müller, Baumeister *et al.,* 1999). In a further refinement of this technique, individual membrane proteins have been imaged, addressed, and unfolded, thereby revealing the individuality of their unfolding pathway (Oesterhelt *et al.,* 2000). In principle, these experiments are simple, and as soon as the sample is imaged at submolecular resolution, the force spectroscopy data can be recorded and correlated. In the future, such experiments will provide much more detailed insights into protein–protein interactions, into interactions between secondary structural elements of different kinds, and into factors that determine the structural and functional stability of membrane proteins (Haltia and Freire, 1995; White and Wimley, 1999).

Acknowledgments

This work was supported by the Swiss National Foundation for Scientific Research (NFP). We are grateful to Dr. Shirley Müller for the critical reading and discussing of the manuscript.

References

Abrahams, J. P., Leslie, A. G. W., Lutter, R., and Walker, J. E. (1994). Structure at 2.8 angstrom resolution of F_1-ATPase from bovine heart mitochondria. *Nature* **370,** 621–628.

Bauer, P.-J., Dencher, N. A., and Heyn, M. P. (1976). Evidence for chromophore-chromophore interactions in the purple membrane from reconstitution experiments of the chromophore-free membrane. *Biophys. Struct. Mech.* **2,** 79–92.

Baumeister, W., Barth, M., Hegerl, R., Guckenberger, R., Hahn, M., and Saxton, W. O. (1986). Three-dimensional structure of the regular surface layer (HPI layer) of *Deinococcus radiodurans. J. Mol. Biol.* **187,** 241–253.

Baumeister, W., and Hegerl, R. (1986). Can S-layers make bacterial connexons? *FEMS Microbiol. Lett.* **36,** 119–125.

Baumeister, W., Wildhaber, I., and Engelhardt, H. (1988). Bacterial surface proteins: Some structural, functional and evolutionary aspects. *Biophys. Chem.* **29,** 39–49.

Becher, B., and Cassim, J. Y. (1977). Effects of bleaching and regeneration on the purple membrane structure of *Halobacterium halobium. Biophys. J.* **19,** 285–297.

Belrhali, H., Nollert, P., Royant, A., Menzel, C., Rosenbusch, J. P., Landau, E. M., and Pebay-Peyroula, E. (1999). Protein, lipid and water organization in bacteriorhodopsin crystals: a molecular view of the purple membrane at 1.9 A resolution. *Struct. Fold. Des.* **7,** 909–917.

Bezanilla, M., Drake, B., Nudler, E., Kashlev, M., Hansma, P. K., and Hansma, H. G. (1994). Motion and enzymatic degradation of DNA in the atomic force microscope. *Biophys. J.* **67,** 2454–2459.

Bianchet, M. A., Hullihen, J., Pedersen, P. L., and Amzel, L. M. (1998). The 2.8-Å structure of rat liver F_1-ATPase: configuration of a critical intermediate in ATP synthesis/hydrolysis. *Proc. Natl. Acad. Sci. U.S.A.* **95,** 11,065–11,070.

Binnig, G., Quate, C. F., and Gerber, C. (1986). Atomic force microscope. *Phys. Rev. Lett.* **56,** 930–933.

Bretaudiere, J.-P., and Frank, J. (1986). Reconstitution of molecule images analysed by correspondence analysis: a tool for structural interpretation. *J. Microsc.* **144,** 1–14.

Brown, D. A., and London, E. (1998). Functions of lipid rafts in biological membranes. *Annu. Rev. Cell. Dev. Biol.* **14,** 111–136.

Butt, H.-J. (1991). Measuring electrostatic, van der Waals, and hydration forces in electrolyte solutions with an atomic force microscope. *Biophys. J.* **60,** 1438–1444.

Butt, H.-J. (1992). Measuring local surface charge densities in electrolyte solutions with a scanning force microscope. *Biophys. J.* **63,** 578–582.

Butt, H.-J., Jaschke, M., and Ducker, W. (1995). Measuring surface forces in aqueous solution with the atomic force microscope. *Bioelect. Bioenerg.* **38,** 191–201.

Carrion-Vazquez, M., Oberhauser, A. F., Fowler, S. B., Marszalek, P. E., Broedel, S. E., Clarke, J., and Fernandez, J. M. (1999). Mechanical and chemical unfolding of a single protein: a comparison. *Proc. Natl. Acad. Sci. U.S.A.* **96,** 3694–3699.

Cheung, C. L., Hafner, J. H., and Lieber, C. M. (2000). Carbon nanotube atomic force microscopy tips: direct growth by chemical vapor deposition and application to high-resolution imaging. *Proc. Natl. Acad. Sci. U.S.A.* **97,** 3809–3813.

Cowan, S. W., Schirmer, T., Rummel, G., Steiert, M., Ghosh, R., Pauptit, R. A., Jansonius, J. N., and Rosenbusch, J. P. (1992). Crystal structures explain functional properties of two *E. coli* porins. *Nature* **358,** 727–733.

Czajkowsky, D. M., Sheng, S., and Shao, Z. (1998). Staphylococcal α-Hemolysin can form hexamers in phospholipid bilayers. *J. Mol. Biol.* **276,** 325–330.

Dammer, U., Hegner, M., Anselmetti, D., Wagner, P., Dreier, M., Huber, W., and Güntherodt, H. J. (1996). Specific antigen/antibody interactions measured by force microscopy. *Biophys. J.* **70,** 2437–2441.

Dammer, U., Popescu, O., Wagner, P., Anselmetti, D., Güntherodt, H. J., and Misevic, G. N. (1995). Binding strength between cell adhesion proteoglycans measured by atomic force microscopy. *Science* **267,** 1173–1175.

Delcour, A. (1997). Function and modulation of bacterial porins: insights from electrophysiology. *FEMS Microbiol. Lett.* **151,** 115–123.

Delcour, A. H., Adler, J., Kung, C., and Martinac, B. (1992). Membrane-derived oligosaccharides (MDOs) promote closing of an *E. coli* channel. *FEBS Lett.* **304,** 216–220.

dela Vega, A. L., and Delcour, A. H. (1995). Cadaverine induces closing of *E. coli* porins. *EMBO J.* **14,** 6058–6065.

Dencher, N. A., Dresselhaus, D., Zaccai, G., and Büldt, G. (1989). Structural changes in bacteriorhodopsin during proton translocation revealed by neutron diffraction. *Proc. Natl. Acad. Sci. U.S.A.* **86,** 7876–7879.

Dmitriev, O. Y., Jones, P. C., and Fillingame, R. H. (1999). Structure of the subunit *c* oligomer in the F_1F_0 ATP synthase: Model derived from solution structure of the monomer and cross-linking in the native enzyme. *Proc. Natl. Acad. Sci. U.S.A.* **96,** 7785–7790.

Drake, B., Prater, C. B., Weisenhorn, A. L., Gould, S. A. C., Albrecht, T. R., Quate, C. F., Cannell, D. S., Hansma, H. G., and Hansma, P. K. (1989). Imaging crystals, polymers, and processes in water with the atomic force microscope. *Science* **243,** 1586–1588.

Dubochet, J., Adrian, M., Chang, J.-J., Homo, J.-C., Lepault, J., McDowall, A. W., and Schultz, P. (1988). Cryo-electron microscopy of vitrified specimens. *Quart. Rev. Biophys.* **21,** 129–228.

Dunlap, D. D., Maggi, A., Soria, M. R., and Monaco, L. (1997). Nanoscopic structure of DNA condensed for gene delivery. *Nucl. Acids Res.* **25,** 3095.

Edman, K., Nollert, P., Royant, A., Belrhali, H., Pebay-Peroula, E., Hajdu, J., Neutze, R., and Landau, E. M. (1999). High-resolution X-ray structure of an early intermediate in the bacteriorhodopsin photocycle. *Nature* **401,** 822–826.

Engel, A., Baumeister, W., and Saxton, W. (1982). Mass mapping of a protein complex with the scanning transmission electron microscope. *Proc. Natl. Acad. Sci. U.S.A.* **79,** 4050–4054.

Engel, A., Lyubchenko, Y., and Müller, D. J. (1999). Atomic force microscopy: A powerful tool to observe biomolecules at work. *Trends Cell Biol.* **9,** 77–80.

Essen, L.-O., Siegert, R., Lehmann, W. D., and Oesterhelt, D. (1998). Lipid patches in membrane protein oligomers: Crystal structure of the bacteriorhodopsin-lipid complex. *Proc. Natl. Acad. Sci. U.S.A.* **95,** 11,673–11,678.

Fisher, T. E., Oberhauser, A. F., Carrion-Vazquez, M., Marszalek, P. E., and Fernandez, J. M. (1999). The study of protein mechanics with the atomic force microscope. *Trends Biochem. Sci.* **24,** 379–384.

Florin, E.-L., Moy, V. T., and Gaub, H. E. (1994). Adhesion forces between individual ligand-receptor pairs. *Science* **264,** 415–417.

Fotiadis, D., Hasler, L., Müller, D. J., Stahlberg, H., Kistler, J., and Engel, A. (2000). Surface tongue-and-groove contours on lens MIP facilitate cell-to-cell adherence. *J. Mol. Biol.* **300,** 779–789.

Fotiadis, D., Müller, D. J., Tsiotis, G., Hasler, L., Tittmann, P., Mini, T., Jenö, P., Gross, H., and Engel, A. (1998). Surface analysis of the photosystem I complex by electron and atomic force microscopy. *J. Mol. Biol.* **283,** 83–94.

Frank, J., Bretaudiere, J.-P., Carazo, J.-M., Veschoor, A., and Wagenknecht, T. (1987). Classification of images of biomolecular assemblies: A study of ribosomes and ribosomal subunits of *Escheria coli. J. Microsc.* **150,** 99–115.

Fritz, M., Radmacher, M., Allersma, M. W., Cleveland, J. P., Stewart, R. J., Hansma, P. K., and Schmidt, C. F. (1995). Imaging microtubles in buffer solution using tapping mode atomic force microscopy. *SPIE* **2384,** 150–157.

Gouaux, J. E., Braha, O., Hobaugh, M. R., Song, L., Cheley, S., Shustak, C., and Bayley, H. (1994). Subunit stoichiometry of staphylococcal alpha-hemolysin in crystals and on membranes: a heptameric transmembrane pore. *Proc. Natl. Acad. Sci. U.S.A.* **91,** 12828–31.

Grigorieff, N., Ceska, T. A., Downing, K. H., Baldwin, J. M., and Henderson, R. (1996). Electron-crystallographic refinement of the structure of bacteriorhodopsin. *J. Mol. Biol.* **259,** 393–421.

Groth, G., and Walker, J. E. (1997). Model of the *c*-subunit oligomer in the membrane domain of F-ATPases. *FEBS Lett.* **410,** 117–123.

Gu, L. Q., Braha, O., Conlan, S., Cheley, S., and Bayley, H. (1999). Stochastic sensing of organic analytes by a pore-forming protein containing a molecular adapter [see comments]. *Nature* **398**(6729), 686–690.

Guthold, M., Zhu, X., Rivetti, C., Yang, G., Thomson, N. H., Kasas, S., Hansma, H. G., Smith, B., Hansma, P. K., and Bustamante, C. (1999). Direct observation of one-dimensional diffusion and transcription by *Escherichia coli* RNA polymerase. *Biophys. J.* **77,** 2284–2294.

Haltia, T., and Freire, E. (1995). Forces and factors that contribute to the structural stability of membrane proteins. *BBA-Bioenerg.* **1228,** 1–27.

Han, W., Lindsay, S. M., Dlakic, M., and Harrington, R. E. (1997). Kinked DNA. *Nature* **386,** 563.

Hansma, P. K., Cleveland, J. P., Radmacher, M., Walters, D. A., Hillner, P. E., Bezanilla, M., Fritz, M., Vie, D., Hansma, H. G., Prater, C. B., Massie, J., Fukunaga, L., Gurley, J., and Elings, V. (1994). Tapping mode atomic force microscopy in liquids. *Appl. Phys. Lett.* **64,** 1738–1740.

Hansma, H. G., Vesenka, J., Siegerist, C., Kelderman, G., Morrett, H., Sinsheimer, R. L., Elings, V., Bustamate, C., and Hansma, P. K. (1992). Reproducible imaging and dissection of plasmid DNA under liquid with the atomic force microscope. *Science* **256,** 1180–1184.

Heinz, W. F., and Hoh, J. H. (1999). Spatially resolved force spectroscopy of biological surfaces using the atomic force microscope. *Nanotechnology* **17,** 143–150.

Helmreich, E. J. M., and Hofmann, K.-P. (1996). Structure and function of proteins in G-protein coupled signal transfer. *Biochem. Biophys. Acta* **1286,** 285–322.

Heymann, J. B., Müller, D. J., Landau, E., Rosenbusch, J., Pebay-Peroulla, E., Büldt, G., and Engel, A. (1999). Charting the surfaces of purple membrane. *J. Struct. Biol.* **128,** 243–249.

Heymann, J. B., Pfeiffer, M., Hildebrandt, V., Fotiadis, D., de Groot, B., Kabak, R., Engel, A., Oesterhelt, D., and Müller, D. J. (2000). Conformations of the rhodopsin third cytoplasmic loop grafted onto bacteriorhodopsin. *Structure* **8,** 643–644.

Hinterdorfer, P., Baumgartner, W., Gruber, H. J., Schilcher, K., and Schindler, H. (1996). Detection and localization of individual antibody-antigen recognition events by atomic force microscopy. *Proc. Natl. Acad. Sci. U.S.A.* **93,** 3477–3481.

Hoh, J. H., Sosinsky, G. E., Revel, J.-P., and Hansma, P. K. (1993). Structure of the extracellular surface of the gap junction by atomic force microscopy. *Biophys. J.* **65,** 149–163.

Israelachvili, J. (1991). "Intermolecular, and Surface Forces," Second Edition edit, Academic Press Limited, London.

Jones, P. C., and Fillingame, R. H. (1998). Genetic fusions of subunit c in the F_o sector of H^+-transporting ATP synthase. *J. Biol. Chem.* **273,** 29,701–29,705.

Jones, P. C., Jiang, W., and Fillingame, R. H. (1998). Arrangement of the multicopy H^+-translocating subunit c in the membrane sector of the *Escherichia coli* F_1F_o ATP synthase. *J. Biol. Chem.* **273,** 17,178–17,185.

Karrasch, S., Dolder, M., Hoh, J., Schabert, F., Ramsden, J., and Engel, A. (1993). Covalent binding of biological samples to solid supports for scanning probe microscopy in buffer solution. *Biophys. J.* **65,** 2437–2446.

Karrasch, S., Hegerl, R., Hoh, J., Baumeister, W., and Engel, A. (1994). Atomic force microscopy produces faithful high-resolution images of protein surfaces in an aqueous environment. *Proc. Natl. Acad. Sci. U.S.A.* **91,** 836–838.

Kasas, S., Thomson, N. H., Smith, B. L., Hansma, H. G., Zhu, X., Guthold, M., Bustamante, C., Kool, E. T., Kashlev, M., and Hansma, P. K. (1997). Escherichia coli RNA ploymerase activity observed using atomic force microscopy. *Biochemistry* **36,** 461–468.

Kataoka, M., Kahn, T. W., Tsujiuchi, Y., Engelman, D. M., and Tokunaga, F. (1992). Bacteriorhodopsin reconstituted from two individual helices and the complementary five-helix fragment is photoactive. *Photochem. Photobiol.* **56,** 895–901.

Kato-Yamada, Y., Noji, H., Yasuda, R., Kinosita, K. J., and Yoshida, M. (1998). Direct observation of the rotation of epsilon subunit in F_1-ATPase. *J. Biol. Chem.* **273,** 19,375–19,377.

Kellenberger, E., Häner, M., and Wurtz, M. (1982). The wrapping phenomenon in airdried and negatively stained preparations. *Ultramicroscopy* **9,** 139–150.

Klebba, P. E., and Newton, S. M. (1998). Mechanisms of solute transport through outer membrane porins: burning down the house. *Curr. Opin. Microbiol.* **1,** 238–47.

Koch, M. H. J., Dencher, N. A., Oesterhelt, D., Plöhn, H.-J., Rapp, G., and Büldt, G. (1991). Time-resolved X-ray diffraction study of structural changes associated with the photocycle of bacteriorhodopsin. *EMBO J.* **10,** 521–526.

Lakey, J. H. (1987). Voltage gating in porin channels. *FEBS Lett.* **211,** 1–4.

Le Dain, A. C., Häse, C. C., Tommassen, J., and Martinac, B. (1996). Porins of *Escherichia coli:* Uniderectional gating by pressure. *EMBO J.* **15,** 3524–3528.

Lee, G. U., Chrisey, L. A., and Colton, R. J. (1994). Direct measurements of the forces between complementary strands of DNA. *Science* **266,** 771–773.

Lee, G. U., Kidwell, D. A., and Colton, R. J. (1994). Sensing discrete streptavidin-biotin interactions with atomic force microscopy. *Langmuir* **10,** 354–357.

Liao, M.-J., Huang, K.-S., and Khorana, H. G. (1984). Regeneration of native bacteriorhodopsin structure from fragments. *J. Biol. Chem.* **259,** 4200–4204.

Lieberman, K., Lewis, A., Fish, G., Shalom, S., Jovin, T. M., Schaper, A., and Cohen, S. R. (1994). Multifunctional, micropipette based force cantilevers for scanned probe microscopy. *Appl. Phys. Lett.* **65,** 648–650.

Luecke, H., Richter, H.-T., and Lanyi, J. K. (1998). Proton transfer pathways in bacteriorhodopsin at 2.3 Angstrom resolution. *Science* **280,** 1934–1937.

Luecke, H., Schobert, B., Richter, H.-T., Certailler, J.-P., and Lanyi, J. K. (1999a). Structural changes in bacteriorhodopsin during ion transport at 2 Angstrom resolution. *Science* **286**, 255–260.

Luecke, H., Schobert, B., Richter, H. T., Cartailler, J. P., and Lanyi, J. K. (1999b). Structure of bacteriorhodopsin at 1.55 A resolution. *J. Mol. Biol.* **291**, 899–911.

Lyubchenko, Y. L., and Shlyakhtenko, L. S. (1997). Direct visualization of supercoiled DNA *in situ* with atomic force microscopy. *Proc. Natl. Acad. Sci. U.S.A.* **94**, 496–501.

Malghani, M. S., Fang, Y., Cheley, S., Bayley, H., and Yang, J. (1999). Heptameric structures of two alpha-hemolysin mutants imaged with in situ atomic force microscopy. *Microsc. Res. Tech.* **44**, 353–6.

Marszalek, P. E., Oberhauser, A. F., Pang, Y. P., and Fernandez, J. M. (1998). Polysaccharide elasticity governed by chair-boat transitions of the glucopyranose ring. *Nature* **396**, 661–664.

Martin, J., Goldie, K. N., Engel, A., and Hartl, F. U. (1994). Topology of the morphological domains of the chaperonin GroEL visualized by immuno-electron microscopy. *Biol. Chem Hoppe-Seyler* **375**, 635–639.

Michel, H., Oesterhelt, D., and Henderson, R. (1980). Orthorhombic two-dimensional crystal form of purple membrane. *Proc. Natl. Acad. Sci. U.S.A.* **77**, 338–342.

Mitsuoka, K., Hirai, T., Murata, K., Miyazawa, A., Kidera, A., Kimura, Y., and Fujiyoshi, Y. (1999). The structure of bacteriorhodopsin at 3.0 A resolution based on electron crystallography: implication of the charge distribution. *J. Mol. Biol.* **286**, 861–882.

Möller, C., Allen, M., Elings, V., Engel, A., and Müller, D. J. (1999). Tapping mode atomic force microscopy produces faithful high-resolution images of protein surfaces. *Biophys. J.* **77**, 1050–1058.

Möller, C., Büldt, G., Dencher, N., Engel, A., and Müller, D. J. (2000). Reversible loss of crystallinity on photobleaching purple membrane in presence of hydroxylamine. *J. Mol. Biol.* **301**, 869–879.

Mou, J., Czajkowsky, D. M., Sheng, S., Ho, R., and Shao, Z. (1996). High resolution surface structure of *E. coli* GroES oligomer by atomic force microscopy. *FEBS Lett.* **381**, 161–164.

Mou, J. X., Yang, J., and Shao, Z. F. (1995). Atomic force microscopy of cholera toxin B-oligomers bound to bilayers of biologically relevant lipids. *J. Mol. Biol.* **248**, 507–512.

Moy, V. T., Florin, E.-L., and Gaub, H. E. (1994). Adhesive forces between ligand and receptor measured by AFM. *Colloids Surf.* **A93**, 343–348.

Müller, D. J., Amrein, M., and Engel, A. (1997). Adsorption of biological molecules to a solid support for scanning probe microscopy. *J. Struct. Biol.* **119**, 172–188.

Müller, D. J., Baumeister, W., and Engel, A. (1996). Conformational change of the hexagonally packed intermediate layer of *Deinococcus radiodurans* imaged by atomic force microscopy. *J. Bacteriol.* **178**, 3025–3030.

Müller, D. J., Baumeister, W., and Engel, A. (1999). Controlled unzipping of a bacterial surface layer with atomic force microscopy. *Proc. Natl. Acad. Sci. U.S.A.* **96**, 13,170–13,174.

Müller, D. J., Büldt, G., and Engel, A. (1995). Force-induced conformational change of bacteriorhodopsin. *J. Mol. Biol.* **249**, 239–243.

Müller, D. J., and Engel, A. (1997). The height of biomolecules measured with the atomic force microscope depends on electrostatic interactions. *Biophys. J.* **73**, 1633–1644.

Müller, D. J., and Engel, A. (1999). pH and voltage induced structural changes of porin OmpF explain channel closure. *J. Mol. Biol.* **285**, 1347–1351.

Müller, D. J., Engel, A., Carrascosa, J., and Veléz, M. (1997). The bacteriophage ø29 head-tail connector imaged at high resolution with atomic force microscopy in buffer solution. *EMBO J.* **16**, 101–107.

Müller, D. J., Fotiadis, D., and Engel, A. (1998). Mapping flexible protein domains at subnanometer resolution with the AFM. *FEBS Lett.* **430**, 105–111.

Müller, D. J., Fotiadis, D., Scheuring, S., Müller, S. A., and Engel, A. (1999). Electrostatically balanced subnanometer imaging of biological specimens by atomic force microscopy. *Biophys. J.* **76**, 1101–1111.

Müller, D. J., Heymann, J. B., Oesterhelt, F., Möller, C., Gaub, H., Büldt, G., and Engel, A. (2000). Atomic force microscopy on native purple membrane. *Biochim. Biophys. Acta* **1460**, 27–38.

Müller, D. J., Sass, H.-J., Müller, S., Büldt, G., and Engel, A. (1999). Surface structures of native bacteriorhodopsin depend on the molecular packing arrangement in the membrane. *J. Mol. Biol.* **285**, 1903–1909.

Müller, D. J., Schabert, F. A., Büldt, G., and Engel, A. (1995). Imaging purple membranes in aqueous solutions at subnanometer resolution by atomic force microscopy. *Biophys. J.* **68**, 1681–1686.

Müller, D. J., Schoenenberger, C. A., Büldt, G., and Engel, A. (1996). Immuno-atomic force microscopy of purple membrane. *Biophys. J.* **70,** 1796–1802.

Müller, D. J., Schoenenberger, C.-A., Schabert, F., and Engel, A. (1997). Structural changes of native membrane proteins monitored at subnanometer resolution with the atomic force microscope. *J. Struct. Biol.* **119,** 149–157.

Nikaido, H. (1989). *Antimicrob. Agents Chemother.* **33,** 1831–1836.

Nikaido, H., and Saier, M. H. (1992). Transport proteins in bacteria: Common themes in their design. *Science* **258,** 936–942.

Nikaido, H., and Vaara, M. (1985). Molecular basis of bacterial outer membrane permeability. *Microbiol. Rev.* **49,** 1–32.

Noji, H., Yasuda, R., Yoshida, M., and Kinosita, K. (1997). Direct observation of the rotation of F_1-ATPase. *Nature* **386,** 299–302.

Oberhauser, A. F., Marszalek, P. E., Erickson, H. P., and Fernandez, J. M. (1998). The molecular elasticity of the extracellular matrix protein tenascin. *Nature* **393,** 181–185.

Oesterhelt, D., Schuhmann, L., and Gruber, H. (1974). Light-dependent reaction of bacteriorhodopsin with hydroxylamine in cell suspensions of *Halobacterium halobium:* Demonstration of an apo-membrane. *FEBS Lett.* **44,** 257–261.

Oesterhelt, F., Oesterhelt, D., Pfeiffer, M., Engel, A., Gaub, H. E., and Müller, D. J. (2000). Unfolding pathways of individual bacteriorhodopsins. *Science* **288,** 143–146.

Pebay-Peyroula, E., Rummel, G., Rosenbusch, J. P., and Landau, E. M. (1997). X-ray structure of bacteriorhodopsin at 2.5 angstroms from microcrystals grown in lipidic cubic phases. *Science* **277,** 1676–1681.

Peters, J., Peters, M., Lottspeich, F., Schäfer, W., and Baumeister, W. (1987). Nucleotide sequence of the gene encoding the *Deinococcus radiodurans* surface protein, derived amino acid sequence, and complementary protein chemical studies. *J. Bacteriol.* **169,** 5216–5223.

Pfeiffer, M., Rink, T., Gerwert, K., Oesterhelt, D., and Steinhoff, H.-J. (1999). Site-directed spin-labelling reveals the orientation of the amino acid side-chains in the E-F loop of bacteriorhodopsin. *J. Mol. Biol.* **287,** 163–171.

Phale, R. S., Schirmer, T., Prilipov, A., Lou, K.-L., Hardmeyer, A., and Rosenbusch, J. (1997). Voltage gating of *Escherichia coli* porin channels: Role of the constriction loop. *Proc. Natl. Acad. Sci. U.S.A.* **94,** 6741–6745.

Pralle, A., Keller, P., Florin, E. L., Simons, K., and Horber, J. K. (2000). Sphingolipid-cholesterol rafts diffuse as small entities in the plasma membrane of mammalian cells. *J. Cell. Biol.* **148,** 997–1008.

Putman, C. A. J., Vanderwerf, K. O., Degrooth, B. G., Vanhulst, N. F., and Greve, J. (1994). Tapping mode atomic force microscopy in liquid. *Appl. Phys. Lett.* **64,** 2454–2456.

Rachel, R., Jakubowski, U., Tietz, H., Hegerl, R., and Baumeister, W. (1986). Projected structure of the surface protein of *Deinococcus radiodurans* determined to 8 Å resolution by cryomicroscopy. *Ultramicroscopy* **20,** 305–316.

Radmacher, M., Fritz, M., and Hansma, P. K. (1995). Imaging soft samples with the atomic force microscope: Gelatin in water and propanol. *Biophys. J.* **69,** 264–270.

Rastogi, V. K., and Girvin, M. E. (1999). Structural changes linked to proton translocation by subunit c of the ATP synthase. *Nature* **402,** 263–268.

Rief, M., Clausen-Schaumann, H., and Gaub, H. (1999). Sequence-dependent mechanics of single DNA molecules. *Nat. Struct. Biol.* **6,** 346–349.

Rief, M., Gautel, M., Oesterhelt, F., Fernandez, J. M., and Gaub, H. E. (1997). Reversible unfolding of individual titin immunoglobulin domains by AFM. *Science* **276,** 1109–1112.

Rief, M., Oesterhelt, F., Heymann, B., and Gaub, H. E. (1997). Single molecule force spectroscopy on polysaccharides by AFM. *Science* **275,** 1295–1298.

Rotsch, C., and Radmacher, M. (1997). Mapping local electrostatic forces with the atomic force microscope. *Langmuir* **13,** 2825–2832.

Sabbert, D., Engelbrecht, S., and Junge, W. (1996). Intersubunit rotation in active F-ATPase. *Nature* **381,** 623–625.

Sato, H., Takeda, K., Tani, K., Hino, T., Okada, T., Nakasako, M., Kamiya, N., and Kouyama, T. (1999). Specific lipid-protein interactions in a novel honeycomb lattice structure of bacteriorhodopsin. *Acta Crystallogr. D Biol. Crystallogr.* **55,** 1251–1256.

Schabert, F. A., Henn, C., and Engel, A. (1995). Native *Escherichia coli* OmpF porin surfaces probed by atomic force microscopy. *Science* **268,** 92–94.

Schäffer, T. E., Cleveland, J. P., Ohnesorge, F., Walters, D. A., and Hansma, P. K. (1996). Studies of vibrating atomic force microscope cantilevers in liquid. *J. Appl. Phys.* **80,** 3622–3627.

Scheuring, S., Müller, D. J., Ringler, P., Heymann, J. B., and Engel, A. (1999). Imaging streptavidin 2D crystals on biotinylated lipid monolayers at high resolution with the atomic force microscopy. *J. Microsc.* **193,** 28–35.

Scheuring, S., Ringler, P., Borgina, M., Stahlberg, H., Müller, D. J., Agre, P., and Engel, A. (1999). High resolution topographs of the *Escherichia coli* waterchannel aquaporin Z. *EMBO J.* **18,** 4981–4987.

Schindler, H., and Rosenbusch, J. P. (1978). Matrix protein from *Escherichia coli* outer membranes forms voltage-controlled channels in lipid bilayers. *Proc. Natl. Acad. Sci. U.S.A.* **75,** 3751–3755.

Schindler, H., and Rosenbusch, J. P. (1981). Matrix protein in planar membranes: Clusters of channels in a native environment and their functional assembly. *Proc. Natl. Acad. Sci. U.S.A.* **78,** 2302–2306.

Schirmer, T. (1998). General and specific porins from bacterial outer membranes. *J. Stuct. Biol.* **121,** 101–109.

Schirmer, T., and Phale, P. S. (1999). Brownian dynamics simulation of ion flow through porin channels. *J. Mol. Biol.* **294,** 1159–1167.

Schulz, G. (1993). Bacterial porins: structure and function. *Curr. Opin. Cell Biol.* **5,** 701–707.

Schurmann, G., Noell, W., Staufer, U., and de Rooij, N. F. (2000). Microfabrication of a combined AFM-SNOM sensor. *Ultramicroscopy* **82,** 33–38.

Seelert, H., Poetsch, A., Dencher, N. A., Engel, A., Stahlberg, H., and Müller, D. J. (2000). Proton powered turbine of a plant motor. *Nature* **405,** 418–419.

Shao, Z., Mou, J., Czajkowsky, D. M., Yang, J., and Yuan, J.-Y. (1996). Biological atomic force microscopy: what is achieved and what is needed. *Adv. Phys.* **45,** 1–86.

Shlyakhtenko, L. S., Potaman, V. N., Sinden, R. R., and Lyubchenko, Y. L. (1998). Structure and dynamics of supercoil-stabilized DNA cruciforms. *J. Mol. Biol.* **280,** 61–72.

Sigrist, H., Wenger, R. H., Kislig, E., and Wüthrich, M. (1988). Refolding of bacteriorhodopsin. *Eur. J. Biochem.* **177,** 125–133.

Simons, K., and Ikonen, E. (1997). Functional rafts in cell membranes. *Nature* **387,** 569–572.

Sleytr, U. B. (1997). Basic and applied S-layer research: an overview. *FEMS Microbiol. Rev.* **20,** 5–12.

Sleytr, U. B., Messner, P., Pum, D., and Sára, M. (1993). Crystalline bacterial cell surface layers: general principles and application potential. *J. Appl. Bacteriol. Symp. Suppl.* **74,** 21S–32S.

Smith, S. B., Cui, Y., and Bustamante, C. (1996). Overstretching B-DNA: The elastic response of individual double-stranded and single-stranded DNA molecules. *Science* **271,** 795–798.

Song, L., Hobaugh, M. R., Shustak, C., Cheley, S., Bayley, H., and Gouaux, J. E. (1996). Structure of staphylococcal alpha-hemolysin, a heptameric transmembrane pore. *Science* **274,** 1859–1866.

Stock, D., Leslie, A. G., and Walker, J. E. (1999). Molecular architecture of the rotary motor in ATP synthase. *Science* **286,** 1700–1705.

Subramaniam, S., Lindahl, M., Bullough, P., Faruqi, A. R., Tittor, J., Oesterhelt, D., Brown, L., Lanyi, J., and Henderson, R. (1999). Protein conformational changes in the bacteriorhodopsin photocycle. *J. Mol. Biol.* **287,** 145–161.

Thompson, B. G., Murray, R. G. E., and Boyce, J. F. (1982). The association of the surface array and the outer membrane of *Deinococcus radiodurans*. *Can. J. Microbiol.* **28,** 1081–1088.

Todt, J. C., Rocque, W. J., and McGroarty, E. J. (1992). Effects of pH on bacterial porin function. *Biochemistry* **31,** 10,471–10,478.

van Heel, M. (1984). Multivariate statistical classification of noise images (randomly orientated biological macromolecules). *Ultramicroscopy* **13,** 165–184.

Viani, M. B., Schäfer, T. E., Chand, A., Rief, M., Gaub, H., and Hansma, P. K. (1999). Small cantilevers for force spectroscopy of single molecules. *J. Appl. Phys.* **86,** 2258–2262.

Walz, T., Tittmann, P., Fuchs, K. H., Müller, D. J., Smith, B. L., Agre, P., Gross, H., and Engel, A. (1996). Surface topographies at subnanometer-resolution reveal asymmetry and sidedness of aquaporin-1. *J. Mol. Biol.* **264,** 907–918.

Weihs, T. P., Nawaz, Z., Jarvis, S. P., and Pethica, J. B. (1991). Limits of imaging resolution for atomic force microscopy of molecules. *Apl. Phys. Lett.* **59,** 3536–3538.

Weiss, M. S., Abele, U., Weckesser, J., Welte, W., Schlitz, E., and Schulz, G. E. (1991). Molecular architecture and electrostatic properties of a bacterial porin. *Science* **254,** 1627–1630.

White, S. H., and Wimley, W. C. (1999). Membrane protein folding and stability: Physical principles. *Annu. Rev. Biophys. Biomol. Struct.* **28,** 319–365.

Wilkens, S., Dahlquist, F. W., McIntosh, L. P., Donaldson, L. W., and Capaldi, R. A. (1995). Structural features of the epsilon subunit of the *Escherichia coli* ATP synthase determined by NMR spectroscopy. *Nat. Struct. Biol.* **2,** 961–967.

Wilkens, S., Dunn, S. D., Chandler, J., Dahlquist, F. W., and Capaldi, R. A. (1997). Solution structure of the N-terminal domain of the delta subunit of the *E. coli* ATPsynthase. *Nat. Struct. Biol.* **4,** 198–201.

Yang, J., Tamm, L. K., Tillack, T. W., and Shao, Z. (1993). New approach for atomic force microscopy of membrane proteins. *J. Mol. Biol.* **229,** 286–290.

Zhong, Q., Inniss, D., Kjoller, K., and Elings, V. B. (1993). Fractured polymer/silica fiber surface studied by tapping mode atomic force microscopy. *Surf. Sci. Lett.* **290,** L688–692.

CHAPTER 14

Single-Molecule Force Measurements

Aileen Chen and Vincent T. Moy

Department of Physiology and Biophysics
University of Miami School of Medicine
Miami, Florida 33136

I. Introduction

In the past several years the atomic force microscope (AFM) has emerged as a powerful tool for measuring the dynamic strength of intermolecular bonds (Moy *et al.*, 1994; Lee, Chrisey *et al.*, 1994; Florin *et al.*, 1994; Lee, Kidwell *et al.*, 1994). The unbinding properties of various ligand–receptor systems including avidin/biotin, antibody/antigen, and *p*-selectin/carbohydrate pairs (Merkel *et al.*, 1999; Hinterdorfer *et al.*, 1996; Fritz *et al.*, 1998) have been characterized by force spectroscopy. These experiments are used to resolve interactions down to a single ligand–receptor pair under well-defined conditions. AFM has also been applied to molecules under more native environments. Here, the binding strength of ligand–receptor interactions has been measured between AFM tips functionalized with specific ligand and receptors on the surface of living cells. In addition to revealing the adhesive forces of the individual receptor–ligand bond (Gad *et al.*, 1997; Lehenkari and Horton, 1999), changes in receptor group dynamics could also be monitored (Chen and Moy, 2000). This review provides basic methodology for acquiring ligand–receptor force measurements, followed by a discussion on some recent applications and findings from these techniques.

II. Experimental Design

A. AFM Schematics

The basic mechanics behind how AFM force measurements are acquired is relatively simple. In brief, a cantilever tip is coated with the ligand or receptor of interest and is then brought into contact with the surface to which its binding partner is attached. Once contact is made, the flexible cantilever bends away from the surface. The deflection of the cantilever is detected optically by changes in the path of a laser beam reflected off its upper surface to a position-sensing detector. Upon retraction, adhesion events are registered by the downward deflection of the cantilever. Once the last bond ruptures, tension on the cantilever is relieved and deflection returns to the resting level. Figure 1A shows a typical force scan following the described sequence of events during approach, contact, and retraction of the cantilever. For an excellent review on the interpretation and potential artifacts of AFM force curves see Heinz and Hoh (1999).

Modification of the typical AFM design used for imaging can improve the quality of the signal acquired in the AFM force measurements. The schematics of the AFM constructed for force measurements in our lab is shown in Fig. 1B. To achieve the sensitivity required for force scans, the mechanisms for lateral and vertical scans were decoupled. This eliminated potential mechanical and electrical noise from the tube piezo-electric translator used in most commercial AFMs. The sample sat on an x-y stage that could be moved with respect to a cantilever mounted on a separate stacked piezo translator (Physik Instrumente, Model P-821.10). Visualization of the sample was through a $20\times$ microscope objective coupled with a CCD camera and positioned directly beneath the sample chamber. To vertically approach and retract the cantilever to and from the

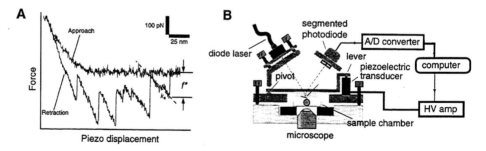

Fig. 1 (A) Force versus displacement curves of the interaction between a streptavidin-functionalized tip and a biotinylated agarose bead. The measurement recorded the force on the AFM cantilever on approach and retraction of the cantilever from the agarose bead. f^* is the rupture force. k_s is the slope of the force versus displacement curve. (B) Schematic representation of the atomic force apparatus used to measure the strength of ligand–receptor bonds.

sample we used a piezo translator. Since piezo translators have an inherent hysteresis that can lead to positioning error, it was necessary to have a position sensor in the translator that would allow for a corrective electronic feedback loop (Physik Instrumente, Model E-810.10).

The deflection of the cantilever was monitored by focusing a laser beam from a 3-mW diode laser (Oz Optics; em. 680 nm) on the upper surface of the cantilever. Changes in the reflection path of the laser were monitored by a two-segment photo-diode (UDT Sensors; Model SPOT-2D). The differential signal from the photo-diodes was then digitized by a data acquisition system equipped with an 18-bit optically isolated analog-to-digital converter (Instrutech Corp., Port Washington, NY). Control of the piezo-electric translator and timing of the measurements were through custom software. Further reduction of mechanical vibration and temperature fluctuation was achieved by suspending the entire apparatus by bungee cords inside of a large evacuated refrigerator.

B. Cantilever Preparation

1. Selecting the Cantilever

Cantilever selection is based on a compromise between maximizing force measurement sensitivity and minimizing thermal noise. In choosing an appropriate cantilever for experiments it should be noted that cantilevers with lower spring constants have higher sensitivity. However, thermal fluctuation of more sensitive tips can degrade the signal-to-noise ratio. For ligand–receptor force measurements we use triangle-shaped unsharpened gold-coated silicon–nitride cantilever tips that have spring constants ranging from 10 to 50 mN/m.

2. Functionalizing the Tip

To acquire direct force measurements of ligand–receptor pairs, it is first necessary to immobilize the ligand on the AFM tip. The techniques commonly used involve either passive chemiadsorption (Moy et al., 1994; Lehenkari and Horton, 1999) or covalent coupling of the ligand to the tip via an extended linker (Hinterdorfer et al., 1996). The linker between the tip and the ligand lends greater mobility and access to receptors on the surface being probed. The following outlines a method for functionalizing tips with the biotin–avidin system as a general linkage for biotinylated ligand (Fig. 2). There are several advantages to this method including the fact the avidin/biotin system has been well characterized and is of high affinity, and the bottom layer of biotin–BSA may help to mask any electrical charges on the cantilever tip that could lead to nonspecific binding.

3. Materials: AFM Cantilevers (MLCT-AUHW; Thermomicroscopes, Sunnyvale, CA), Biotin–BSA (A-6043; Sigma, St.Louis, MO), Neutravidin (cat.#: 31000; Pierce, Rockford, IL)

1. Soak cantilever for 5 min in acetone and then UV irradiate for 15 min.

2. Incubate cantilever in a 50-μl drop of biotin–BSA (0.5 mg/ml in 0.1 M sodium bicarbonate, pH 8.3) overnight at 37°C in a humidified incubator.

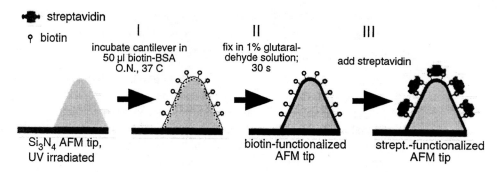

Fig. 2 Schematics for the functionalization of AFM tips with streptavidin.

3. Wash cantilever three times in PBS (pH 7.4) to remove unbound protein. (Note: At this point cantilevers can be stored in PBS at 4°C for up to a week.)

4. Incubate cantilever in a 50-μl drop of neutravidin (0.5 mg/ml in 0.01 M PBS, pH 7.4) for 10 min at room temperature.

5. Wash cantilever three times in PBS.

6. Incubate cantilever in a 50-μl drop of biotinylated ligand (e.g., biotinylated con-canavalin A; 0.5 mg/ml in PBS, pH 7.4) for 10 min at room temperature.

7. Wash cantilever three times in PBS before use.

Note:

1. It is important that biotin–BSA adsorption takes place at pH 8.3 or higher, as the basic conditions seem to facilitate BSA adsorption to the cantilever.

2. Kits for biotinylating the ligand of interest are readily available from Pierce Chemical Company.

4. Calibrating the Tip

To translate the deflection of the cantilever, x, to units of force, F, it is necessary to determine the spring constant of the cantilever, k (i.e., $F = kx$). Presently, there are several techniques for calibrating tips, including theoretical methods that provide an approximation of k (Sader, 1995). Theory, however, does not take into account the effect of coating tips and/or manufacturer variability from cantilever to cantilever. Thus, it is necessary to determine k using empirical methods. Such methods involve the measurement of cantilever bend with application of a constant known force (Senden and Ducker, 1994) or measurements of cantilever's resonant frequency (Hutter and Bechhoefer, 1993). The following is a brief overview, based on Hutter and Bechhoefer (1993), of the method we use for calibrating cantilevers.

The cantilever tip can be treated as a simple harmonic oscillator whose power spectrum of thermal fluctuation can be used to derive the spring constant. In brief, the cantilever is raised several microns from the surface and its natural frequency of vibration (resonant frequency) is monitored for 2–3 s. Since each vibration mode of the cantilever receives the thermal energy commensurated to one degree of freedom, $\frac{1}{2}k_BT$, the measured variance of the deflection $\langle x^2 \rangle$ can be used to calculate the spring constant (i.e., $\frac{1}{2}k_BT = \frac{1}{2}C\langle x^2 \rangle$, where k_B and T are Boltzmann's constant and temperature, respectively). To separate deflections belonging to the basic (and predominant) mode of vibration from other deflections or noise in the recording system, the power spectral density of the temperature-induced deflection is determined, and only the spectral component corresponding to the basal mode of vibration is used to estimate the spring constant. Using this approach, the spring constants of cantilevers can be calibrated in either air or in solution.

C. Sample Preparation

A prerequisite for any AFM measurement is the immobilization of the sample. Numerous methods exist for immobilizing ligand to a variety of different substrates (Hermanson *et al.*, 1992). Some commonly used ligands for affinity chromatography can be found already attached to agarose beads (e.g., biotin or D-mannose).

In choosing a substrate for carrying out the measurements, it should be noted that on receptors attached to elastic elements there is an increased probability for receptor–ligand interactions compared to receptors bound directly to a flat surface. Since an elastic substrate (e.g., agarose beads) will conform to the shape of the cantilever tip, there is greater surface area contact and hence a higher probability for receptor–ligand interactions (Moy *et al.*, 1994). The probability of binding can also be enhanced by attaching receptors to tethers that allow for a wider range of lateral motion and, therefore, more encounters between ligand and receptor (Hinterdorfer *et al.*, 1996). In addition, tethers can provide latitude for proper reorientation of the molecule during stretching so that the external force being applied to the molecule is perpendicular to the surface. Some of the common substrates that have been used for AFM force measurements include derivatized agarose beads, dextran-coated surfaces (Rief *et al.*, 1997), and cells (Lehenkari and Horton, 1999). The following is the protocol that we use for the preparation of biotinylated agarose beads for avidin–biotin rupture force measurements.

1. Materials: 35-mm Plastic Petri Dishes, Neutravidin (Sigma), PBS (pH 7.4), Biotinylated Agarose Beads (Sigma)

    1. Prepare neutravidin-coated dishes for immobilizing biotinylated agarose beads.

    (a) Place a 100-μl drop of neutravidin (0.05 mg/ml in sodium bicarbonate, pH 9.6) on the bottom of a 35-mm plastic Petri dish.

    (b) Incubate overnight at 37°C in a humidified incubator.

    (c) Rinse the dish three times in PBS just before adding beads.

    2. Add 100 μl of biotinylated agarose beads to 1.5 ml PBS.

    3. Centrifuge beads at 10,000g for 10 s and remove supernatant.

4. Repeat wash twice.

5. After removing supernatant from last wash, resuspend beads in 0.01% BSA in PBS.

6. Add 100 μl of washed beads to neutravidin-coated plate. Beads should adhere to the dish almost immediately.

III. Applications

A. Probing Molecular Landscapes of Streptavidin–Biotin Unbinding

Direct force measurements of the unbinding strength of single streptavidin/biotin pairs opened the way for examining the molecular determinants of ligand–receptor unbinding (Merkel *et al.*, 1999). As predicted by the Bell Model, the rupture force of the individual streptavidin–biotin bonds increased with increasing loading rate (Bell, 1978; Evans and Ritchie, 1997). Figure 3 shows the dynamic response of streptavidin and a streptavidin mutant, W120F, in which the tryptophan residue at position 120 was replaced by a phenylalanine. As shown, the mutation altered the force spectrum of the streptavidin–avidin interaction. The analysis of these force spectra provides a direct approach for probing the landscape of the streptavidin–biotin unbinding (Yuan *et al.*, 2000).

The streptavidin–biotin force measurements were carried out as described in earlier sections with a streptavidin-functionalized tip and an agarose bead. To ensure that single bonds were being measured, the frequency of adhesion events was reduced to 30% by restricting indentation depth and/or by adding excess free biotin. By simple statistical analysis based on the Poisson distribution, one is thus ensured that 80% of the measured events are due to the rupture of single-molecule pairs (Merkel *et al.*, 1999). The loading rate of the measurement is dependent on both the elasticity of the system and the speed

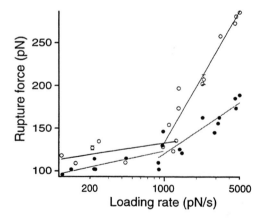

Fig. 3 Loading rate dependence of the rupture force in the unbinding of the streptavidin–biotin (o) and W120F-biotin (•). The force measurements revealed two loading regimes in the unbinding of the complexes. Both regimes in force spectra were fitted to the Bell model (Yuan *et al.*, 2000).

at which the cantilever is retracted. The loading rate was varied by pulling the molecules apart at different cantilever retraction rates. The elasticity of the system was determined by measuring the slope of the retract trace.

B. Probing Adhesion Receptors on Cells

We applied AFM toward measuring ligand–receptor unbinding on the surface of living and fixed cells to examine the effect of receptor crosslinking on receptor/ligand unbinding strength (Chen and Moy, 2000). For these experiments, the cantilever tip was functionalized with biotinylated concanavalin A (C2272, Sigma) using the methods outlined earlier. Measurements were carried out at room temperature in glucose-free RPMI supplemented with 0.01% BSA and 0.01 mM $MnCl_2$. Glucose was eliminated from the culture medium to prevent potential competitive binding with the Con A-functionalized tip and Con A receptors on the cell. $MnCl_2$ was a source of Mn^{2+}, a necessary cofactor for Con A binding. BSA was added not only to reduce nonspecific binding but also to provide a permissive environment for adhesion of cultured NIH-3T3 fibroblast to the bottom of an uncoated plastic tissue culture dish. Measurements were carried out on both unfixed and lightly fixed cells to determine if crosslinking of Con A receptors would have an effect on receptor unbinding strength. A minimal applied force of 250 pN was used in these measurements and the scan speed was maintained at 1 μm/s.

Compared to measurements on agarose beads, unfixed cells had much longer regions of stretch before final separation between the tip and membrane (compare Fig. 1A and 4A). Typical distances spanned 500 nm. Thus, receptors seemed to be anchored to cell tethers. Rupture force measurements revealed a stronger rupture force for chemically

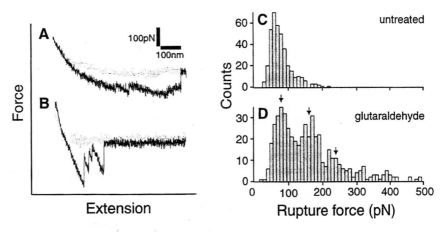

Fig. 4 Force versus extension curves acquired from Con A-functionalized AFM tips interacting with Con A receptors on the surface of NIH-3T3 cells that were (A) not fixed and (B) fixed with glutaraldehyde. Histograms of rupture force between Con A-functionalized AFM tips and Con A receptors on (C) untreated cells and (D) glutaraldehyde-fixed cells. Arrows in (D) indicate quantized peaks at 80, 160, and 240 pN following fixation of cells in glutaraldehyde.

fixed cells (173 ± 6.1 pN) compared to unfixed cells (86 ± 2.6 pN) (Figs. 4C and 4D). Moreover, differences in cell compliance were readily apparent from the slope of the retract trace as the tip pulled on the surface of the cell. Force histograms revealed multiple quantal peaks that were absent in the unfixed cell histograms (Fig. 4D), suggesting that much of the increase in rupture force was due to a shift toward cooperative binding of cells. In addition a shift in the first peak indicated that changes in loading rate resulting from changes in cell elasticity could also lend to the increase in rupture force following fixation.

Acknowledgments

This work was supported by grants from the American Cancer Society and the NIH (1 R29 GM55611-01) to VTM.

References

Bell, G. I. (1978). Models for the specific adhesion of cells to cells. *Science* **200**, 618–627.

Chen, A., and Moy, V. T. (2000). Cross-linking of cell surface receptors enhances cooperativity of molecular adhesion. *Biophys J.* **78**, 2814–2820.

Evans, E., and Ritchie, K. (1997). Dynamic strength of molecular adhesion bonds. *Biophys. J.* **72**, 1541–1555.

Florin, E. L., Moy, V. T., and Gaub, H. E. (1994). Adhesion forces between individual ligand-receptor pairs. *Science* **264**, 415–417.

Fritz, J., Katopodis, A. G., Kolbinger, F., and Anselmetti, D. (1998). Force-mediated kinetics of single P-selectin/ligand complexes observed by atomic force microscopy. *Proc. Natl. Acad. Sci. U.S.A.* **95**, 12,283–12,288.

Gad, M., Itoh, A., and Ikai, A. (1997). Mapping cell wall polysaccharides of living microbial cells using atomic force microscopy. *Cell Biol. Int.* **21**, 697–706.

Heinz, W. F., and Hoh, J. H. (1999). Spatially resolved force spectroscopy of biological surfaces using the atomic force microscope. *Trends Biotechnol.* **17**, 143–150.

Hermanson, G. T., Mallia, A. K., and Smith, P. K. (1992). "Immobilized Affinity Ligand Techniques." Academic Press, San Diego, CA.

Hinterdorfer, P., Baumgartner, W., Gruber, H. J., Schilcher, K., and Schindler, H. (1996). Detcetion and localization of individual antibody-antigen recognition events by atomic force microscopy. *Proc. Natl. Acad. Sci. U.S.A.* **93**, 3477–3481.

Hutter, J. L., and Bechhoefer, J. (1993). Calibration of atomic-force microscope tips. *Rev. Sci. Instrum.* **64**, 1868–1873.

Lee, G. U., Chrisey, L. A., and Colton, R. J. (1994). Direct measurement of the forces between complementary strands of DNA. *Science* **266**, 771–773.

Lee, G. U., Kidwell, D. A., and Colton, R. J. (1994). Sensing discrete streptavidin-biotin interactions with AFM. *Langmuir* **10**, 354–361.

Lehenkari, P. P., and Horton, M. A. (1999). Single integrin molecule adhesion forces in intact cells measured by atomic force microscopy. *Biochem. Biophys. Res. Commun.* **259**, 645–650.

Merkel, R., Nassoy, P., Leung, A., Ritchie, K., and Evans, E. (1999). Energy landscapes of receptor-ligand bonds explored with dynamic force spectroscopy [see comments]. *Nature* **397**, 50–53.

Moy, V. T., Florin, E.-L., and Gaub, H. E. (1994). Adhesive forces between ligand and receptor measured by AFM. *Colloids Surf.* **93**, 343–348.

Rief, M., Oesterhelt, F., Heymann, B., and Gaub, H. E. (1997). Single molecule force spectroscopy on polysaccharides by atomic force microscopy. *Science* **275**, 1295–1297.

Sader, J. E. (1995). Parallel beam approximation for V-shaped atomic force micrscope cantilevers. *Rev. Sci. Instrum.* **66,** 4583–4587.

Senden, T. J., and Ducker, W. A. (1994). Experimental determination of spring constants in atomic force microscopy. *Langmuir* **10,** 1003–1004.

Yuan, C., Chen, A., Kolb, P., and Moy, V. T. (2000). Energy landscape of the streptavidin-biotin complexes measured by atomic force microscopy. *Biochemistry,* **39,** 10219–10223.

CHAPTER 15

Forced Unfolding of Single Proteins

S. M. Altmann and P.-F. Lenne

European Molecular Biology Laboratory
Cell Biology and Biophysics Programme
Meyerhofstrasse 1
69117 Heidelberg, Germany

METHODS IN CELL BIOLOGY, VOL. 68
Copyright 2002, Elsevier Science (USA). All rights reserved.
0091-679X/02 $35.00

I. Introduction

The present chapter is focused on forced unfolding of proteins by atomic force microscopy (AFM). Protein folding remains one of the most fascinating mechanisms of biology. Different approaches can be used to understand this complex mechanism. During the last years, exciting advances have been made trough new detailed experimental and theoretical studies [for a review, see Brockwell *et al.* (2000)]. Forced unfolding is among the new experimental techniques promising new insights into the energy landscapes of protein folding processes.

AFM provides experimenters with the means to manipulate single molecules under physiological conditions. This powerful new tool can produce the forces necessary either to rupture ligand–receptor bonds (Florin *et al.*, 1994) or to stretch DNA. More recently, the AFM has been applied to unfold proteins (Rief *et al.*, 1997). For a review, see Fisher *et al.* (1999).

The unfolding of proteins by applying a force to single proteins attached between a surface and an AFM tip complements more classical techniques using either temperature or chemicals as denaturants. This approach provides single-molecule information that has not been available previously. It is of particular interest for proteins that are under mechanical stress in living cells as, for example, the muscle protein titin or the cytoskeleton protein spectrin. In this chapter we will present the method of forced unfolding and illustrate its use to probe the mechanical properties of a single-spectrin domain.

A. General Scheme

The general scheme for forced unfolding of protein resembles a "fishing" experience. Proteins attached on a surface are picked up with the silicon–nitride tip of a flexible cantilever (Fig. 1). The probability of fishing one or more molecules depends not only

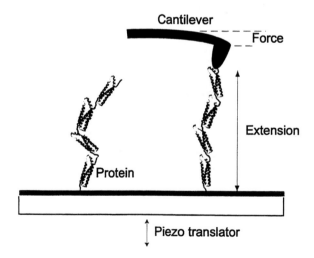

Fig. 1   Experimental scheme for protein fishing.

on the density of proteins on the surface but also on the interactions between the tip and the protein (See Section II,B). Once a protein is picked up, it can be stretched to more than 10 times its folded length (depending on its folded structure) reaching almost its total contour length.

The extension of the elastic, already unfolded part of the protein produces a restoring force that bends the cantilever. This bending, and therefore the force, can be measured with the high precision of the AFM. With proper sample preparation and well-adapted instrumental techniques, single-molecule unfolding processes generate a signature, i.e., a force–distance profile, which can be clearly distinguished from background noise and other events not related to unfolding processes (see Sections III,B and IV.A).

II. The Biological System

A. Spectrin Proteins

Spectrin is a member of a large family of actin-binding proteins. These are able to crosslink actin filaments into loose networks or tight bundles. This property makes the members of the spectrin family scaffolds for both cytoplasmic and membrane assemblies. Forming a two-dimensional network in red blood cells, spectrin molecules are assumed to provide the cell with special elastic features (Elgsaeter et al., 1986). This ability of the spectrin molecule to contract and expand has been attributed to the modular structure made of repeats, initially identified by Speicher and Marchesi (1984). Moreover, this ability seems to be a key element for structures, also containing spectrin, that are regularly subjected to mechanical stresses in cellular complexes ranging from muscle Z bands to stereocilia. Recent experiments have also demonstrated that spectrin acts as a protein accumulator that traps and stabilizes proteins at specific points on cell membranes (Hammarlund et al., 2000; Moorthy et al., 2000; Dubreuil et al., 2000).

The basic constituent of spectrin chains is the repeat which typically has 106 amino acids and is made of three antiparallel α-helices separated by two loops, folded into a left-handed coiled-coil (Fig. 2) (Pascual et al., 1997; Djinovic-Carugo et al., 1999; Grum et al., 1999). Grum et al. (1999) proposed a model for the flexibility of spectrin, based on structural data.

It is rather difficult to deduce the mechanical response of the molecule under stress from structural data, since the energy landscape of proteins is unknown. Mechanical properties of proteins must be measured directly, because they depend on a particular pathway along a preferred direction through the energy landscape. This can be done by AFM.

Rief et al. (1999) studied the natural α-spectrin chain by AFM, which is composed of homologous but not identical domains. When the chain is stretched, it is not possible to know which domains unfold first. Hence, the study of engineered constructs consisting of identical domains provides more insight into the mechanical stability and the unfolding features of the spectrin repeat (Lenne et al., 2000).

Fig. 2 Structure of the spectrin repeat: a left-handed antiparallel triple-helical coiled-coil (Protein Data Bank ID 1AJ3).

B. Protein Engineering

Protein engineering is required to

1. Fix the protein to both a surface and the tip.

2. Construct polyproteins to amplify the features of unfolding of one domain. These can be handled much more easily than single-domain proteins.

3. Construct mutants for a detailed study of the relation between structure and mechanical stability.

The application of AFM for protein unfolding has so far been restricted to a small group of proteins. Most of published works were focused on natural or engineered proteins, organized in linear arrays of globular domains. This is the case for natural spectrin, titin, and fibronectin. To our knowledge, only two works dealt with nonmodular proteins, namely, the bacteriorhodopsin (Oesterhelt *et al.,* 2000) and the HPI protomer (Muller *et al.,* 1999).

Nonmodular proteins are difficult to handle in the AFM, as it is not easy to attach the proteins at one end on a surface and at the other to the tip without affecting their structure.

With modular proteins, even if the protein is not so ideally attached, the elements of the chain, which are not directly attached to a solid surface, span the gap between the surface and the tip and can contain one or few domains that lend themselves to forced unfolding. In this case the force traces will exhibit a sawtooth-like pattern that gives a fingerprint of the modular protein.

The changes introduced upon adsorption of proteins to a surface are still poorly understood. This process can be at least partially controlled by engineering specific ends. The thiol group of cysteine allows proteins to be attached specifically onto a gold-coated surface. The spectrin clones were therefore fused with a COOH Cys2 tag for immobilization purposes. A cysteine residue could be as well engineered at the other (properly oriented) terminus of the protein to enable specific attachment of the gold-coated AFM tip to the cysteine (Oesterhelt *et al.*, 2000). But the same trick cannot be used for the surface and the tip at the same time.

Oesterhelt *et al.* (2000) used two-dimensional (2D) crystals of proteins that fix the orientation the proteins. The specific interactions of the supporting lipid layer results in the ordering of the proteins. In the case of bacteriorhodopsin, the proteins form 2D crystals spontaneously. In the 2D lattice, the protein has a well-defined direction and only a part of the protein is accessible to the tip. But it is a quite particular case of a protein that forms large 2D crystal domains easily. Protein engineering also allows constructing modular proteins from ones the are not naturally modular. Yang *et al.* (2000) used an original strategy to polymerize lysosyme proteins by solid-state synthesis.

A system that would guarantee uniform orientation of the molecules is still needed. To preserve the native states of the protein, few or no specific interactions are required. A good candidate is the *N*-nitrilo-triacetic acid (NTA)/His tag system, which is widely used in molecular biology to isolate and purify histidine–tagged fusion proteins. Here the histidine tag acts as a high-affinity recognition site for the NTA chelator. Schmitt *et al.* (2000) have shown that the binding forces between histidine–peptide and NTA chelator are in the 50-pN range. The strength of such bonds is too small to prevent detachment of the molecule before the total unfolding of a protein.

C. Preparation of Samples

One can use different surfaces to attach proteins. Glass and mica are suitable to non-specifically adsorb proteins [compare Norde *et al.* (1986)]. It is preferable though to use specific interactions to immobilize the protein and to forbid the detachment of it during stretching. We used engineered proteins (see Section II,C) with a cysteine residue at one end, which can form a specific bond with the gold-coated surfaces. The proteins were suspended in PBS or another suitable buffer at a concentration ranging from 10 to 100 μg/ml. To prepare the working buffers, it is recommendable to use ultra-pure water rather than double-distilled water. This guarantees that the salt concentrations, which are very important for the adsorption characteristics, are well defined. A small drop (20–50 μl) of the protein solution is deposited on the surface. Proteins were absorbed during 10 min, and samples were washed with PBS (10 times with 100 μl) afterwards.

Osterhelt *et al.* (2000) proposed an alternative method, based on 2D protein crystallization (see Section II,C).

D. Results

To get a better understanding of the mechanical properties of the spectrin repeat, we constructed four identical repeats recombinant proteins consisting of multiple repeats of the same domain. The unfolding features in the motif are thus multiplied and can be compared within the frame of one repeat (Carrion-Vazquez *et al.*, 1999). We chose the 16th repeat from the α chain of the chicken brain spectrin, because its tertiary structure has been resolved by NMR and X-rays. Polyproteins consisting of 4 repeats were cloned and expressed. Details can be found in Lenne *et al.* (2000). We will refer to these as $(R16)_4$ in the following.

Using AFM, we have shown that the single-spectrin domain can unfold either in an all-or-none fashion or in a step-like fashion. The force–extension patterns are indeed compatible with at least one intermediate between a folded and a totally unfolded state. As documented in the Fig. 3, we detected two populations of events: elongations of 32 and 15 nm (± 3 nm). The value of 32 nm corresponds to to the total unfolding of a domain: it is obtained by considering that the polypeptide chain is extended to 90% of maximum (0.34 nm per residue), and a folded domain has 106 residues and is 4 nm

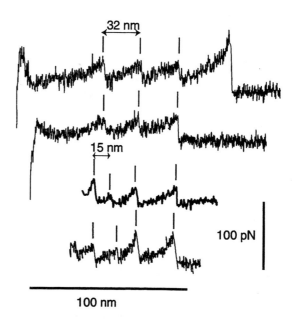

Fig. 3 Unfolding traces of a spectrin construct made of four identical tandem domains $(R16)_4$. As the proteins are picked up at random, the maximum extension varies from one curve to the other but never exceeds the total contour length of the protein. In the two lower curves, partial unfoldings are detected.

long (32 nm ~106 × 0.34–4). As two short elongation events (15 nm) are consecutive in most cases, we think that the unfolding of one domain results from two transitions, each producing a similar elongation. The corresponding unfolding forces were ranged from 25 to 80 pN depending on the pulling speed.

III. Forced Unfolding

A. Atomic Force Microscopy: Force Spectroscopy Mode

AFM principally allows manipulation of the protein in the subnanometer range and can generate and measure forces up to several nanonewtons with piconewton precision. Most commercial AFM apparatus have a force spectroscopy mode. However, the possible settings are in general restricted. We describe here general and necessary features of an AFM to allow high-resolution forced unfolding measurements. More generally these features are those required for force measurements.

First, the noise level of the detection system should be low enough to allow small forces to be detected. So far, forced unfolding studies have been carried out on proteins of high-mechanical stability, such as, e.g., titin, requiring decidedly more than 100-pN unfolding forces. To extend these studies to a very large number of proteins that might be less stable as spectrin, efforts must be made to construct an AFM with a high ratio of signal over noise.

To study the dependence of speed on unfolding forces, a feedback with different speeds for retract is required. The speeds that are generally used range from 0.1 to 10 nm/ms. To manipulate chains of different lengths a tuneable relative retract position with respect to the surface as well as to any intermediate position is necessary. The details of our instrument, which allows such precise control, are described in Section VII,A. Levers were calibrated by thermal fluctuation analysis. The values were cross-checked by stretching DNA as described in Section VII,B.

B. Force Curves

Generally the surface is brought into contact with the tip and is retracted away from the surface. When a protein attaches to the tip, it is pulled away from the surface and stretched. During extension, the protein undergoes structural transitions that lead at last to the total stretching of the protein. These transitions are of two kinds: (i) elongation of the polypeptide chain that requires an increasing force and (ii) unfolding events that produce a relaxation of the chain (and drop of the force) under mechanical stress.

Figure 3 shows few force curves recorded during the unfolding of a spectrin molecule. The last peak corresponds to the rupture of the bond between the protein and the tip. Detachment from the surface is highly improbable as the protein is strongly attached by cysteine-gold linkage but this cannot be excluded. This peak is generally of higher amplitude than the previous ones because the strength of the bond formed by nonspecific adsorption with its large entropic contribution is typically much larger than the unfolding forces measured for the proteins that have been the object of forced unfolding studies thus far.

This eliminates the possibility of successive detachment from the surface. In such a situation, one observes a large first force peak with consecutively decreasing force peaks which are due to the gradual loss of interactions between the protein(s) and the surface (Hemmerle *et al.*, 1999).

Position and amplitude of force peaks were measured. The distance between peaks reflects the gain of distance after unfolding and stretching of a folded structure in the protein. An unfolding event occurring at a given extension is thus specified by both the amplitude of the force peak at this extension and the distance from the same peak to the next one. The collected data may be presented in histograms of force and distances. These are fingerprints of the protein that could reveal different populations of unfolding events.

C. Force–Clamp Experiments

In the AFM available so far, the force acting on the lever in contact drives the feedback. When this latter is retracted and a protein is unfolded, this force drops to values of a few piconewtons and fewer on which a stable force feedback cannot be run. Problems with low-frequency noise, such as, e.g., drift, are therefore the most prominent among the various reasons why so far exclusively dynamic force spectroscopy has been possible in the low-force regime, which is particularly important for single-protein studies. With the multiple sensor stabilization system (MSS system, Section VII,A) of our AFMs, we are now able to unfold proteins in a new way that will for the first time give direct access to the natural lifetime of the folded state. By retracting the tip in small steps with stalling periods at constant distances to the surface in between, one can extend the unfolded polymer content just far enough that the force transduced to the next folded domain is not yet large enough to lead to an unfolding event that is dominated by the exponential dependency on the applied force.

Because the extension of the polymer is kept at constant length during these stall periods, the force applied to the folded domain is also constant (see Section IV,C). This means that during a following period of constant distance and therefore constant length one can measure the stability of the folded domain as a function of constant force. This has not been possible in dynamic force spectroscopy. Using the MSS system, one can now control the distance and therefore the applied force with the necessary temporal (from milliseconds to hours) as well as spatial resolution (sub-nanometer) and therefore high-force resolution ($\sim$10 pN), depending only on the force constant of the lever, if the noise level of the detection system is low enough.

It is of great importance to notice that it is only during such segments of the stretching process where force applied to the protein is constant that the induced unfolding can start from an equilibrium state, as the external force is only kept constant here.

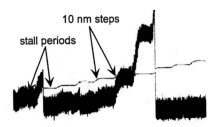

Fig. 4 Forced unfolding of spectrin with 10-nm ramped steps.

In Fig. 5 we have extracted those parts from the MSS–force curve in Fig. 4 that shows the force on the cantilever only while it is being moved because the D-lever setpoint is being increased during these periods. This recreates the standard force curve as it is known from dynamic force spectroscopy. The important difference is that some of the unfolding events occurred after the distance was kept constant for some time. The large difference in the noise on the curves is due to the fact that only the curve in Fig. 5 was low-pass-filtered as it is commonly done.

In Fig. 6 we have now extracted those parts from the MSS–force curve in Fig. 4 that show the force on the cantilever while the D-lever's setpoint is kept constant, and therefore the extension length of the polymer content between the M-lever's tip and the surface is kept constant. As can be seen in Fig. 6, this leads to well-timed segments during which the force applied to the unfolded segments of the protein is kept constant for constant distance, as is expected from the worm-like chain model. In this particular graph, unfolding happens at constant force during the second constant distance/force segment.

D. Refolding

AFM is able to measure refolding of protein domains. After stretching the protein, it can be relaxed by bringing the substrate close to the lever again. The protein can then be extended again, and force curves will again display unfolding events if refolding were

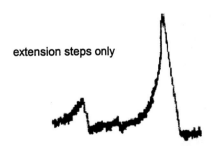

Fig. 5 The MSS–force curve after extraction of only the dynamic force curve segments.

stall periods only

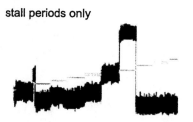

Fig. 6 The MSS–force curve after extraction of the constant force segments.

to occur. This demonstrates the recovery of unfolded domains during relaxation. We estimated the refolding rate in the range of 1 s^{-1} for spectrin domain.

Refolding experiments are very delicate and especially so for the small spectrin domain. We would like to insist on the fact that this cannot be done easily with the usual atomic force microscope (AFM). Indeed after few cycles the extension of the molecule is not known with good accuracy anymore as drift may occur in between.

We propose to process refolding with a multiple-pulse protocol (Fig. 7). Once the protein is attached onto the tip it is necessary to prevent it from adhesion into the surface by pressing the tip on the surface. The zero-length point is defined as the point where the AFM cantilever contacts the surface.

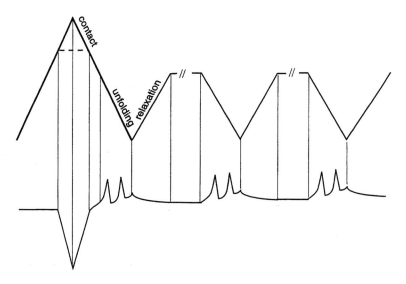

Fig. 7 Schematic procedure for refolding. Signal fed to the piezo-actuator (upper trace) and force trace (lower trace). The first approach brings the tip into contact with the surface and one single protein can then be picked up. The protein is stretched up to a position before it breaks off from the tip. The protein is then relaxed, but the sample is not brought again into contact to prevent any interaction with other proteins on the surface. The cycle can be repeated many times.

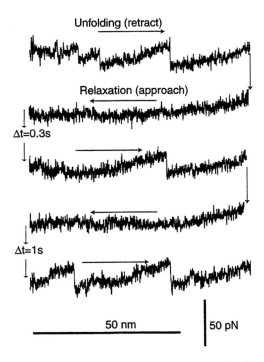

Fig. 8 Cycles of unfolding–refolding of a single molecule of spectrin.

Figure 8 shows few cycles of unfolding–relaxation. After reaching the extended state and before rupturing the bond between the protein and the tip, the protein was relaxed to an extension of 20 nm. This demonstrates that refolding occurs on a time scale of seconds and is therefore significantly slower than the time scale for forced unfolding.

IV. Analysis

A. Sorting Data

The complexity of the fishing process, i.e., of the different types of interactions between the tip, the protein, and the surface, leads to very different situations in one and the same experiment: the tip can pick up none, one, or few molecules. Molecules can also lay down on the surface and be detached gradually, leading to artifacts as already mentioned earlier. Whenever only one protein is picked up, this will occur randomly at any position along the chain. All these contributions lead to very different force curves carrying very different information that needs to be sorted out. It is necessary to make a careful analysis of the force curves. The statistical nature of the fishing process unavoidably leads to an experimental situation where most of the force curves have to be discarded due to various artifacts.

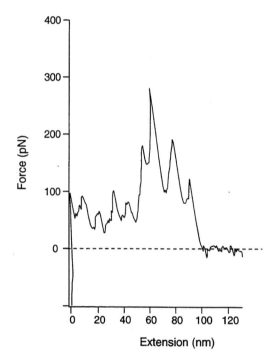

Fig. 9 Multiple pickup lead to force curves presenting high forces and large off-set.

B. Multiple Pickup

One such artifact is the simultaneous pickup of multiple proteins. This creates a situation of multiple parallel and therefore additive springs between tip and surface and leads to higher forces and steep slopes to each peak (Fig. 9). Any force curves displaying an off-set in the force indicate multiple parallel springs, i.e., multiple-molecule pickup. These must be discarded from analysis to ensure single-molecule manipulation.

We propose to analyze the magnitude and distances between successive force peaks. The latter is known with nanometer or better accuracy. So far researchers in the field have sorted their curves by using the criterion of distance. Misspellings led to the exclusion of the curves from the data sets. We sorted our data by keeping curves where there were none of the artifacts cited earlier, such as large adhesion peaks, high-force peaks indicative of multiple pickups, but we did not use the criterion of distance. We should like to emphasize that the distribution of the contour length increments that has been used in the majority of work is based on fitting of force curves by the worm-like chain model. This model relates the force to the extension of the chain with two free parameters: the persistence length and the contour length of the chain (see Section IV,C). Fits of our force curves based on the WLC model led to quite scattered persistence length. Because these fits did not always match well at the rupture point (force peaks indicating unfolding in a single trace), we preferred to show the elongation distribution, i.e., the distribution of peak spacings that can be determined with a resolution better than 1 nm. The advantage of this procedure is that it is only a relative measurement which can be carried out anywhere on the force curve and independent of prior processes, whereas the criterion of distance uses absolute measurements depending strongly on the initial starting point of the single-molecule unfolding process.

C. Fitting Procedures

To keep a polymer at a certain extension at constant temperature, it is necessary to apply a force. The work done by stretching the polymer chain goes into the reduction of the conformational entropy. In other words, the polypeptide chain acts as an entropic spring. This explains the ascending part of force curves. Once a folded structure has unfolded, an extra-length of the unfolded part of the chain is available for stretching.

Each extension segment between force peaks on the sawtooth-like pattern of the curve is well described by the WLC model. It relates the force F of the stretched chain to its extension x using two characteristic parameters: the contour length of the chain L_c and its persistence length L_p. Marko and Siggia (Bustamente *et al.*, 1994) have proposed the following interpolation formula that is commonly used:

$$F = \frac{k_B T}{L_p}\left[\frac{1}{4(1-x/L_c)^2} - \frac{1}{4} + \frac{x}{L_c}\right].$$

This expression, improved by Bouchiat *et al.* (1999), can be useful for fitting experimental data with better accuracy; for example,

$$F = \frac{k_B T}{L_p}\left[\frac{1}{4(1-x/L_c)^2} - \frac{1}{4} + \frac{x}{L_c} + \sum_{i=2}^{i\leq 7} a_i\left(\frac{x}{L_c}\right)^i\right],$$

with $a_2 = -0.5164228$, $a_3 = -2.737418$, $a_4 = 16.07497$, $a_5 = -38.87607$, $a_6 = 39.499944$, and $a_7 = -14.17718$.

The application of these formulas to the force curves generally yields persistence lengths equal to a few times the amino acid size that is 0.38 nm. This is somewhat surprising, since the original model of Krafty and Porod (Fixman and Kovac, 1973) leads asymptotically to the WLC only when the number of monomers per persistence length is large. As Bouchiat *et al.* (1999) mentioned, the concept of effective persistence length would be more appropriated in the case of protein stretching.

Fits of the spectrin unfolding force curve are shown in Fig. 10. Four peaks were fitted with Bouchiat *et al.*'s formula (1999). We fixed the persistence at $L_p = 0.58$ nm and the contour length was kept free. L_p was fixed to the average value obtained from the fits of 20 force peaks. It must be noted that some deviation from entropic behavior is evident in some curves.

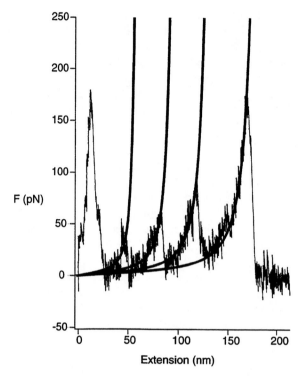

Fig. 10   Fits of force curves by the WLC model, with $L_p = 0.58$ nm and increasing contour lengths.

V. Models

A. Questioning Unfolding Pathways

The folding energy landscape of protein is very complex. It is often represented as a funnel through which the protein follows its pathway from unfolded to folded states (Onuchic *et al.*, 1997). The AFM may prove to be a very powerful tool to probe this energy landscape.

Theoretical work (Klimov and Thirumalai, 1999) has shown that forced unfolding could reveal folding processes. Proteins that fold in one step should be unfolded cooperatively, whereas those that fold in two or more steps should do so by the formation of intermediates. Our results on spectrin show that different pathways may be followed during unfolding.

Comparison between experiments and simulations can provide clues on unfolding pathways. Marszalek *et al.* (1999) showed features that are compatible with intermediates. They found an abrupt extension of the titin domain by 7 Å before the first unfolding event. This fast initial extension before a full unfolding event is considered by the authors to produce a reversible "unfolding intermediate." Steered molecular dynamics

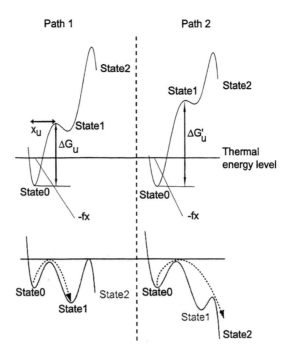

Fig. 11 Modeling unfolding into two states. Conceptual pathways in the energy landscape traversed under force. By adding a mechanical potential, $-fx$, the external force tilts the landscape and lowers the barriers. When the first barrier becomes lower than the thermal energy level, it can be crossed. In the model, a difference in relative height of the first to the second barrier along two possible paths determines whether state 1 is accessible.

simulations show that the rupture of a pair of hydrogen bonds near the amino terminus of the protein domain causes an extension of about 6 Å, which is in good agreement with their observations. Disruption of these hydrogen bonds by site-directed mutagenesis eliminates the unfolding intermediate.

As demonstrated by this work, the combination of experiments and simulations could lead to a better understanding of the unfolding pathways.

The work that is provided by the external force drives the protein from the folded to the unfolded states. The free energy ΔG is discounted by the mechanical energy $F \cdot x$. This results in a tilt of the energy profile as shown in Fig. 11. Energy barriers are effectively lowered so that the system can cross them by thermal energy excitation.

B. A Single-Parameter Model for Forced Unfolding Using Three States

To explain our data, we refrain from an elaborate and potentially wrong modeling based on a large number of parameters and rather propose a model consisting of only two pathways for the protein unfolding under mechanical stress through the energy landscape. These pathways are defined by the direction of pulling, where each pathway may be interpreted as to represent a projection of a large volume of configuration space onto one of two preferred tracks. Paci and Karplus (1999) found two sets of unfolding pathways for fibronectin-type 3 modules by molecular dynamics simulations, which are analogous to those that we propose. The same authors have recently detected stable intermediates during unfolding of a single-spectrin domain (Paci and Karplus, 2000).

Both tracks follow similar directions in real space, as defined by the direction of the applied force; however, according to our model, they may lead to two different unfolded states starting from the native folded state (Fig. 11). The intermediate state is conceptually available along both pathways, but each pathway by itself either leads from the native *state 0* to the partially unfolded *state 1* or the completely unfolded *state 2*. The relative difference of the height of the free energy barrier along either of the two pathways determines whether state 1 or state 2 is attained. The advantage of this strongly simplified modeling is that only one free parameter is needed for differentiating between the two averaged pathways. At the same time, we find this approach to be in good agreement with our data from the experiment and from Monte Carlo (MC) simulations (see following).

In the native folded state, the protein is in state 0 at the bottom of the potential well. The directional mechanical stress applied by the AFM tip not only decreases the barrier height to thermally activated unfolding but also reduces the options of the protein to those of following either *path 1* or *path 2* during unfolding. The protein will follow only one path leading to a bimodal probability distribution with 35 and 65% probabilities for path 1 and path 2, respectively, according to our experimental data.

The external stretching force reduces the effective energy barriers so that the system can cross them by thermal activation (Evans and Ritchie, 1997). As the applied force increases, the height of the energy landscape is reduced linearly along the generalized reaction coordinate. Along path 1 this reduction will lower the free energy barrier of the partially unfolded state below the thermal energy level and thereby grant access to this state. The remaining free energy difference of the totally unfolded state is too large, such that this state cannot be reached. The protein will therefore unfold only partially. Along path 2 the forced reduction will simultaneously lower the barriers of both states 1 and 2 below the thermal energy level, such that the barrier height of the intermediate state is still the one dominating the kinetics of the unfolding pathway. Along this path though, the barrier to the totally unfolded state is now lower than the barrier of the intermediate state, and the protein will unfold completely and at most stay only intermittently in the intermediate state, because the thermal energy will drive it immediately into the completely unfolded state. A similar concept has been proposed by Merkel *et al.* (1999) to explain the rupture of the streptavidin–biotin bond.

Since the free-energy barrier for the intermediate state is higher along path 2 and also closer to the total unfolding barrier, a higher force is needed on average to reach the completely unfolded state than to reach the only partially unfolded state. The difference in free energy for state 1 along path 1 is lower than that along path 2. Because the height of the barrier to complete unfolding in state 2 is roughly the same along the very similar directions through the conformational space of the protein starting in the native folded conformation, state 1 will be accessible to the protein well before state 2. The thermal energy will allow the protein to unfold partially into state 1, while state 2 is still hidden behind a barrier that cannot be overcome by thermal activation. Because the free-energy barrier to state 1 is lower along path 1, the average force needed to reach this state is lower than that for state 2.

C. Monte Carlo Simulations

We have included these two scenarios in a simple MC simulation (See Section VII,D) by testing the reaction kinetics simultaneously for the short and long elongation events. The kinetics can be characterized by two parameters: the width of the first barrier and an effective "attempt" frequency, which includes the barrier height as a multiplicative exponential factor, normalized by the thermal energy. The width of the first barrier was kept the same for both scenarios, while the attempt frequency was adjusted to agree with the relative difference in barrier height. Figures 12a to 12c show the force and elongation histograms obtained from 5000 consecutive runs of a Monte Carlo simulation. The simulations reproduced well the general features of the experimental data with a barrier width of 0.4 nm and an attempt frequency of 0.5 Hz along path 1 and 0.05 Hz along path 2 (corresponding to about a 2-kT difference in barrier heights). The selected pathway guides the folded domain either to a state where it is totally unfolded or to a state where it is partially folded.

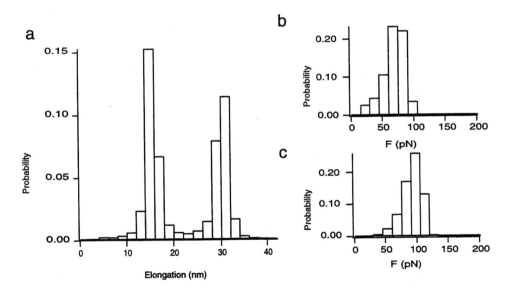

Fig. 12 Probability histograms of elongation (a) and unfolding forces for short elongation (b) and long elongation events (c). These were obtained by 5000 Monte Carlo simulations of unfolding of four domains placed in series. By testing the two reaction kinetics simultaneously associated with the two different pathways, short and long elongation events were allowed. A barrier width of 0.4 nm and an attempt frequency of 0.5 Hz along path 1 and 0.05 Hz along path 2 fitted best to the experimental data.

VI. Conclusion and Prospects

Although the molecular complexity of unfolding pathways can be very high, force spectroscopy of properly engineered single proteins can provide important clues to energy landscapes on time scales from milliseconds to seconds and larger (the stability of the instrument permitting). In the future, we expect an increasing contribution by forced unfolding measurements to the understanding of protein folding as, on the one hand, proteins can be engineered to systematically perturb the unfolding pathways imposed by the real-space directionality and, on the other hand, instrument developments as, e.g., outlined in this chapter, will enable new types of measurements.

This combination will be able to provide a more detailed understanding of the link between mechanical stability and folding features of proteins. The comparison of experimental results to increasingly available simulation results could offer a deeper insight into unfolding pathways. In particular, as we have shown, this technique can reveal—possibly functionally relevant—intermediates that were not detected thus far by other techniques.

A. Biological Implications

Molecular elasticity is a physicomechanical property that is associated with a number of proteins in both the muscle and the cytoskeleton. These new details about unfolding of single domains revealed by precise AFM measurements show that force spectroscopy can be used to not only determine forces that stabilize protein structures but also analyze the energy landscape and the transition probabilities between different conformational states. Applications of force spectroscopy on single molecules may thus lead to a better understanding about molecular biophysics in both the muscle and the cytoskeleton.

VII. Appendices

A. The Double-Sensor-Stabilized AFM

We have designed a special AFM with a unique local stabilization system, which lends ultrahigh positional accuracy to forced unfolding measurements. It is made of two crossed, i.e., independent, optical detection systems (Figs. 13 and 14). The exact details of this double-detection system will be described elsewhere.

In a normal AFM one lever is used simultaneously for the feedback that drives the displacement of the piezo-tube and the force measurement. This technique is being used with great success in various imaging applications of the AFM, but it has several serious drawbacks in forced unfolding applications. Because the AFM only knows where the surface is (as long as the lever is in sensory contact with it), the absolute distance between the free position of the lever and the surface is not known when the lever is not in contact with the surface. While the lever is out of contact, drifts and other low-frequency noise changing the distance between the tip and the sample cannot be detected. On approaching

Fig. 13   Our double-sensor atomic force microscope has two full crossed optical detection units.

or retracting from the sample one controls the extension of the piezo only and not the distance between tip and sample. Compare the lower part of Fig. 15. This is why, in conventional dynamic force spectroscopy, force curves must be done fast enough, such that absolute and relative measurements on force curves can be carried out by basing them on the distance the piezo surface traveled during this time according to the voltages applied to it.

Fig. 14 The two optical detection units are used to simultaneously detect the deflection signals from two levers on the same substrate independently. (See Color Plate.)

The stabilization system we used in our instrument is based on the idea of using two levers simultaneously. This allowed us to split the feedback control system off from the one for the force measurements. In our double-sensor-stabilized AFM, (DSS-AFM) two sensors mounted side by side with a distance of few hundred microns on the same substrate carrier and slightly tilted with respect to the sample surface are used for this purpose. Because of the latter, one of the two sensors will make contact with the surface before the other. The distance between the second sensor and the sample surface can then actively be controlled with subnanometer resolution. Thus, measurement and distance controls are split up between the two levers. The first lever will detect all the noise that would change the distance between sensor array and sample. This can be controlled by a fast feedback. We eliminated all drift between the tip and sample by using a fast integral feedback.

The second lever signal can be used to carry out measurements at any distance, measured by the first lever, from the surface from milliseconds to hours at forces determined only by the sensitivity of the detection system (the thermal noise amplitude of the second cantilever, which was about 10 pN in our case) and by the average statistical error (which comes out to be about 4 at 30 pN according to the Gaussian law of error propagation), if we assume angstrom resolution and 10% uncertainty in the determination of the force constant of the cantilever by one of the calibration methods listed in the following.

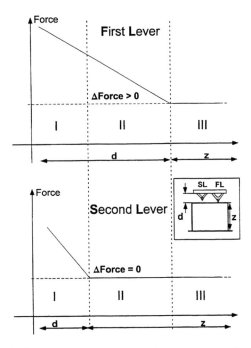

Fig. 15 Force curve representation of the principle of the double sensor stabilization system in our AFM. After the first lever has contacted the surface, the distance between the second lever and the surface can be controlled with the typical subnanometer resolution.

The distance between the sample and the second tip, which is used to do the actual unfolding experiments, can be controlled with the subnanometer resolution typical for AFM by selecting the proper setpoint, i.e., the normal force, of the first lever (Fig. 15).

With this control one can do what is called "force clamping" a protein between the lever and the surface. As is shown in the Fig. 16, as long as the first lever is in contact with the surface, it is possible to stop the retract of the second lever at any time (e.g., t_{01}) for any duration ($t_{02} - t_{01}$) while keeping the distance (d_0) and therefore the force (F_0) constant.

B. Data Acquisition and Evaluation Techniques

Data were acquired by a 32-bit PCI-M-I/O-16E-4 acquisition card (National Instruments) with 16 single-ended analogue inputs with a 12-bit resolution. The maximal speed of acquisition was 500 kSamples/s for single-channel acquisition. For most measurements, we recorded multiple channels at 100 kSamples/s.

Force curves were recorded and saved on a Computer with a 266-MHz Intel Pentium II CPU with 128-MByte RAM. We implemented programs for digitally controlling the instrument and storing the data acquired with either Labview (National Instruments)

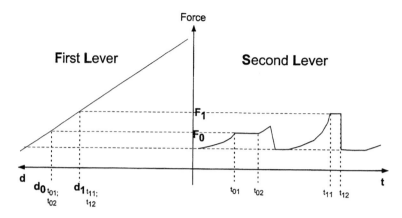

Fig. 16 Force clamp based on the double sensor stabilization system. The force can be kept constant by simply keeping the distance constant.

or Igor Pro using NI-DAQ tools (Wavemetrics). All force curves from an experiment were continuously recorded using FIFO-buffering and saved without prior sorting. After each experiment, data were separated, scaled, and sorted. A box-smoothing window (of variable width depending on the acquisition rate) or a 2-kHz low-pass filter was applied to reduce the laser noise and thermal noise on the force signals. This low-pass filtering does not alter the curves acquired with Hz-scanning. Positions and force were then analyzed manually with Igor Pro.

C. Calibration

1. Thermal

The analysis of the thermal fluctuations of the vibrating lever gives access to the stiffness of the latter. It is based on the equipartition theorem and may be performed as described in (Florin *et al.*, 1995). However, this approach requires that no additional noise is added to the thermal noise. This would lead to an overestimation of the displacement of the lever and hence to an underestimation of the measured stiffness.

2. The B–S Transition of λ-Phage DNA

When a single λ-digest DNA molecule is stretched it goes into a highly cooperative conformation transition (B–S). The well-pronounced transition force plateau provides a valuable method for lever calibration. This plateau is observed at 65 pN at room temperature of 20°C (see Fig. 17) (Rief, Clausen-Schaumann *et al.*, 1999; Clausen-Schaumann *et al.*, 2000). Practically, λ-BstE digest DNA was used (Sigma) (see (Rief, Clausen-Schaumann *et al.*, 1999) for methods). The procedure is easy and at the same time constitutes a test for correct setup for force measurements.

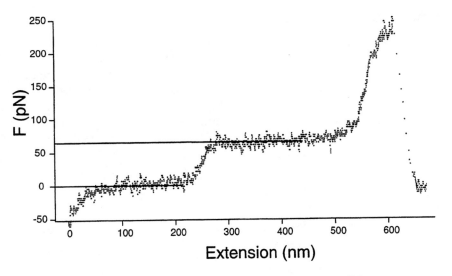

Fig. 17 The B–S transition of λ-phage DNA measured by AFM.

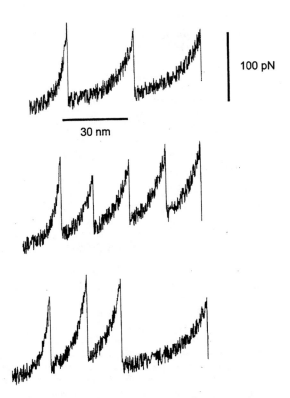

Fig. 18 Monte Carlo simulations of spectrin extension.

D. Monte Carlo Simulations

Monte Carlo simulations (Fig. 18) were set up analogously to Rief *et al.* (1998) by combining the WLC model to calculate the force with the kinetics governed by a two-state model. The method was developed further to a three-state model with a choice of two pathways. The unfolding rate ν_u of a folded structure is the product of a natural vibration ν_0 and the likelihood of reaching the transition state with an energy barrier ΔG_u discounting by mechanical energy $F \cdot x_u$, where x_u is the width of the activation barrier (Bell, 1978), $\nu_u(F) = \nu_0 \exp(-(\Delta G_u - F \cdot x_u)/k_B T) = \nu_{\text{eff}} \exp(F \cdot x_u/k_B T)$ ($k_B T = 4.1$ pN $\cdot$ nm at room temperature). The effective frequency ν_{eff} represents the number of attempts to cross the barrier of width x_u.

References

Bell, G. I. (1978). Models for the specific adhesions of cells to cells. *Science* **200**, 618–627.

Bouchiat, S. M., Wang, M. D., Allemand, J.-F., Strick, T., Block, S. M., and Croquette, V. (1999). Estimating the persistence length of a worm-like chain molecule from force-extension measurements. *Biophys. J.* **76**, 409–413.

Brockwell, D. J., Smith, D. A., and Radford, S. E. (2000). Protein folding mechanisms: New methods and emerging ideas. *Curr. Opin. Struct. Biol.* **10**, 16–25.

Bustamante, C., Marko, J. F., Siggia, E. D., and Smith, S. (1994). Entropic elasticity of λ-phage DNA. *Science* **265**, 1599–1600.

Carrion-Vazquez, M., Oberhauser, A. F., Fowler, S. B., Marszalek, P. E., Broedel, S. E., Clarke, J., and Fernandez, J. M. (1999). Mechanical and chemical unfolding of a single protein : A comparison. *Proc. Natl. Acad. Sci. U.S.A.* **96**, 3694–3699.

Clausen-Schaumann, H., Rief, M., Tolksdorf, C., and Gaub, H. E. (2000). Mechanical stability of single DNA molecules. *Biophys. J.* **78**, 1997–2007.

Djinovic-Carugo, K., Young, P., Gautel, M., and Saraste, M. (1999). Structure of the alpha-actinin rod: Molecular basis for cross-linking of actin filaments. *Cell* **98**, 537–546.

Dubreuil, R. R., Wang, P., Dahl, S., Lee, J., and Goldstein, L. S. (2000). Drosophila beta spectrin functions independently of alpha spectrin to polarize the Na,K ATPase in epithelial cells. *J. Cell. Biol.* **149**, 647–656.

Elgsaeter, A., Stokke, B. T., Mikkelsen, A., and Branton, D. (1986). The molecular basis of erythrocyte shape. *Science* **234**, 1217–1223.

Evans, E., and Ritchie, K. (1997). Dynamic strength of molecular adhesion bonds. *Biophys. J.* **72**, 1541–1555.

Fisher, T. E., Oberhauser, A. F., Carrion-Vazquez, M., Marszalek, P. E., and Fernandez, J. M. (1999). The study of protein mechanics with the atomic force microscope. *Trends Biochem. Sci.* **24**, 379–384.

Fixman, M., and Kovac, J. (1973). Polymer conformational statistics. III. Modified Gaussian models of stiff chains. *J. Chem. Phys.* **56**, 1564–1568.

Florin, E.-L., Moy, V. T., and Gaub, H. E. (1994). Adhesive forces between individual ligand receptor pairs. *Science* **264**, 415–417.

Florin, E.-L., Rief, M., Lehmann, H., Ludwig, M., Dornmair, C., Moy, V. T., and Gaub, H. E. (1995). Sensing specific molecular interactions with the atomic force microscope. *Biosens. Bioelectron.* **10**, 895–901.

Grum, V. L., Li, D., MacDonald, R. I., and Mondragon, A. (1999). Structures of two repeats of spectrin suggest models of flexibility. *Cell* **98**, 523–535.

Hammarlund, M., Davis, W. S., and Jorgensen, E. M. (2000). Mutations in beta-spectrin disrupt axon outgrowth and sarcomere structure. *J. Cell. Biol.* **149**, 931–942.

Hemmerle, J., Altmann, S. M., Maaloum, M., Horber, J. K., Heinrich, L., Voegel, J. C., and Schaaf, P. (1999). Direct observation of the anchoring process during the adsorption of fibrinogen on a solid surface by force-spectroscopy mode atomic force microscopy. *Proc. Natl. Acad. Sci. U.S.A.* **96**, 6705–6710.

Klimov, D. K., and Thirumalai, D. (1999). Stretching single-domain proteins: Phase diagram and kinetics of force-induced unfolding. *Proc. Natl. Acad. Sci. U.S.A.* **96**, 6166–6170.

Lenne, P.-F., Raae, A. J., Altmann, S. M., Saraste, M., and Hörber, J. K. H. (2000). States and transitions during forced unfolding of a single spectrin repeat. *FEBS Lett.* **476,** 124–128.

Marszalek, P. E., Lu, H., Li, H., Carrion-Vazquez, M., Oberhauser, A. F., Schulten, K., and Fernandez, J. M. (1999). Mechanical unfolding intermediates in titin modules. *Nature* **402,** 100–103.

Merkel, R., Nassoy, P., Leung, A., Ritchie, K., and Evans, E. (1999). Energy landscapes of receptor-ligand bonds explored with dynamic force spectroscopy. *Nature* **397,** 50–53.

Moorthy, S., Chen, L., and Bennett, V. (2000). Caenorhabditis elegans beta-G spectrin is dispensable for establishment of epithelial polarity, but essential for muscular and neuronal function. *J. Cell. Biol.* **149,** 915–930.

Muller, D. J., Baumeister, W., and Engel, A. (1999). Controlled unzipping of a bacterial surface layer with atomic force microscopy. *Proc. Natl. Acad. Sci. U.S.A.* **96,** 13,170–13,174.

Norde, W., Macritchie, F., Nowicka, G., and Lyklema, J. (1986). Protein Adsorption At Solid Liquid Interfaces-Reversibility and Conformation Aspects. *J. Coll. Interf. Sci.* **112,** 447–456.

Oesterhelt, F., Oesterhelt, D., Pfeiffer, M., Engel, A., Gaub, H. E., and Muller, D. J. (2000). Unfolding pathways of individual bacteriorhodopsins. *Science* **288,** 143–146.

Onuchic, J. N., Luthey-Schulten, Z., and Wolynes, P. G. (1997). Theory of protein folding: The energy landscape perspective. *Annu. Rev. Phys. Chem.* **48,** 545–600.

Paci, E., and Karplus, M. (1999). Forced unfolding of fibronectin type 3 modules: An analysis by biased molecular dynamics simulations. *J. Mol. Biol.* **288,** 441–459.

Paci, E., and Karplus, M. (2000). Unfolding proteins by external forces and temperature: The importance of topology and energetics. *Proc. Natl. Acad. Sci. U.S.A.* **97,** 6521–6526.

Pascual, J., Pfuhl, M., Walther, D., Saraste, M., and Nilges, M. (1997). Solution structure of the spectrin repeat: A left-handed antiparallel triple-helical coiled-coil. *J. Mol. Biol.* **273,** 740–751.

Rief, M., Gautel, M., Oesterhelt, F., Fernandez, J. M., and Gaub, H. (1997). Reversible unfolding of individual titin immunoglobulin domains by AFM. *Science* **276,** 1109–1112.

Rief, M., Fernandez, J. M., and Gaub, H. E. (1998). Elasticity coupled two-level systems as a model for biopolymer extensibility. *Phys. Rev. Lett.* **81,** 4764–4767.

Rief, M., Clausen-Schaumann, H., and Gaub, H. E. (1999). Sequence-dependent mechanics of single DNA molecules. *Nat. Struct. Biol.* **6,** 346–349.

Rief, M., Pascual, J., Saraste, M., and Gaub, H. (1999). Single molecule force spectroscopy of spectrin repeats: Low unfolding forces in helix bundles. *J. Mol. Biol.* **286,** 553–561.

Schmitt, L., Ludwig, M., Gaub, H. E., and Tampe, R. (2000). A metal-chelating microscopy tip as a new toolbox for single-molecule experiments by atomic force microscopy. *Biophys. J.* **78,** 3275–3285.

Speicher, D. W., and Marchesi, V. T. (1984). Erythrocyte spectrin is comprised of many homologous triple helical segments. *Nature* **311,** 177–180.

Yang, G., Cecconi, C., Baase, W. A., Vetter, I. R., Breyer, W. A., Haack, J. A., Matthews, B. W., Dahlquist, F. W., and Bustamante, C. (2000). Solid-state synthesis and mechanical unfolding of polymers of T4 lysozyme. *Proc. Natl. Acad. Sci. U.S.A.* **97,** 139–144.

CHAPTER 16

Developments in Dynamic Force Microscopy and Spectroscopy

A. D. L. Humphris and M. J. Miles

H. H. Wills Physics Laboratory
University of Bristol
Tyndall Avenue
Bristol, BS8 1TL
United Kingdom

I. Introduction

The advantages of atomic force microscopy (AFM) for the study of biological specimens are unique and are discussed in many of the accompanying chapters. The ability to image at molecular resolution in three dimensions in any environment appropriate to biology, including aqueous buffers and growth media, means that the observed structures are close to those of biological relevance. The specimen must, of course, be immobilized on a surface for a time scale that is comparable to the time to scan the image. This is typically about 1 min, but can be reduced to currently a few seconds for imaging small and flat areas. The use of staining or coating to increase image contrast is not usually necessary with AFM, and so the specimen is free to change with time. This allows processes to be followed *in situ*. There are exciting developments to dramatically increase AFM scan rates and these will be of great value in the study of biomolecular processes.

To obtain a three-dimensional topographic image of the specimen surface, the force between the tip and the specimen is usually maintained at a constant preset value by moving the specimen (or the tip) toward or away from the tip (or specimen). The most important parameter to control in AFM imaging of delicate biological specimens is the force applied by the probe to the specimen to avoid either distortion of the structure during imaging or, worse still, permanent damage to the specimen. An additional benefit of using lower forces is that the strength of tethering the molecule to the surface can be less, which again results in less distortion of the biomolecular structure.

The AFM can be operated in various modes. The highest resolution has been achieved in contact mode (see, for example, Baker *et al.*, 2000), but this is only possible on sufficiently rigid specimens, as the lateral force on the specimen as the tip scans in contact over the surface leads in many cases to specimen damage. Reduction of the normal force applied by the tip to the specimen alleviates this lateral deformation. When operating in air under ambient conditions, a thin layer of water exists on the specimen and tip surfaces. Furthermore, at sufficiently high relative humidity, capillary condensation may occur between the tip and the specimen resulting in a neck of water forming as the tip approaches the specimen. The resultant surface-tension force will act to pull the tip of the probe into the specimen, increasing the normal force to about 50 nN. The lateral force on the specimen is proportionally increased. This capillary force can be virtually eliminated by working in a liquid environment so that the liquid interface is moved above both tip and cantilever. The normal force in this case is typically about 1 nN, so that lateral force, although much reduced, is still sufficient to cause damage to many biological specimens. It has been estimated that the normal force should be less than 100 pN to avoid damage to most biological structures. Working in an aqueous environment, for example, may have other advantages in terms of simulating physiological conditions appropriate to the particular biological system.

The use of "tapping" or intermittent-contact mode of AFM reduces the lateral force applied to the specimen by the tip, which spends considerably less time in contact with the specimen as it scans over the specimen surface. The adsorbed water layer still plays an important role in this process, as the tip may either oscillate within the water layer if driven at low amplitudes or move in and out of the layer on each cycle at higher amplitudes. In "tapping" mode operation, the cantilever is oscillated at or close to its resonant frequency either by a small piezo-electric transducer at the fixed end of the cantilever or by an oscillating magnetic field, in which case the cantilever must be coated with a magnetic material. Again, it is often important to work in a liquid environment not only to avoid instabilities caused by the presence of the water layer but also because it is desirable to work in physiologically relevant conditions. It is therefore necessary to oscillate the cantilever in the liquid environment. This can be achieved by either driving the cantilever with an oscillating magnetic field as in the case of air operation or acoustically coupling the oscillation of a piezo-electric transducer through the liquid and the liquid cell to the cantilever. This latter method is more complicated to analyze, as the the transfer function of the cell and the liquid dominates the frequency spectrum.

The damping of the cantilever oscillation by the surrounding liquid leads to a change in the quality factor (Q) of the cantilever from, typically, 50–300 in air to between 1 and 5 in water. This has several serious consequences. First, the energy and the force required to drive the cantilever are much greater than those for the same amplitude in air. The force to excite the cantilever at resonance is given by

$$\text{Force} = (kA)/Q, \qquad\qquad [1]$$

where k is the spring constant of the cantilever and A is the amplitude of oscillation. The forces applied to the specimen in intermittent contact mode in liquid are, in fact, about an order of magnitude greater than those in simple contact mode in liquid, although the lateral forces are less. This may account for the observed decrease in resolution compared to that in contact mode imaging.

The resonant peak in the frequency spectrum is now so broad that any shift in frequency due to the interaction between the tip and the specimen is essentially undetectable. Indeed, there is a decrease in resonant frequency as the probe approaches the surface due to an increase in the effective mass of the cantilever, which results from the hydrodynamic interaction of the probe with the liquid and the surface. The probe essentially drags liquid with it increasing its effective mass by between 10 and 50 times. This shift in frequency is not usually noticed in the conventional setup, owing to the great width of the resonant peak in liquid. At high values of quality factor Q, the cantilever oscillation during intermittent contact can be regarded as harmonic due to the small effect of the relatively small amount of energy lost in each contact of the surface compared to the energy stored in the cantilever oscillating at resonance. However, at low values of Q experienced in conventional tapping or intermittent contact mode in liquid, the energy stored in the oscillating cantilever is much less so that intermittent contact with the surface results in a nonsinusoidal motion of the cantilever; that is, the oscillation becomes anharmonic and higher modes become significant.

The decrease in the quality factor of the cantilever in liquid has the consequence that greater forces are applied to the specimen during the tapping process, which results in distortion and lower resolution imaging of the specimen. It also gives incorrect values for the heights of the specimens. Higher values of Q in liquid bring many benefits to imaging. The simplest method of increasing the value of Q is by designing the cantilever to present a low area in the direction of motion. An alternative method, applicable to any cantilever, uses an active feedback system based on the time-dependent displacement of the cantilever (Anczykowski et al., 1998; Tamayo et al., 2000). With this method it is possible to increase the effective quality factor of the cantilever by up to three orders to magnitude in liquid. This means that the forces imposed by the tip on the specimen can be reduced, in practice by up to two orders of magnitude, that is, to about 10 pN. Several examples of imaging soft, delicate specimens with this active Q control in operation will be described later, but first a brief description of the technique will be given.

II. Active Q Control

A. Background

The technique requires the addition of electronics that provide an additional drive signal to the drive piezo or driving magnetic field. The input for the electronics is derived from the displacement of the cantilever as measured by the split photo-diode. This additional drive term can be used to essentially cancel the damping term due to the liquid in the equation of motion of the cantilever. In practice, such an electronics unit can be easily used with conventional commercial SPMs to increase the effective Q value of cantilevers in liquid (see, for example, *infinitesima* Ltd, Bristol, UK)

To see how this works, consider

$$m\frac{d^2z}{dt^2} + \gamma\frac{dz}{dt} + kz = F_0 e^{i\omega t} + F_{int}(z),\qquad [2]$$

the equation of motion of a damped oscillator. Where z is the displacement of the cantilever, m is the effective mass, of the cantilever, γ is the damping constant, and k is the spring constant. The second term on the left-hand side represents the damping force due to the motion of the cantilever through the liquid environment. This force depends on the velocity of the cantilever and the damping constant γ. The right-hand side of the equation represents the sum of the time-dependent driving force and the DC interaction force between the tip and the specimen. The resonant frequency of the cantilever is given by $\omega_0 = (k/m)^{1/2}$ and the quality factor by $Q = \frac{m\omega_0}{\gamma}$. To increase the effective Q of the cantilever in liquid, a further driving term is added to the right-hand side of Eq. [2]:

$$m\frac{d^2z}{dt^2} + \gamma\frac{dz}{dt} + kz = F_0 e^{i\omega t} + F_{int}(z) + Ge^{i\pi/2}z(t).\qquad [3]$$

This term represents a positive feedback of the cantilever displacement with variable gain (G) and phase shifted by $\pi/2$ so as to be in phase with the velocity rather than the displacement of the cantilever. Equation [3] can be rewritten as

$$m\frac{d^2z}{dt^2} + \gamma_{eff}\frac{dz}{dt} + kz = F_0 e^{i\omega t} + F_{int}(z),\qquad [4]$$

in terms of an effective damping constant, γ_{eff}, where $\gamma_{eff} = \gamma - G/\omega$ and $Q_{eff} = \frac{m\omega_0}{\gamma_{eff}}$. The effective damping constant, γ_{eff}, resulting from the feedback loop may be increased or decreased depending on the value and sign of the gain G.

B. Practical Implementation of Active Q Control

In practice, information about the position and velocity of the tip or end of the cantilever is most readily obtained from the difference output of the split-domain photodiode used in association with an optical lever in most conventional AFM heads. This photo-diode signal is proportional to the displacement of the cantilever. For control over the effective value of the quality factor Q of the cantilever, this signal is amplified by a factor G and

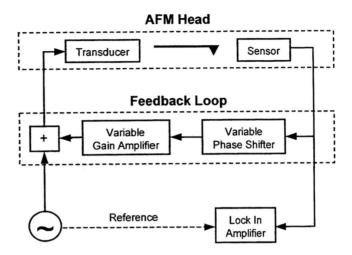

Fig. 1 A schematic diagram of a conventional AFM head with the addition of the feedback loop.

the phase of this oscillating signal shifted by $\pi/2$ so as to bring the signal in phase with the velocity of the cantilever. In practice, the actual shift required to bring the displacement in phase with the velocity will not be exactly $\pi/2$ owing to other phase shifts in the system, particularly in the electronics. The resulting signal is then added to the drive signal for the piezo-voltage or magnetic-field coil current. This feedback arrangement is shown in relation to the cantilever drive and detector in Fig. 1. Another practical consideration is that the effective mass of the cantilever is increased near the specimen surface owing to the hydrodynamic interaction, so that the resonant frequency will be lower, requiring the cantilever to be tuned near the surface, This effect can be neglected for liquid tapping operation without feedback-enhanced Q because the great breadth of the damped resonant peak results in this shift in resonant frequency being practically unimportant.

We are concerned here with the use of positive feedback to increase the effective value of the Q for the primary purpose of increasing the force sensitivity. It should be mentioned that negative feedback to decrease the value of Q has been used in situations where its value is particulary high, such as in vacuum, to decrease the energy stored in the cantilever and to increase its response time to allow scan rates to be increased.

The dramatic increase in effective Q of the fundamental resonance of a cantilever in a liquid environment as a result of the active feedback is shown in Fig. 2. This effect is demonstrated using both the magnetic and the acoustic methods of driving the cantilever. The initial breadth of the cantilever resonance in water is clearly seen in Fig. 2a, where the cantilever is driven magnetically and the frequency–amplitude plot shows a broad (asymmetric) resonant peak having a value of Q of about 1.6. In the case of the same cantilever driven acoustically (Fig. 2b), other resonances associated with the liquid cell dominate the spectrum making it impossible to identify the cantilever resonance.

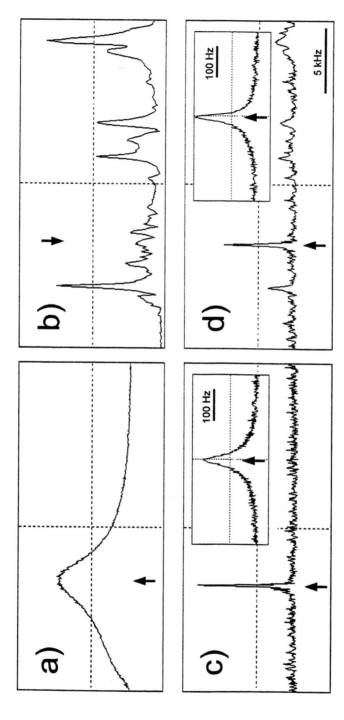

Fig. 2 Frequency spectra of the system in conventional operation (a and b) and with quality factor enhancement (c and d). Acoustic (b and d) and magnetic (a and c) methods were used to excite the cantilever. Insets show magnified region around resonance peak. All data sets were obtained with the same cantilever with a nominal spring constant of 0.4 N/m. Arrow marks 16.05 kHz.

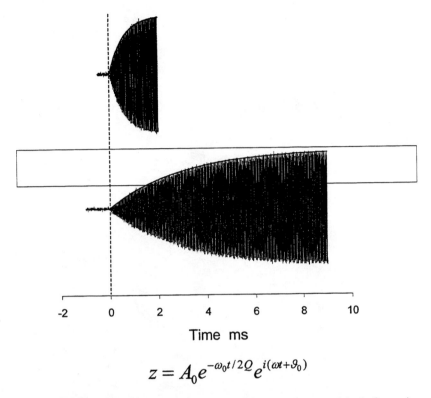

$$z = A_0 e^{-\omega_0 t / 2Q} e^{i(\omega t + \vartheta_0)}$$

Fig. 3 Transient response of cantilever after turn on of drive signal at time $t = 0$ in air. Comparison of a conventionally driven cantilever (top trace) and quality enhanced (lower trace) with a true quality factor of 58 and effective quality factor of 291, respectively. Frequency spectra were recorded and fitted with the harmonic oscillator to estimated the effective quality factors. Data sets are overlaid with the predicted transient response (envelope line) using these fitted quality factors. The same cantilever was used for both measurements with a resonant frequency of 32.1 kHz and nominal spring constant 0.1 N/m.

With the active-Q feedback enabled, the resonant peak in the magnetically driven case (Fig. 2c) is seen to have considerably sharpened and now has an effective Q value of 280. A similar result is also seen for the accoustically driven cantilever (Fig. 2d). The effective value of Q is dependent on the value of the feedback gain G. The response of the cantilever is in most respects as if the true value of Q has been increased; for example, the transient response time of the cantilever also increases. Figure 3 shows the measured transient response of the cantilever (in air) without (upper) and with (lower) the active-Q feedback enabled. The envelope of the increasing oscillation amplitudes with time agrees well with the curve calculated for a harmonic oscillator having the same values of Q as the experimental resonant peaks corresponding to the two cases, without and with active-Q feedback. This is further evidence that the cantilever is behaving as if its Q value were actually greater. A disadvantage of the increased values of Q is the corresponding increase in the response time of the cantilever that results in an upper limit to scan rates. In practice, this is not a serious problem as the values of effective Q obtained in liquid with the feedback are of the same order as the values of Q of the natural resonance in air, and so scan speeds are similar to those of tapping mode in air, though somewhat lower owing to the lower resonant frequency of the liquid tapping cantilevers.

III. Application of Active Q-Control AFM

A. Imaging

Some of the most difficult surfaces to image by AFM are the surfaces of swollen gels and unfixed cells. These specimens are very soft and highly deformable and so provide a severe test of the active-Q technique. Figure 4 shows liquid tapping mode images of

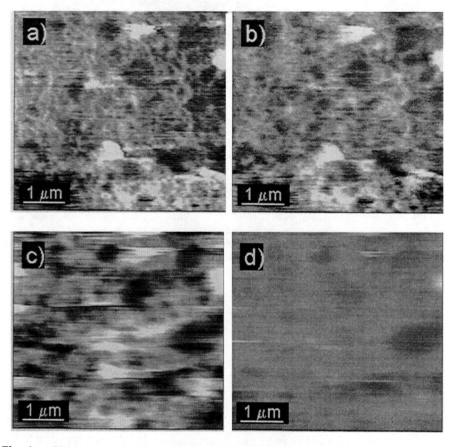

Fig. 4 A 30% isotactic polystyrene (iPS)/decahydronapthalene (dekalin) gel under dekalin, imaged with (a and b) and without (c and d) Q control . Topography (a and c) is displayed with a z range of 80 nm and phase contrast (b and d) with a range of 5 degrees. (See Color Plate.)

an isotactic polystyrene/dekalin gel imaged in dekalin. The gel was not allowed to dry out at the surface between its formation and the AFM imaging. Figures 4a and 4c are topographic images of the same area of the gel using the same cantilever recorded with and without active-Q enabled, respectively. Two significant improvements are clearly seen in a comparison of these images. First, the resolution in Fig. 4a is considerably higher than that in Fig. 4c; second, the deformation of the surface evident in the streaking of the images is considerably reduced in Fig. 4a compared to that in Fig. 4c. This is a result of the greater force sensitivity and thus of the ability to use lower imaging forces with active Q enabled in Fig. 4a. An even more dramatic improvement is seen in the phase images. The detailed phase image with active-Q enabled in Fig. 4b stands in stark contrast to the almost featureless phase image of Fig. 4d. This results from the increased phase sensitivity associated with the "sharper" resonant peak when active-Q is enabled. A similar result is obtained from a 1% agarose gel in water (Fig. 5) which was imaged under water and again never allowed to dry at the surface. The resolution of the topographic images (Figs. 5a and 5c) and the contrast of the phase images (Figs. 5b and 5d) are superior with the Q enhancement on Figs. 5b and 5d. Line profiles across the images illustrate the higher spatial frequencies contained in the image with active-Q enabled compared to the image without, and similarly the profiles across the phase images show the greater phase contrast also obtained with the active-Q feedback activated.

The difficulties encountered in imaging living cells are similar to those experienced in imaging swollen gels. The thickness of cells combined with their low stiffness results in large deformations, so that the tip can distort the structure locally to such an extent that lateral forces again become a problem and result in streaks in the image. The value in using active-Q for imaging unfixed cells is apparent from the images presented in Fig. 6. The topographic image with active-Q enabled shown in Fig. 6 a reveals little distortion of the cell as a result of the imaging process. Figure 6b is the corresponding active-Q phase image. This image shows high-resolution contrast of various structures of the cell, and in particular the cytoskeleton can be seen. This structure is usually seen when large forces are applied by the tip to the cell surface to deform the cell surface sufficiently to sense the more rigid structures beneath. However, in this case the forces applied are relatively low, and the structure is revealed through the dramatically increased sensitivity of the phase signal associated with the higher Q resonance.

The lower imaging forces facilitated by active-Q tapping mode imaging also have benefits for imaging single isolated molecules. The lower forces result in less compression of single molecules, and this results in molecular heights closer to the values that might be expected for the molecular structure. The molecules are also narrower in these images, so that the lower deformation also results in improved molecular resolution. Figure 7 is a comparison of the same doubled-stranded DNA molecule imaged under butanol with (Fig. 7b) and without (Fig. 7a) active-Q control enabled. The lower forces also allow the use of weaker adsorption to immobilize the molecule during imaging which results in a further reduction in molecular distortion.

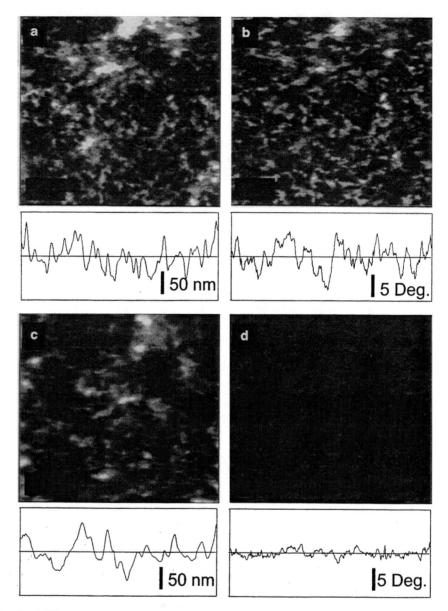

Fig. 5 A 1% agarose gel under water, imaged with an effective quality factor of 300 (a and b) and with conventional tapping mode (c and d). Topography (a and c) is displayed with a z range of 150 nm and phase contrast (b and d) with a range of 5 degrees. (See Color Plate.)

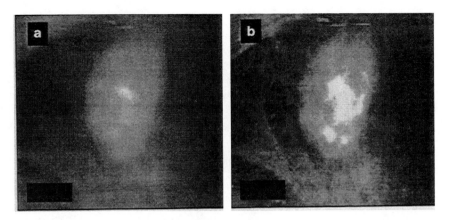

Fig. 6 A living rat kidney cell imaged in buffer with an effective quality factor of 300, height (a) z range 2.5 μm and phase (b) range 60°. (See Color Plate.)

B. Dynamic Force Spectroscopy

1. Background

The measurement of force as a function of extension of a single molecule using AFM force-sensing techniques has been named force spectroscopy. Some of the measurements already reported involved the forces associated with the unfolding of protein structures (Lenne *et al.*, 2000; Oesterhelt *et al.*, 2000; Rief *et al.*, 1998, 1999) and it is hoped that information on the nature of the folded structure can be extracted from these force measurements. Such molecular processes are dynamic and energetic and exhibit both conservative and dissipative forces. The use of the active-Q technique allows the measurement of the complex meachanical properties of such a process in the appropriate buffer environment (Humphris *et al.*, 2000). The increase in the quality factor to over 300 in liquid facilitates the tracking of the resonant frequency and the separation of conservative and dissipative forces. Tracking the resonant frequency via a phase-locked loop is the approach that has been used here. Such information also allows the measurement of the effective viscosity of a single molecule. The implementation of dynamic force spectroscopy not only increases the sensitivity of the measurement but also decouples the conservative (elastic) and dissipative (viscous) components of the force associated with the molecular extension. One realization of the dynamic force spectroscopy experiment is shown in Fig. 8. In the standard method of force spectroscopy, a molecule bound at one end to the AFM tip and at the other end to a fixed surface is extended by the displacement of the tip away from the surface. In the dynamic version of the experiment, a small vertical oscillation is superimposed on the displacement of the tip, and the force response to both the DC stretching and the AC displacement is recorded. The elastic force of the molecule acting, on the tip acts as a spring changing the resonant frequency of the cantilever by a factor $(1 - f'/k)^{1/2}$, where f' is the force gradient of the interaction and k the spring constant of the cantilever. The dissipative force changes the effective damping constant of the cantilever (γ), and, therefore, the effective quality factor of the

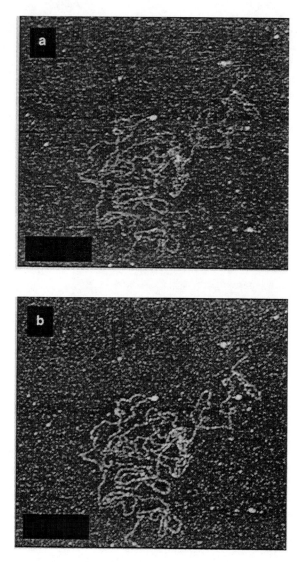

Fig. 7 DNA imaged under butanol with (b) and without (a) Q control, z range 3 nm. (See Color Plate.)

cantilever, $Q = m\omega_0/\gamma$, where ω_0 is the free resonant frequency of the cantilever. The change in Q is reflected in the change of the amplitude at resonance, given by $A = QF/k$, where F is the excitation force. The damping in liquid, resulting in a decrease in Q by 3 or more orders of magnitude, prevents the detection of changes in resonant frequency and quality factor with high enough sensitivity to observe molecular processes. The low Q is a result of the damping being dominated by the hydrodynamic interaction of the cantilever with the surrounding liquid, such that the much smaller molecular

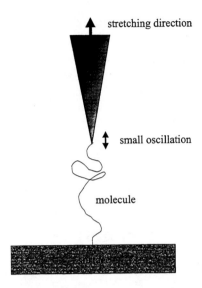

Fig. 8 A schematic representation of the dynamic force spectroscopy technique, showing a molecule being stretched by the linear motion of the AFM tip, but with a small oscillating displacement being added.

elastic and viscous forces are unobservable. The active-Q feedback technique effectively counteracts the hydrodynamic damping and increases the effective Q of the cantilever to provide sufficient sensitivity to measure molecular processes during molecular stretching.

Transient measurements (Fig. 3) have shown the motion of the cantilever to be essentially harmonic in nature, and so it is straightforward to calculate the elastic force gradient, f', from the observed frequency shift, $\Delta\omega$, using the nominal spring constant of the cantilever, k:

$$f' = k\left(1 - \left(\frac{\omega + \Delta\omega}{\omega_o}\right)^2\right). \tag{5}$$

The damping constant γ is related to the observed change in the oscillation amplitude by

$$\gamma \approx \frac{k}{\omega_0 Q_{\text{eff}}} \frac{A_0}{A}. \tag{6}$$

The intrinsic damping of the molecule, γ_{m}, can be evaluated by subtracting the damping constant of the free system γ_{sys}; for example,

$$\gamma_{\text{m}} = \gamma_{\text{eff}} - \gamma_{\text{sys}} = \frac{k}{Q_{\text{eff}}}\left(\frac{A_0}{\omega A} - \frac{1}{\omega_0}\right). \tag{7}$$

2. Applications of Dynamic Force Spectroscopy

As an illustration of the application of active-Q force spectroscopy, measurements on two polysaccharides, dextran and methyl cellulose, are shown in Fig. 9. The trace shown in Fig. 9a is the conservative (elastic) component, Fig. 9b is the dissipative component related to the effective damping constant of the dextran molecule, and Fig. 9c is the total force derivative derived from the DC deflection of the cantilever. These extension curves show a transition that has been previously reported for the dextran molecule using conventional DC force spectroscopy and which has been explained by the chair-to-boat transition of the glucose ring on molecular extension (Marszalek *et al.*, 1998). Dynamic differential force spectroscopy enables the separation of the elastic and dissipative components of the force during the transition. The inflection in the DC force

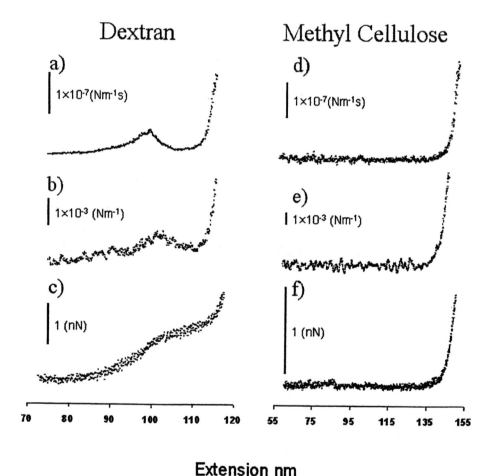

Fig. 9 Dynamic force spectroscopy measurements of a dextran molecule and a molecule of methyl cellulose. (a) and (d) are the dynamic elastic force gradient plots; (b) and (e) are the dynamic dissipative force gradient plots; (c) and (f) are the DC force plots.

measurements (Fig. 9c) is seen to correspond to a broad maximum in the differential elastic force response (Fig. 9a). This chair-to-boat transition is known not to occur for the polysaccharide methyl cellulose, and the corresponding dynamic force spectroscopy data shown in Figs. 9d and 9e do not show the features associated with the transition seen for dextran.

IV. Transverse Dynamic Force Techniques

A. Transverse Dynamic Force Spectroscopy

Force spectroscopy in a conventional AFM has some undesirable aspects. For example, as the molecule tethered between the cantilever tip and a suitable surface is stretched, the force on the cantilever increases; therefore the amount that the cantilever is bent increases (Fig. 10). This in turn, unfortunately, has the effect of changing the separation of the ends

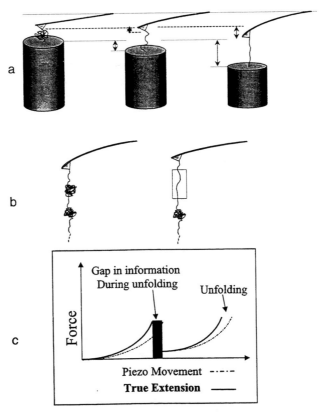

Fig. 10 A series of diagrams illustrating some undesirable aspects of conventional force spectroscopy. (a) A diagram illustrating the bending of a cantilever during the stretching of a molecule, resulting in an end-to-end separation being different from the displacement of the piezo transducer. (b) and (c) Illustration of the sudden jump in displacement of the cantilever tip during an unfolding process, and the consequent loss of information.

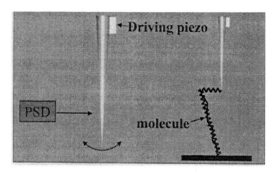

Fig. 11   A diagram showing the transverse dynamic force arrangement with vertically mounted probe. The stretched molecule provides part of the horizontal restoring force on the cantilever.

of the molecule so that there is not true control over the extension or the loading rate on the molecule. There is another, perhaps more serious, drawback in using conventional AFM for force spectroscopy. This involves the energy stored in the bent cantilever during the observation of a molecular process such as unfolding. The bent cantilever stores elastic energy that is released when the barrier to unfolding is overcome, causing a sudden jump in tip position. Thus, no information is available on the unfolding process itself, potentially the most interesting aspect of the extension process. This sudden jump could be avoided if the force transducer had infinite stiffness parallel to the extension direction, and this would also restore true extension and loading rates. Transverse dynamic force microscopy (TDFM) essentially provides a rigid force transducer as it employs a cantilever mounted vertically to the specimen surface (Fig. 11). The TDFM cantilever is made to oscillate laterally, usually by a few nanometres, at or near its resonant frequency, by either a dither piezo or by a magnetic field (Antognozzi *et al.*, 2000, 2001). The oscillations of the end of the probe are monitored by a position-sensitive detector. One version of the TDFM, known as the shear-force microscope (ShFM), is used to control the position of the optical fiber probe in scanning near-field optical microscopy (SNOM) (Betzig *et al.*, 1992).

In TDFM, only the tip of the cantilever need be in the liquid, and so the damping is much less than that in conventional AFM. The active-Q technique can be used to increase the effective Q of the TDFM probe in liquid, but this is usually unnecessary.

With a molecule attached between the TDFM cantilever tip and the surface, an additional restoring force component on the cantilever due to the molecule exists (Fig. 11b). Similarly, any dissipative component of the molecular force during extension will contribute to the damping of the cantilever. As was the case with dynamic force spectroscopy measurements with a conventional AFM, the additional restoring force shifts the resonant frequency to higher values and the dissipative force causes a reduction in amplitude of oscillation at resonance. As with the conventional AFM active-Q force spectroscopy, the resonant frequency was again tracked. Quantitative values of both elastic and dissipative components of the force exerted by the molecule during extension can be derived from a continuum mechanics model of the oscillating probe. TDFM force spectroscopy is a method that provides a rigid force transducer for controlled extension and loading rates, a means of avoiding "blind spots" in the data due to cantilever "snap back," and also simultaneous information on the elastic and dissipative components of the molecular force during extension. Figure 11 shows the basic TDFM arrangement and its application to force spectroscopy.

B. Application of Transverse Dynamic Force Spectroscopy and Microscopy

Results from the extension of a dextran molecule are shown in Fig. 12. The transition during extension of the dextran described here is also observed in measurements of the elastic (Fig. 12a) and dissipative (Fig. 12b) components recorded with transverse dynamic force spectroscopy.

TDFM is essentially a non contact force microscopy and so is of particular importance in imaging easily deformable structures such as unfixed, living cells. Figure 13 shows a TDFM image of a living rat kidney cell in growth medium. The image is similar to those obtained using active-Q tapping mode, but the TDFM images were recorded without the tip contacting the surface of the cells. TDFM clearly has great potential for imaging very soft specimens in liquid, but the current scan rates are rather

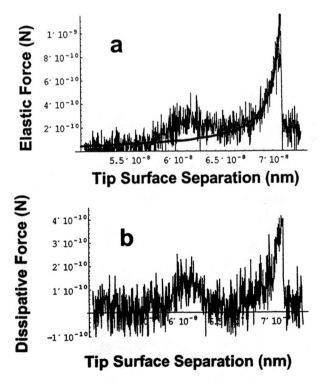

Fig. 12 Elastic (a) and dissipative (b) transverse dynamic force spectra of a dextran molecule.

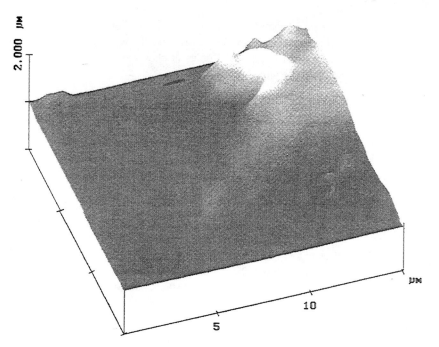

Fig. 13 A TDFM image of a living rat kidney cell imaged in growth medium. (See Color Plate.)

low, being limited by the resonant frequency of the probes and the associated lock-in techniques.

V. Conclusions

The increase in the effective Q of the cantilever applied to the liquid tapping mode is an important advance that will allow soft and delicate specimens to be imaged with less deformation, higher resolution, and greater phase contrast. Some examples of these advantages have been presented here. TDFM provides an alternative, essentially non-contact, image technique, but is significantly slower in imaging. Both techniques provide dynamic mechanical property measurements of single molecules with the high rigidity of the TDFM probe offering further advantages, particularly when following sudden transitions such as a protein unfolding process.

References

Anczykowski, B., Cleveland, J. P., Kruger, D., Elings, V., and Fuchs, H. (1998). Analysis of the interaction mechanisms in dynamic mode SFM by means of experimental data and computer simulation. *Appl. Phys.* A **66,** S885–S889.

Antognozzi, M., Haschke, H., and Miles, M. J. (2000). A new method to measure the oscillation of a cylindrical cantilever: The laser reflection detection system. *Rev. Sci. Instrum.* **71**, 1689–1694.

Antognozzi, M., Humphris, A. D. L., and Miles, M. J. (2001). Observation of molecular layering in a confined water film and study of the layers viscoelastic properties. *Appl. Phys. Lett.* **78**, 300–303.

Baker, A. A., Helbert, W., Sugiyama, J., and Miles, M. J. (2000). New insight into cellulose structure by atomic-force microscopy shows the I-alpha crystal phase at near-atomic resolution. *Biophys. J.* **79**, 1139–1145.

Betzig, E., Finn, P. L., and Weiner, J. S. (1992). Combined shear force and near-field scanning optical microscopy. *Appl. Phys. Lett.* **60**, 2484–2486.

Humphris, A. D. L., Tamayo, J., and Miles, M. J. (2000). Active quality factor control in liquids for force spectroscopy. *Langmuir* **16**, 7891–7894.

Lenne, P. F., Raae, A. J., Altmann, S. M., Saraste, M., and Hoerber, J. K. H. (2000). Stales and transitions during forced unfolding of a single spectrin repeat. *FEBS Lett.* **476**, 124–128.

Marszalek, P. E., Oberhauser, A. F., Pang, Y. P., and Fernandez, J. M. (1998). Polysaccharide elasticity governed by chair-boat transitions of the glucopyranose ring. *Nature* **396**, 661–664.

Oesterhelt, F., Oesterhelt, D., Pfeiffer, M., Engel, A., Gaub, H. E., and Mueller, D. J. (2000). Unfolding pathways of individual bacteriothodopsins. *Science* **288**, 143–146.

Rief, M., Gautel, M., Schemmel, A., and Gaub, H. E. (1998). The mechanical stability of immunoglobulin and fibronectin III domains in the muscle protein titin measured by atomic force microscopy. *Biophys. J.* **75**, 3008–3014.

Rief, M., Pascual, J., Saraste, M., and Gaub, H. E. (1999). Single molecule force spectroscopy of spectrin repeats: Low unfolding forces in helix bundles. *J. Mol. Biol.* **286**, 553–561.

Tamayo, J., Humphris, A. D. L., and Miles, M. J. (2000). Piconewton regime dynamic force microscopy in liquid. *Appl. Phys. Lett.* **77**, 582–584.

CHAPTER 17

Scanning Force Microscopy Studies on the Structure and Dynamics of Single DNA Molecules

Giampaolo Zuccheri and Bruno Samorì

Department of Biochemistry
University of Bologna
40126 Bologna, Italy

METHODS IN CELL BIOLOGY, VOL. 68

I. Introduction

The study of DNA is crucial. Gerd Binnig, the Nobel Prize laureate and inventor of both the scanning tunneling microscope (STM) (Binnig *et al.,* 1982a,b) and the scanning force microscope (SFM; atomic force microscope, AFM) (Binnig *et al.,* 1986), was the scientist who also initiated this field of research in 1984 by experimenting on DNA with the STM (Binnig, 1992; Binnig and Rohrer, 1984). Soon after its invention, the SFM was used for most of the DNA research that was begun with the STM because of its ability to image nonconducting specimens, as DNA is normally considered,[1] and because its images could be interpreted straightforwardly in terms of surface topography.

The late Albrecht Weisenhorn and his collaborators, the pioneers of SFM imaging of DNA (Weisenhorn *et al.,* 1990, 1991) prepared the groundwork for both the studies at the beginning of the 1990s and the *annus mirabilis* for this field. In 1992, many papers appeared, marking the emergence of DNA imaging with the SFM as a field of research. These papers focused on the methodologies used and stated that imaging had arrived, finally, to the point of being termed "reproducible." The turning point for making DNA imaging reliable was the control of adsorption of DNA molecules from a solution onto a surface adequate for SFM imaging. Since 1992, research groups found solutions for this problem and led SFM studies on the structure and function of DNA.

Due to the importance of DNA and the many themes of research related to both its structure and function, the studies on DNA led by SFM users soon became varied and tackled many different problems. The utility of the scanning force microscope for structural biology soon became evident. In the past few years, researchers in the very young field of nanotechnology are viewing DNA molecules, because of their peculiar physical

---

[1]Very interestingly, under particular conditions DNA has been found to be conductive and to function as a molecular wire (Aich *et al.,* 1999).

characteristics, as construction bricks for nanostructures. It is certain that the SFM will play a major role in the construction, function, and control of these new devices. In this chapter, we present a mixed historical and conceptual review of the methods used with the scanning force microscope to analyze the structure of nucleic acids, trying to provide the reader with the elements for evaluating the most proper technique for his/her own application.

II. The Control of Adsorption of DNA on Surfaces

Adsorption of DNA on surfaces has been, and still is to some extent, the primary concern of researchers in this field for many years. In the early 1990s, the stress was placed on the reproducibility and the efficiency of adsorption. A stable adsorption is necessary to prevent the scanning probe from sweeping the molecules around on the surface. Nowadays, attention is paid not only to the modulation and the specificity of the adhesion of DNA but also to the possibility of using the mildest conditions, so that fragile structures or complexes will not be altered or disrupted either in the deposition on the surface or in any other part of the experiment.

A. Electrochemical Adsorption of DNA

Discharging DNA onto a gold electrode in a controlled fashion was one of the first methods employed to adsorb DNA on a flat surface that could be suitable for imaging (Lindsay *et al.*, 1992). The development of STM (often fostered in the same laboratories that made the transition to SFM) made possible methodologies for preparing flat surfaces of gold (such as flat facelets on a small gold ball). This deposition method is reversible (simply by inverting the potential) and suitable for STM experiments but did not encounter much success since it both required a complex instrumental setup and produced data whose biological relevance could be questioned.

B. The Inheritance of the Electron Microscope

In many cases, the early results of SFM of DNA are due to the experience of electron microscopists. To cite one example, Yang and co-workers (Yang and Shao, 1993; Yang *et al.*, 1992) deposited the copolymer of DNA with cytochrome C on carbon-coated mica and obtained images of a quality comparable to other protocols.

Other protocols for DNA spreading have been modified by electron microscopy. The deposition assisted by benzyldimethylalkylammonium chloride (BAC) or similar cationic detergents has proved successful (Pietrasanta *et al.*, 1994; Schaper *et al.*, 1993), but the presence of the detergent might raise doubts over the possible alteration of structurally relevant information on DNA or DNA–protein complexes.

The data coming from conventional and cryo-electron microscopy experiments on DNA are always fruitfully compared to those produced by the SFM. Both techniques have their pros and cons.

C. Dehydration of Solutions of DNA on Mica

The most straightforward method for spreading DNA on a surface is to dry a dilute solution of molecules on a suitable substrate. To promote the adsorption of polyelectrolytes, the substrate must be highly hydrophilic. To be amenable to SFM analysis, the deposition substrate must be very flat and should not provide any structural features that could be mistaken for the specimen: ideally, it should be featureless and very clean. The need for a very reproducible substrate led to muscovite mica, which soon became the most frequently used surface for the adsorption of biological macromolecules for the SFM. This phyllosilicate can be easily cleaved to provide the microscopists with atomically flat and highly hydrophilic surfaces. In principle, every cleavage should expose a surface which is chemically equivalent to all the others.

The choice of the solution conditions for deposition is crucial. While the pH must be buffered to ensure both the chemical stability of the DNA and the biological relevance of its structure, the chemicals used for the buffering present a serious problem at the moment of dehydration, since they crystallize and precipitate on the surface as their concentration is increased by dehydration. One solution was to dissolve the DNA in ultrapure deionized water, checking that the pH was not too low, and to dehydrate a drop of such a solution on freshly cleaved mica (Samorì et al., 1993). The drawbacks of such a protocol were the scarce efficiency of the adsorption and the fear that DNA would assume undesired conformations in the extremely low-ionic-strength medium. A clever means of avoiding these drawbacks was to use a low concentration of ammonium acetate as a buffer (Henderson, 1992; Thundat, Allison et al., 1992; Thundat, Warmack et al., 1992). This salt is volatile and it should evaporate completely in water, leaving only the DNA on the surface. The method is reported as very efficient (using DNA solutions as dilute as 0.3 μg/ml), but it does not allow a fine control of the ionic strength of the solution as the water evaporates.

Alternatively, some researchers dried specimens that contained nonvolatile buffers (such as Tris) and sometimes resorted to rinsing the spreads after dehydration (Murray et al., 1993). A very high-electrolyte concentration near the surface could explain the very condensed DNA shapes sometimes shown by the SFM images obtained with this method.

D. Silane Treatment of Mica

Yuri Lyubchenko and his collaborators obtained a very strong adhesion of DNA to mica by treating the surface with a silane bearing an amino group, most commonly amino-propyl triethoxy-silane (APTES) (Lyubchenko, Gall et al., 1992; Lyubchenko, Jacobs et al., 1992). The amino group is sometimes methylated after binding the silane to mica (Lyubchenko, Gall et al., 1992). In water, the positively charged ammonium groups bind the negatively charged DNA (and RNA) and the adsorption allows SFM imaging. Methylation of the amino group on the surface makes the positive charge independent of solution pH and greatly increases the efficiency of DNA binding (Lyubchenko, Gall et al., 1992). There are several available protocols for surface functionalization with APTES: the freshly cleaved mica can be exposed to either APTES vapors (Lyubchenko,

Gall *et al.*, 1992; Lyubchenko, Jacobs *et al.*, 1992) or to acetone (Karrasch *et al.*, 1993) toluene, dimethylformamide (Lyubchenko, Gall *et al.*, 1992), and water solutions (Feng *et al.*, 2000). Some researchers prefer to use partly hydrolyzed silane solutions (Fang and Hoh, 1998). The surface of mica functionalized in such a fashion is referred to as AP-mica. The efficiency of AP-mica in adsorbing DNA is out of the question, as supported by the abundant evidence presented in the past (see, for instance, Feng *et al.*, 2000; Lyubchenko and Shlyakhtenko, 1997; Oussatcheva *et al.*, 1999; Shlyakhtenko *et al.*, 2000). AP-mica can bind DNA under a variety of conditions, from 0 to 60°C and under different ionic strengths, and from very dilute DNA solutions (down to 0.01 μg/ml for lambda DNA). In principle, it should be possible to modulate the strength of adsorption by reducing the density of the aminopropyl groups on mica, but the experiments reported in the literature all show that DNA is very strongly attached to the surface: at times too strongly. As will be discussed in the following, presently, one of the main concerns in SFM of DNA is both the control and the modulation of adhesion. Too strong an adhesion that cannot be modulated will prevent surface equilibration of the DNA molecules (Rivetti *et al.*, 1996) and seriously limit the possibility of performing dynamic studies on the adsorbed molecules. As shown by Lyubchenko and Shlyakhtenko (1997) AP-mica is very effective even for SFM in aqueous environments, and its adhesion is so strong and fast that molecules are frozen in the state they probably were in, in solution, but so strong that the residual mobility of the long DNA molecules is very small, as the authors demonstrate in their time-lapse experiments. Silane treatment could prove very useful to image the dynamics of very small molecules, whose mobility on the surface would be too high, unless their adsorption was very strong (Shlyakhtenko *et al.*, 2000).

It must also be noted that most surface treatment operations are somewhat tricky and often produce "dirty" surfaces, as not all the experimental parameters can be under strict control. APTES treatment is not an exception, and impurities in the reactants, partial polymerization or inhomogeneous treatments, can make a discontinuous surface, where only portions of the surface area are flat enough to be amenable to SFM analysis (Fang and Hoh, 1998; Lyubchenko, Gall *et al.*, 1992).

One of the main concerns in using surfaces treated with APTES is its evidenced ability to condense DNA (Lyubchenko, Gall *et al.*, 1992). Plasmids adsorbed from low-salt solutions can form rods and condensed amorphous structures (our unpublished data; Fang and Hoh, 1998). In previously reported experiments, the treated surfaces were carefully rinsed to ensure that unbound silanes were removed: the authors proposed that the DNA molecules could recruit mobile bound silanes that then cause the molecules to condense. Heat or vacuum treatments should reduce the presence of mobile silanes and have proved efficient in eliminating DNA condensation on treated silicon substrates. This post-treatment is strongly advisable when preparing surfaces for DNA adsorption.

E. Multivalent Cation Activation of Mica and the Control of Electrostatic Adsorption of DNA

In 1992 and later, another reliable method to adsorb DNA molecules to mica was employed by several groups (Bustamante *et al.*, 1992; Feng *et al.*, 2000; Hansma *et al.*,

1993; Hansma, Vesenka et al., 1992; Vesenka et al., 1992). It was found that substituting the monovalent cations of mica with divalent cations from a solution (usually magnesium) would make it bind the DNA molecules to which it was later exposed. These methods are said to activate the mica for the adsorption. Glow discharge was also used to make the magnesium-substituted surface more hydrophilic, even though it was considered unnecessary (Hansma, Vesenka et al., 1992). To reduce the effects of the capillary forces that lead to DNA damage, the images were produced under dry atmosphere using electron beam-deposited "supertips" (EBD tips) as probes (Keller and Chih-Chung, 1992), even though standard commercial probes can also be used. (Rabke et al., 1994). A trivalent cation pretreatment of mica has also been shown to work (Weisenhorn et al., 1990).

It was also evidenced that the magnesium pretreatment was not necessarily a distinct step from DNA adsorption (Hansma et al., 1993; Rabke et al., 1994), and solutions containing both DNA and magnesium salt could be deposited on freshly cleaved mica. The authors employing this protocol at the beginning would also dry the salt solution on mica, thus promoting adsorption; and only later they would (at times) rinse the solution to do away with the excess salt that could cause DNA structural artifacts (Henderson, 1992; Zenhausern, Adrian et al., 1992; Zenhausern, Adrian, Ten Heggeler-Bordier, Eng et al., 1992; Zenhausern, Eng et al., 1992).

Rabke and co-workers made an effort to understand magnesium-promoted adsorption in depth (Rabke et al., 1994). They challenged the hypothesis that magnesium would substitute potassium ions on mica, by measuring very little change in the quantity of the present potassium in the various phases of the surface preparation and DNA spreading. If an ion exchange takes place, it must be on a very limited number of sites.

It must be noted that these protocols of adsorption generally employ relatively high concentrations of DNA (especially those that deposit DNA on pretreated surfaces) ranging from 10 to 100 μg/ml of plasmids.

1. Soluble Cations Promote the Adsorption of DNA

The realization that a millimolar concentration of multivalent cations in solution with the DNA is enough to promote the adsorption of DNA to untreated freshly cleaved mica has really changed the way DNA is imaged with the SFM. The adsorption survives both a thorough rinsing step with deionized water and a subsequent drying of the salt-free surface (Vesenka et al., 1993).

As a rule of thumb, multivalent cations in solution promote the adsorption of DNA to mica, while monovalent cations decrease it (Bezanilla et al., 1995). As studied experimentally by Rivetti and co-workers (1996), in magnesium-mediated DNA adsorption, diffusion transports the DNA to the surface where it is irreversibly adsorbed. Under proper conditions, the adsorbed molecules still maintain two-dimensional diffusional freedom on the surface up to the moment the layer of water on the surface is dried.

Hansma and Laney studied the effect of the type and concentration of the cation on DNA deposition (Hansma and Laney, 1996). Many divalent inorganic cations have proved effective in promoting the adsorption of DNA on mica, if the adsorbed layer of DNA was dried prior to imaging. If, on the other hand, DNA is imaged in solution without first dehydrating the solution, the researchers were able to produce, with only a limited number of cations, stable images with good resolution of DNA molecules. The authors conclude that Ni(II), Co(II), and Zn(II) are suitable for solution imaging of DNA, either if used directly in solution with the DNA or if either Ni(II) or Co(II) is used for pretreating the mica surface later exposed to a Mg(II) containing DNA solution. Hansma and Laney also report that in Mg(II) solutions (those otherwise most commonly employed) DNA does not bind strongly enough to mica to permit imaging without dehydration. In our own experience, protocols for imaging DNA in solution are not yet completely reproducible. We and others have reported successful experiments where DNA molecules were imaged on untreated mica after being deposited from a solution containing Mg(II) as the only divalent cation. The solution has never been dehydrated (Rippe, Mücke et al., 1997; van Noort et al., 1998; Zuccheri et al., 1998). It seems that the success of the experiment is a complex function of the experimental conditions and mostly hinges on the quality of the probe. Our success rate is higher when employing EBD probes, probably due to their hydrophobic nature, even though we also obtained images with commercial Si_3N_4 probes, especially if the buffer solution was substituted with deionized water after initial deposition and imaging (unpublished results). Also, other groups have reported that commercially available tips can perform well in imaging DNA in solutions containing Mg(II) (van Noort et al., 1998).

Even though certain metal cations are more efficient in depositing DNA on mica, they could affect the structure of the nucleic acids by substituting the counterions they physiologically bear; furthermore, this could be poisonous for proteins, practically limiting the scope of the SFM protocols that employ them. It has been shown that, quite interestingly, Zn ions cause kinking in DNA (Han, Dlakic et al., 1997; Han, Lindsay et al., 1997); furthermore, Zn(II) is poorly soluble in water. Divalent cations can alter the secondary structure of DNA and they have been known to affect the contour length of DNA molecule images (Fang et al., 1998), probably by altering the winding of the DNA helix (Xu and Bremer, 1997).

The optimization of protocols for depositing DNA on untreated mica from solutions containing inorganic cations have made the observations of DNA routine. If a researcher has a pure sample of DNA, he/she can certainly deposit it on untreated mica and produce high-quality images by following these protocols.

2. Thermodynamic Equilibration of DNA on Mica

The structural analysis of DNA molecules requires a careful control of adhesion. As pointed out by Rivetti and co-workers (1996) many protocols that produce efficient DNA deposition on surfaces also lead to kinetic trapping of the molecules on the mica substrate. Such trapped molecules, depending on the path they followed for landing on the surface, display a global conformation that should represent the projection of the shape they had in solution. On the other hand, if the conditions are properly adjusted, adsorption would not completely hinder two-dimensional diffusion on the surface; thus molecules can equilibrate thermodynamically and assume global conformations that are directed by their physicochemical nature (which can be studied from their imaging). Many protocols for pretreatment of mica or direct deposition of DNA molecules can lead to a strong adsorption and trapping (several are listed in Rivetti's very important paper) and should be avoided. If the deposition protocol only exploits the ionic-exchange properties of mica, it seems sufficient to always keep a millimolar concentration of Mg(II) in solution to avoid trapping. Other treatments that significantly alter the nature of mica establishing covalent bonds with it or with DNA must be evaluated on an individual basis analyzing the statistical global shape of the adsorbed molecules with the methods described by Rivetti and co-workers (1996).

If the structural features of interest in DNA are more local, or are studied with methods that analyze local spatial parameters of the adsorbed DNA molecules, then the issue of surface equilibration is expected to be less serious. Many fast local dynamics must compose to change the global shape of a molecule. The local structural features (like bending angles) could be faithfully responding to the molecular properties even if the global shape is still not (see also Section VI,C). A stronger deposition protocol could provide thermodynamically meaningful data on a smaller scale than others or for shorter DNA molecules: for instance, on a comparable time scale, long DNA molecules diffuse considerably less on the surface of AP-mica than shorter molecules (Lyubchenko and Shlyakhtenko, 1997). (Shlyakhtenko et al., 2000). To harvest the greatest amount of information from SFM experiments, researchers should employ deposition methods that can be tailored not only to their experiments but also to the other boundary conditions. In the next sections, we will try to demonstrate one of our possible solutions for this problem based only on the electrostatics of untreated mica; however, the strength of many deposition techniques could certainly be modulated for particular applications.

F. Other Charged Molecules in Solution Can Promote DNA Adsorption

Inorganic cations are not the only effective electrostatic means for adsorbing DNA on mica. More complex organic charged molecules also proved effective. As mentioned earlier, Achim Schaper and collaborators, modifying protocols they had already used for electron microscopy, successfully employed quaternary ammonium salts (benzyldimethylalkylammonium chloride) to spread DNA (Pietrasanta et al., 1994; Schaper et al., 1993). The concentration of the detergent used in the reported experiments does not seem to affect protein–DNA interactions. A recent paper by Fang and Hoh (1999) demonstrates that soluble silanes bearing two or three amino groups are effective in promoting the adsorption of a water solution of DNA to freshly cleaved mica. The proposed method seems interesting, but the ability of the silanes to condense the DNA must be kept under control, unless one wants to study DNA condensation itself.

The interaction of these "artificial" complex molecules with biological macromolecules might be difficult to predict, and it would be advisable to reduce the factors that could lead both to artifacts and to a difficult interpretation of the obtained data.

G. Other Surface Treatments

It sometimes appears that the limitation for all possible surface treatments is only the fantasy of researchers. With the fine control of surface chemistry presently attainable, there are many possible modifications that would allow DNA depositions to be amenable to SFM structural analyses. We will mention only one more: cationic bilayers have been placed on the substrate and the DNA adsorbed onto them has been imaged at unprecedented high resolution with the SFM (Mou et al., 1995). The authors report a successful measure of DNA pitch and evidence for the visualization of DNA helix chirality. With such a high resolution, very local structural variations of the DNA could be directly visualized as the molecule interacts with proteins or other ligands.

H. Covalent DNA Anchoring for SFM Imaging

Since the beginning of force microscopy of DNA, researchers proposed to covalently immobilize DNA molecules on a surface. Since gold was already in use as a STM substrate, the immobilization of DNA on ultraflat gold via gold–sulfur bonds was immediately realized (Hegner et al., 1993). Thiols bind readily to clean gold in aqueous solutions, and a thiol group is easily introduced into the DNA molecules by means of modified nucleotides (for example, by using thiolated primers in PCR). Other methods have been developed in the following years, due to the developments of other techniques (for instance DNA chip technology) to bind DNA reliably to surfaces. These methods often employed self-assembled monolayers (SAM) of thiols on gold to present reactive groups to the solutions so that DNA molecules could attach reliably and controllably. Thiolated single-stranded oligonucleotides can be bound to a gold surface within a SAM, so that they can bind double-stranded DNA equipped with a single-stranded "capture tail" complementary to the surface-bound one, with the result of fishing them out of the solution (Bamdad, 1998).

Very recently, Shlyakhtenko and co-workers (1999) presented a method to photocrosslink DNA to a properly modified mica surface. The proposed method is reported to create homogeneous spreads of DNA on clean substrates, and thus is amenable to SFM measures (even though the authors reported that DNA appears unexpectedly kinked when deposited on this functionalized surface).

As in the previously mentioned studies by the group of Yuri Lyubchenko, other groups employ some silanes to covalently attach nucleic acids to surfaces (glass or silicon): these methods could be exploited for tailoring SFM protocols. Kumar and co-workers synthesized silanized nucleic acids, for example, by coupling a mercaptosilane with a thiolated oligonucleotide and then attaching them to the surface (Kumar et al., 2000). A variety of heterobifunctional crosslinkers are commercially available to help in binding oligonucleotides at surfaces, such as those utilized by Strother and co-workers (2000). Methods are available to functionalize silicon surfaces and make them expose methyl, chlorine, ester, acid, and even N-hydroxysuccinimidyl ester groups that could be used for the attachment of properly prepared DNA or oligonucleotides (Strother et al., 2000; Wagner et al., 1997, and references therein). The main (and well-evident) difference between this class of methods and that presented here is that the latter is designed to specifically adsorb properly functionalized molecules. Often, the protocols also reduce the non specific adsorption of other molecules in solution (Bamdad, 1998). Such careful design makes the methods generally complex, involving multistep functionalizations, and works on a properly designed and modified target molecule (with the exception of that presented by Shlyakhtenko and co-workers (1999)). For the general application in biology, a deposition method usually cannot modify the molecules it is interested in, because it is practically too difficult not to alter their properties. Due to the widespread interest of DNA attachment to surfaces for many techniques and uses (DNA chip technology, only to mention one), it is certain that these methods will continue to evolve. The researcher will soon be offered a variety of methods to employ for the solution of any research problem.

III. Air Imaging of DNA: Which Present, Which Future?

Imaging dehydrated specimens with the microscope operating in air is the easiest of the SFM operations. This is, indeed, one of the reasons for the popularity of the microscope and one of the advantages in its use with respect to the electron microscope, which must always be used under high vacuum. In our experience, any new user can master operations in ambient air reasonably well after a few hours of lecture in class and some hours of hands-on experience with the microscope. In the last few years, the stability of the microscopes operating, for instance, in tapping mode (see following) and the constant quality of the commercial probes make imaging in air significantly easier than it was in the past.

A. The Humidity Issue and SFM under Organic Solvents

The control of the interaction forces between the probe and the specimen is of fundamental importance in SFM, especially in *contact-mode* SFM, the first to be employed during the years. In contact mode, the probe and the specimen are always in contact, while the probe is dragged along the surface of the specimen. If the interaction forces are not minimal, the shear forces generated by the motion are sufficient not only to seriously damage the soft biological macromolecules irreversibly but also to produce bad quality images.

Among the possible interactions between the sample and the probe, the most relevant one in air imaging by contact mode is due to the hydration of the surfaces exposed to humid air. The layer of water normally present on any hydrophilic surface exposed to air creates a meniscus around the SFM probe, and this causes the onset of very high capillary forces that pull the probe toward the specimen.

What is commonly thought in SFM is that the sharper the probe, the better is the resolution of the images it produces. This concept takes a more subtle meaning in the presence of capillary forces. Here, a probe with a larger surface entering the hydration layer would imply the onset of higher capillary forces (Israelachvili, 1992). This argument in favor of very sharp probes could be balanced by the consideration that the residual forces that attract the probe toward the specimen would produce higher pressures if they were applied on the smaller area upon contact with a sharper probe and consequently be more disruptive for a soft sample. The researchers preferred to use EBD probes in an attempt to both reduce the effect of capillary forces and improve the resolution of the early images of DNA. An EBD probe is more hydrophobic than an unmodified Si_3N_4 probe: this could help in reducing the capillary forces. The reduction of the source of these forces took two possible paths. Some authors decided to reduce, by imaging while keeping the microscope with the mounted sample in dry nitrogen, the relative humidity of the environment in which the probe and the specimen were (Bustamante *et al.*, 1992; Thundat, Allison *et al.*, 1992; Vesenka *et al.*, 1992). Other authors eliminated the air–water interface at which the meniscus would form, by submerging the sample and the probe in propanol (Hansma *et al.*, 1993; Hansma, Sinsheimer *et al.*, 1992; Hansma, Vesenka *et al.*, 1992; Murray *et al.*, 1993; Samorì *et al.*, 1993). For DNA, this operation also has the side effect of improving the adhesion of the molecules to the surface, since DNA is insoluble in propanol. At the time this protocol was used, a stronger adsorption was certainly desired. The images obtained under propanol can still rival newer techniques as far as resolution is concerned (Hansma *et al.*, 1995).

Several authors studied the effect of both the relative humidity and the applied force on imaging of DNA with the SFM (Bustamante *et al.*, 1992; Ji *et al.*, 1998; Thundat, Allison *et al.*, 1992; Thundat, Warmack *et al.*, 1992; Thundat *et al.*, 1993; Vesenka *et al.*, 1992; Vesenka *et al.*, 1993; Yang *et al.*, 1996; Yang and Shao, 1993) and they generally found that a lower imaging force at the lowest possible relative humidity was the desired working condition.

B. AC Modes

The emergence of tapping mode SFM, in which the cantilever is continuously oscillated and in which it contacts the specimen only intermittently, has certainly represented a great improvement for SFM imaging of DNA (Hansma *et al.*, 1995). The lateral motion of the probe takes place almost totally when the probe is not in contact with the sample, so shear forces are virtually eliminated. With this new technique, the control of ambient humidity is still helpful but not normally necessary to produce high-quality images. Specimens can be scanned repeatedly in air without any recognizable damage. The instruments operating in tapping mode are generally very stable, and imaging has become easier and quicker.

C. In Search of Sharper Probes

A constant theme of research in SFM has always been the search of sharper or more specific probes. The report of the many advances of the field and the many varieties of probes available on the market is beyond the scope of this paper. We simply want to highlight two important landmarks in the improvement of the probe quality. The first major advance was the invention of EBD tips with an end radius of curvature on the order of 10 nm or less for probes (Keller and Chih-Chung, 1992). A second major advance was completed recently with the study of protocols attaching single carbon-nanotubes on the commercial SFM probes (Cheung *et al.*, 2000; Hafner *et al.*, 1999) [after other researchers previously resorted to manually attaching them to the apex of the probes (Dai *et al.*, 1996; Wong, Harper *et al.*, 1998)]. With single or multiwalled nanotubes pointing out from the probe, imaging tips with an end radius of 2–3 nm are within reach; furthermore, nanotubes can be chemically functionalized to give probes the ability either to measure specific properties or to manipulate molecules (Wong, Joselevich *et al.*, 1998).

D. The Interaction of DNA with Proteins and Other Ligands

Among the many research applications of the SFM in air, the structural studies of protein–DNA complexes are certainly some of the most remarkable. The scanning force microscope can visualize the complexes directly and under physiologically relevant conditions. Studies of protein–DNA complexes performed on single molecules of DNA demonstrated the advantages of distinctly showing specific versus nonspecific complexes and comparing the structural features of the two kinds of complexes. While other methods provide some information on the structure of specific complexes (X-ray diffraction, gel electrophoresis, and many others), one should rely on microscopic techniques on single molecules to learn something about the structural features of nonspecific complexes and the nature of the DNA–protein interaction (Erie *et al.*, 1994; Schepartz, 1995). A review of DNA–protein interaction studies with the SFM is beyond the scope of this chapter, but it seems useful to mention some of the important achievements. Bending angles and the location of binding proteins on long DNAs are easily measured (Bustamante and Rivetti, 1996; Jeltsch, 1998). The wrapping of DNA around proteins can be implied indirectly from comparisons to the length of uncomplexed DNA (Rivetti *et al.*, 1999). Often dimers or multimers of the binding protein can be distinguished also by means of volume measurements (Wyman *et al.*, 1997); looping and other unusual structures can be evidenced (Rippe, Guthold *et al.*, 1997). Obviously, protein-induced structural changes that interest the DNA can also be visualized when the individual proteins are too small to be seen on the DNA (Dame *et al.*, 2000).

Small DNA ligands cannot be seen with the SFM under normal conditions. Often the SFM analysis needs to resort to the study of the structural alterations that the ligand induces, as if the structure were a reporter of the binding or the activity (Coury *et al.*, 1996, 1997). Sometimes what the researchers are really interested in is the structural alteration that the ligand induces and whether the structure is fundamentally connected to the ligand activity. As an example, the coiling up of DNA plasmids in solution in the presence of ethidium bromide has been followed in real-time on mica (Pope *et al.*, 1999).

Sometimes a bulky tag (like a protein) can be tethered to a small ligand to report its location: as an example, by binding a bulky streptavidin to a biotinylated PNA probe

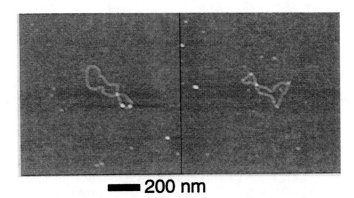

Fig. 1 The complexes between PNA and DNA have been evidenced by tagging a biotinylated PNA with streptavidin, which is seen as a globular object on the DNA strands. The formation of the complex should be sequence specific. The streptavidins bound at the crossovers of the DNA strands could be crosslinking two biotinylated PNA molecules through their multiple binding sites.

(unpublished results) (see Fig. 1) we verified the binding and imaged the location of peptide nucleic acids–DNA duplexes (PNA–DNA) on a large DNA plasmid.

E. A or B? This is the Question: The Secondary Structure of DNA from SFM Data

Since the beginning of SFM of DNA, concerns were raised regarding the structural alterations that dehydrating DNA molecules could imply (Bustamante *et al.*, 1992). Dehydration itself can drive the transition from B- to A-DNA, the average form being present at a reduced relative humidity. In several instances, researchers found that DNA molecules imaged after dehydration were somewhat shorter than expected for B-DNA and they attributed such shortening to a partial B-to-A transition (Bustamante *et al.*, 1992; Hansma *et al.*, 1996; Rivetti and Codeluppi, 2000). In our opinion, the phenomena involved in dehydration of a DNA spread on mica are not completely under control. The extent of such partial transition certainly depends on the time required for the drying step, the total residual humidity on the sample, and the degree of adhesion of DNA on the surface at the moment of dehydration. Even ethanol dehydration proved inefficient in contracting DNA molecules adsorbed on mica under trapping conditions (Fang *et al.*, 1999). Some authors (us, among others) concluded that DNA retained its B structure on mica (Fang *et al.*, 1999; Hansma *et al.*, 1993; Muzzalupo *et al.*, 1995; Rippe, Mücke *et al.*, 1997). Plausibly, very long DNA molecules find more adhesion sites on the mica and are more refractory to dehydration-driven contraction, conserving a higher fraction of B-DNA along their length, even if thoroughly dehydrated (Hansma *et al.*, 1996; Rivetti and Codeluppi, 2000).

Considering the amount of structural results obtained from studying DNA and DNA–protein interactions by operating the microscope in air and the degree of their general accordance with data from other techniques, we would be willing to conclude that data collected on linear DNA in air do not seem to be heavily affected by this structural transition.

F. Solid–State Sizing and Other Possible Futures for the Imaging of Dried Specimens

Although the scope of the alterations induced by sample drying could be limited, it is certainly advisable to collect SFM data from DNA in solution when possible. With the constant development of the technology and the protocols, we expect that SFM imaging in solution will soon be as easy as imaging in air is now. In any case, it is our opinion that there is still a future in SFM imaging of DNA on dehydrated specimens.

The speed, stability, and ease of SFM operations in air make them amenable to automation. With the available instruments, a sample can be scanned automatically to produce hundreds of images in a day. Automatic image processing techniques are beginning to tackle the problem of extracting structural data from the images (Spisz et al., 1998). One of the future roles of SFM imaging of DNA will be that of complementing gel electrophoresis in the sizing of DNA fragments or the measure of DNA damage (Fang et al., 1998; Murakami et al., 2000). Such analysis will possibly be conducted automatically, probably as one step in a multistep synthesis and characterization. The advantage, beyond automation, is that the SFM can work on minute quantities of DNA and represent a new level in sensitivity for DNA sizing (Fang et al., 1998). Another significant advantage of imaging dry specimens (obtained with carefully controlled protocols) is that specimens can be stored for future observations; reactions can be stopped at particular stages to enable kinetic studies that can parallel solution bulk analyses (Hansma et al., 1999).

IV. Imaging DNA in Fluid

One of the main advantages of SFM over other high-resolution imaging techniques (namely, EM) is that it can operate when both the probe and the specimen are submerged in liquid. For the operation of the most widespread commercial microscopes, the liquid needs to be transparent to the laser beam used for reporting the vertical position of the probe on the surface. Less common cantilever and SFM detector types (piezo-resistive cantilevers, for example) could allow imaging to take place in highly optically absorbing or dispersing media.

A. Imaging under Water or Buffers

Even though more technically challenging, imaging was performed under fluid since the emergence of SFM investigations of DNA (Hansma, Vesenka et al., 1992; Samorì et al., 1993). As mentioned previously, submerging the probe and the DNA in propanol (or butanol, or ethanol) increased its adhesion to the substrate and reduced the strong capillary forces that could hamper imaging in air by contact-mode SFM. Although the spatial resolution of the recorded images was very good (Hansma et al., 1995), imaging under organic solvents has been abandoned in the search for more native conditions, such as imaging fully hydrated DNA under water.

In contact-mode imaging, the adhesion of DNA molecules to the surface needs to be strong to prevent the scanning probe from scraping the DNA off the surface. The two main solutions found were imaging DNA on AP-mica (Lyubchenko, Gall *et al.*, 1992) and imaging DNA that had been thoroughly dehydrated on untreated mica (Hansma *et al.*, 1993). As mentioned previously, the attachment of DNA to AP-mica is so strong that it inhibits almost every motion, even though the sample has never been dehydrated. The re-hydration of DNA does not allow it to move on the surface; furthermore, serious doubts can be cast either on the structural relevance of rehydrated DNA molecules or on the possible induced strand damage.

The next major advance was the introduction of tapping-mode SFM to the operations under fluid. This advance, fostered in the labs of both Paul Hansma (Hansma *et al.*, 1994) and Jan Greve (Putman *et al.*, 1994), allowed soft samples to be imaged under liquid without being damaged and without being swept around by the scanning probe, even if it was only weakly attached to the surface.

The most relevant problem for solving imaging under aqueous solutions, which we discussed at length in the previous sections, is finding the proper conditions for DNA adsorption. Further historical or methodological details go beyond the scope of this chapter. The chapter written by Helen Hansma and her collaborators on the techniques of imaging in fluid is certainly a very good source of information on the subject. In the following sections, we will limit ourselves to describing a few contributions to the field in which our lab had a relevant role.

B. The Modulation of DNA Mobility and the Imaging of DNA Dynamics in Solution

The success of experiments of DNA imaging in solution depends on the strength of adhesion of the DNA to the surface, usually mica. The success of the experiments designed to study the dynamics of DNA is mainly dependent on the ability to modulate the adsorption of DNA on mica. A very strongly adsorbed molecule would be imaged faithfully and at high resolution by the SFM probe but would not be interested by any motion and the strong adsorption would likely hinder its interaction with either proteins or other ligands (van Noort *et al.*, 1998; Zuccheri *et al.*, 1998). Molecules that were too mobile would be difficult to image at high resolution, since the speed of the motion of the DNA chains is comparable to that of the scanning probe. Technical improvements are continuously speeding up image collection in solutions and will soon allow the imaging of very fast dynamics with the SFM (Argaman *et al.*, 1997; van Noort *et al.*, 1998).

In our laboratory, we implemented a careful control of electrostatics to modulate the adhesion of DNA on freshly cleaved mica. During the course of our experiments, the specimen was never dehydrated. Surface charge screening has a significant effect on the adhesion of charged polyelectrolytes on mica, and we chose to tinker with it in order to continuously and reversibly change the degree of adhesion of DNA. Once deposited from a solution containing magnesium cations, DNA will stay bound on the mica for a long time, even if the solution in the SFM fluid cell is substituted with low-salt buffers without any magnesium or multivalent cations. It seems likely that the exchange of the bound Mg(II) with the monovalent cations in solutions is very slow, while the change in the charge screening guaranteed by solution cations not bound to the surface is very quick and immediate after a change in the solution environment of the fluid cell (Samorì et al., 1996; Zuccheri et al., 1998). A gravity-driven injection system piped to the fluid cell enables continuous changes in the ionic strength of the medium: as deionized water is injected, the charge screening decreases, but the DNA molecules on the surface do not desorb. On the contrary, they experience an increased electrostatic attraction to the surface. When higher ionic strength buffers are injected, the charge screening is reconstituted, the electrostatic interaction between the DNA molecules and the surface can now be competed by all other cations, and the DNA molecules can diffuse two-dimensionally on the surface shifting between binding sites: they appear more mobile. It is conceivable that the scanning probe could influence such motion promoting local adsorption/desorption reactions, but we expect that its effect will only superimpose to other desorption/adsorption trends. No neat molecular motions in the direction of the probe scan were recorded in our experiments. As shown in Fig. 2, this ability to switch adhesion on and off in a reversible and controlled manner was used to visualize the dynamics of supercoiled DNA molecules in solution (Zuccheri et al., 1998). In this experiment, supercoiled pBR322 molecules were deposited on the surface of freshly cleaved mica from a solution containing DNA at 1 μg/ml, 4 mM HEPES buffer (pH 6.8), and 1 mM MgCl$_2$. Conditions were such that molecules adsorbed on the surface but conserved some diffusional freedom in two dimensions. When the adsorption is made stronger by the reduction of surface charge screening, the SFM images are clearer and DNA strands are observed at high resolution (see 5th and 7th frame). On the other hand, when a higher ionic strength buffer is on the surface, the chain motions make the images more blurred and the resolution poorer, but differential dynamics among DNA chains can be evaluated (for example, 1st to 4th frame in Fig. 2). The dynamics of the supercoiled molecules can be studied from the comparison of frames obtained before and after "on" and "off" states of chain mobility.

Most of the time we used EBD tips that we grew locally in a scanning electron microscope (Keller and Chih-Chung, 1992), but we used commercial Si$_3$N$_4$ probes successfully, as also reported by other groups (van Noort et al., 1998). In our opinion, the chemical nature of the probes (their hydrophobicity), more than the end tip radius, can make an observation successful. EBD tips are expected to be significantly more hydrophobic than commercial Si$_3$N$_4$ ones. The success of the observation of DNA in buffer is strongly dependent on the probe: some probes can only observe DNA deposited in buffer after a water injection, while others can image the DNA chains distinctly in buffer, where the mobility is higher. Obviously, a probe that interacts strongly with DNA can sweep it around while scanning. As we have reported (Zuccheri et al., 1998), we frequently observed transient adsorption states on freshly cleaved mica. With a time scale slower than the charge screening changes induced by the operator (about 10 min), the DNA molecules spontaneously regain their diffusional freedom after a deionized water injection has frozen them on the surface, without any further external change of the bulk of the solution in contact with mica. This mobilization is reversible with a further injection

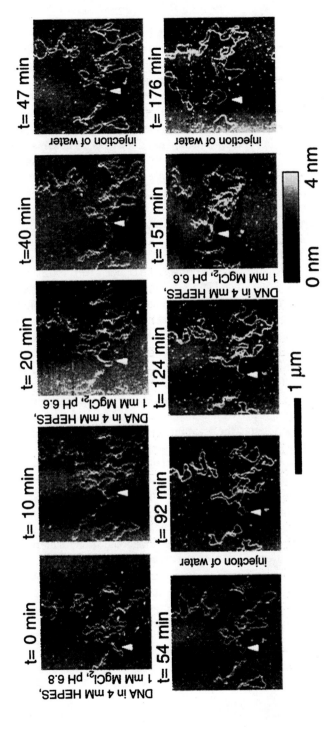

Fig. 2 SFM study of the dynamics of supercoiled DNA in solution. DNA molecules were deposited on freshly cleaved mica and a time-sequence of their images was recorded. The ionic environment was changed as detailed in the figure to modulate the strength of adhesion on the substrate. Changes in the shape, in the size of the loops, and in the adhesion can be clearly seen, especially for the molecules indicated by the arrows. Reproduced with permission from Zuccheri, G., Dame, R. T., Aquila, M., Muzzalupo, I., and Samorì, B. (1998). Conformational fluctuations of supercoiled DNA molecules observed in real time with a scanning force microscope. *Appl. Phys. A* **66**(suppl., pt. 1–2), S585–S589, courtesy of Springer–Verlag.

of deionized water. After any water injection, DNA molecules stay trapped on the surface for a longer time. After a few cycles of spontaneous remobilization/water-induced immobilization, the DNA molecules can stay adsorbed on the surface for either a time longer than the experiment or until an injection of buffer increases the charge screening. We believe that this is not only a manifestation of the cation-exchange properties of mica but also a result of slow release of cations from its basal plane. In the absence of cations in the bulk, there is a slow release of the cations from the mica (K^+) that exchanges with the H^+ present in solution (Nishimura et al., 1995). The departure of K^+ and the diffusion of H^+ could cause an increase in the concentration of K^+ in the layers of solution in contact with the solid–liquid interface. This increase in K^+ concentration might not be too small, due to the small thickness of the layer interested by the adsorption of DNA. After several injections of water, all the mica is H-exchanged and there cannot be any net changes in the electrolyte concentrations, unless the solution is opportunely changed. As also reported by Rivetti et al., (1996) the H^+-exchanged mica is stickier than the freshly cleaved mica for DNA. There is evidence that the kinetics of ion release from mica could be on the observed time scale (Paige et al., 1992).

It could be argued that since DNA was deposited from a solution that contains salts, the remobilization after a water injection could be due to mixing problems in the fluid cell during the injection. Residual salts that have not flowed away with the water could diffuse to the surface and cause the increased charge screening. In our opinion, this is not the case. Injections in the SFM cell always use an abundant volume of fluid: in 10–15 s, several milliliters of deionized water or buffer are flowed through a cell whose volume is around 30 μl. Under these conditions, we believe that the buffer substitution in the cell is complete, although we have never performed accurate measures of the type of flow. SFM imaging of DNA can also be made with a constant flow in the cell, if no vibrations are transmitted to the scanning probe. This is a better technique to allow the probe to function at thermal equilibrium. In experiments performed with a constant buffer or water flow, we never observed the transient adsorption states described previously.

As also shown in Fig. 2, we also tried to implement a pH-based method for the electrostatic control of DNA adsorption on mica, but the results were less encouraging. One of the problems in controlling the adsorption by playing with the ionic strength is that many biological processes and the structure of DNA itself are very sensitive to the ionic strength of the medium. For a change in the ionic strength of the medium, not only the degree of adsorption of DNA on the surface changes. We verified this possibility in the experiment shown in Fig. 3. A covalently closed circular DNA molecule coils up during the experiment because of the increase of the ionic strength of the medium. Later, the increased ionic strength allows the motion of the molecules with a speed that was not compatible with imaging. (Only their "ghosts" are visualized.)

C. Movies of Molecular Motion

With the current technical development in SFM and the fine control of DNA adsorption, it is presently possible to record true movies of the DNA motion on the surface. To compose a movie, a great number of frames are necessary to show gradual changes in

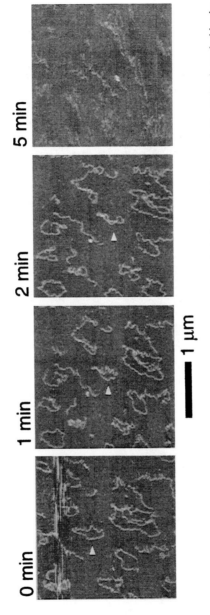

Fig. 3 Time sequence of DNA molecules imaged in fluid with the SFM as the ionic strength is increased. The molecule marked by the arrow coils up during imaging as a result of the change in the ionic environment. In the last frame, the high ionic strength of the medium is responsible for an increased mobility of the molecules on the surface that makes imaging impossible. Reproduced with permission from Zuccheri, G., Dame, R. T., Aquila, M., Muzzalupo, I., and Samori, B. (1998). Conformational fluctuations of supercoiled DNA molecules observed in real time with a scanning force microscope. *Appl. Phys. A* **66**(suppl., pt. 1–2), S585–S589, courtesy of Springer–Verlag.

the shape of the molecules. The frame rate of tapping-mode SFM imaging in fluid is still fairly low, even if there have recently been important technical improvements (Argaman *et al.,* 1997; van Noort *et al.,* 1998, 1999). The current limit in the frame rate seems to be around 10 s/frame, which is a reasonably high scan rate but still much slower than what is attainable with optical microscopy (the other family of microscopy techniques that can record dynamics).

One of the most severe limitations intrinsic to the SFM is that it must operate on specimens adsorbed on a surface. All those interesting biological processes that take place in solution are outside the realm of the SFM. This limitation turns out to be a valuable advantage for the study of DNA. The dynamics that would interest the DNA molecules in solution would have time scales well beyond the possibilities of a SFM experiment. The adsorption on the surface tremendously slows the motion of the DNA chains. If the adsorption sites on the surface are isotropically distributed and they do not impose preferential motion paths to the DNA chains, then the effect of the adsorption is only to slow down the kinetics of the processes. We believe that for the case of DNA adsorbed on mica, the system could be considered as an isotropic distribution of binding sites, and the DNA segments can shift from site to site without a preferential direction determined by the surface structure. The distance between charged sites on mica should be 0.5 nm (Hansma and Laney, 1996; Nishimura *et al.,* 1994, 1995). It seems plausible that this is dense enough to allow a DNA phosphate to shift from one site to another without imposing significant strain on the rest of the molecule.

The quest for thermodynamic data obtained from conformational fluctuations of single molecules in real-time requires the ergodicity of the system (see Section VI,C). A sample constituted of a single molecule is also the maximally homogeneous specimen that can be obtained, and dynamic behaviors that are different from one molecule to the other can be picked up during the observation [as could be the case for RNA polymerase (Davenport *et al.,* 2000)]. Such classes of different behaviors would be lost in a method that deals with a large population of molecules. Processes like DNA transcription have already been followed by SFM (Kasas *et al.,* 1997); transcription pausing and arrest could be studied by SFM in solution on single molecules having both a structural and a kinetic characterization at the same time.

Surface-bound enzymatic systems are very common in biology: true two-dimensional enzyme factories work on biological membranes and take advantage of the reduced dimensionality of the system. It is very likely that a good deal of the future work in scanning force microscopy will be devoted to surface structural biology and enzymology.

D. DNA–Protein Interaction on Stage

Many applications of the SFM in solutions have already appeared in the literature in the last few years. Besides the exciting observation of RNA polymerase moving on a DNA template (Bustamante *et al.,* 1999; Guthold *et al.,* 1999; Kasas *et al.,* 1997), the complexes between DNA and photolyase have been observed (van Noort *et al.,* 1998). The effects on nuclease digestion have been observed in real-time (Bezanilla *et al.,* 1994). The structure of chromatin has been imaged in solution (for example, on glass, in the absence of divalent cations) and SFM studies of the dynamics of chromatin fibers may be on their way (Bustamante *et al.*, 1997; Fritzsche *et al.*, 1995).

V. DNA Manipulation with the SFM: The Controlled Dissection of DNA

To produce an image of a nano-object, the AFM scans a solid probe in its proximity. A probe that touches a macromolecule has the potential of doing much more than simply imaging it. There already are several examples in the literature where peculiar SFM probes, or particular motions applied to the probe, function as nanoscalpels, a nano-brooms, nanotweezers, or other similar tools (Baur *et al.*, 1997; Ramachandran *et al.*, 1998, and references therein).

A. Dissection of Single Molecules

From the very beginning of SFM of DNA, the researchers realized that the DNA was easy to damage if the imaging conditions were not perfect for imaging. By scanning only one line of the sample with an increased imaging force, researchers could cut the DNA double chain and see the incision, often a wide gap in the molecule (Hansma, Vesenka *et al.*, 1992; Henderson, 1992; Vesenka *et al.*, 1992; Yang *et al.*, 1996).

The novel potentiality of the SFM to image DNA dynamics in solution was employed in our laboratory to experiment with DNA dissection in solution (Samorì, 1998). Always working in solution allows the DNA to stay in its native conformation, so that it can still be subjected to enzymatic reactions. SFM-induced dissection can be used for physically mapping the DNA: a section of interest can first be localized with the SFM in its imaging mode (or coupling the SFM with optical or confocal microscopes); then it can be abstracted with opportune techniques and handled as desired. In the experiment presented in Fig. 4, a single hydrated DNA molecule is precisely cut at a single point by increasing the force applied by the scanning probe. The cut produced after the first frame creates a linear double-stranded DNA molecule from a previously circular one. If the imaging conditions are completely under control, the mobility of the DNA molecules on the mica surface can be opportunely tuned (see Section IV,B). As shown in Fig. 4, after being cut, the new linear DNA molecule rearranges on the surface on which it is weakly bound. As time passes, thermal motion drives a conformational transition in which the DNA molecule starts from the initial conformation at zero end-to-end distance and then assumes states with increasing end-to-end distance (see the frames after the first). The mobility on the surface controls the kinetics of the transition, but it is the probe-induced cut that starts it when the operator wants. The same probe is used not only to image the scene before and after the incision but also to make the incision and start the conformational rearrangement. After a survey of a large portion of the surface, the operator chooses which molecule he/she wants to subject to the strand cut; the other molecule in the time sequence (the coiled one in the lower part of the frames) was not subjected

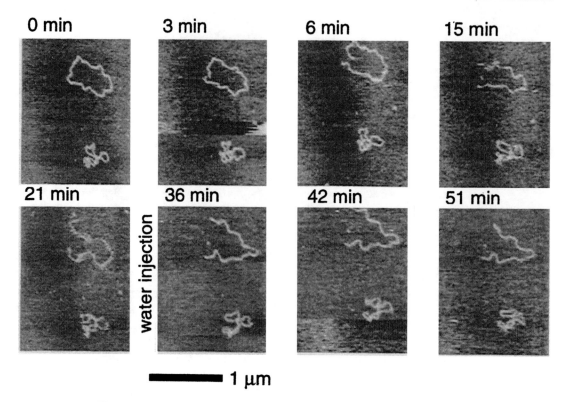

Fig. 4 Single-molecule microsurgery performed with the SFM probe on a DNA molecule in fluid. The probe induces a single cut in the molecule on the top of the first frame at $t = 0$. The molecules are adsorbed under conditions that do not prevent them from diffusing two-dimensionally; thus, the newly created linear molecule starts opening after the cut. The supercoiled molecule in the bottom part of the time-sequence frames had not been cut and it maintains its shape unaltered, except for diffusional motions. Reproduced with permission from Samorì, B. (1998). Stretching, tearing, and dissecting single molecules of DNA. *Angew. Chem. Int. Ed.* **37**(16), 2198–2200, courtesy of Wiley–VCH.

to any dissection and it maintained its shape during the entire lapse of the sequence. Since it shares the same portion of surface of the relaxed molecule that is cut, the coiled molecule experiences the same conditions of adhesion on the surface; thus, it appears blurred when the other molecule is also blurred (since they both are weakly bound).

The success rate of these experiments is still low. Often the increase of applied force that is necessary to cut a DNA molecule is enough to ruin the SFM probe, for instance, by making it pick up something from the surface. The resulting loss of resolution is sometimes enough to prevent further imaging.

The reported experiment is really the inverse path of a DNA cyclization experiment (Crothers *et al.*, 1992), and many useful pieces of information could be gathered on DNA by the careful study of the kinetics of the process. The advantage with respect to solution cyclization experiments is that the high-resolution structure of the molecules can be studied at the same time, and the sequence elements responsible for the conformational change can be easily pinpointed. A similar experiment was modeled for circular oligopeptides (Steinberg, 1994).

B. The Ablation of Fragments of DNA

A further step forward on the road to single-molecule analysis and DNA nanotechnology is the possibility of using the SFM tip as either a nanopipette or a nano-tweezer and abstracting one carefully selected piece of DNA from a population of available fragments that have been adsorbed on a surface. The chosen fragment could already be the result of previous nanotechnological operations, such as a probe-induced cuts. The fragment withdrawn from the surface could then be subjected to PCR or other molecular biological techniques, after having been solubilized again. This single-molecule biochemistry is not the optimistic forecast for the future improvement of probe technologies. Xu and Ikai (1998) already obtained a fine control over the adhesion of DNA to the SFM probe, so that adhesion could be "switched on and off " to image DNA and then attach it strongly to the tip and remove it from the surface. Xu and Ikai then successfully subjected the resuspended DNA molecule to PCR to amplify it.

These types of experiments were also performed on chromosomes (Heckl, 1998; Thalhammer et al., 1997), from which the probe could scrape fragments of DNA from locations chosen during the imaging phase of the experiment (also with the aid of other imaging techniques) (Thalhammer et al., 1997).

VI. The Study of DNA Conformations and Mechanics: Curvature and Flexibility

Many biophysical techniques are presently available to study the physical chemistry of nucleic acids. Among the EM studies of DNA, a certain number of studies were obviously devoted to the investigation of DNA's physical characteristics, such as elasticity or intrinsic curvature (Bettini et al., 1980; Frontali et al., 1979, 1988; Joanicot and Revet, 1987; Muzard et al., 1990). The emergence of the SFM gave new impulse to these studies due to the simplicity of preparing DNA spreads for the SFM and to the higher biological relevance of observing a partially or completely hydrated specimen (with respect to the vacuum-dehydrated ones) without the need of any staining or any coating that could alter the properties of the nucleic acids. In the past few years, cryo-electron microscopy (cryo-EM) has become a new and interesting method for biophysical chemistry measurements. Some very interesting studies were performed, taking advantage of the cryo-EM three-dimensional imaging of macromolecules in solution to study nucleic acids (Adrian et al., 1990; Bednar et al., 1995; Dubochet et al., 1992; Dustin et al., 1991; Furrer et al., 1997). Some concerns have been raised in the past about the validity of cryo-EM data: some authors fear that the very low temperatures used in freezing can force the molecules into unusual conformations, that the evaporation of the thin layer of solution in the moments preceding vitrification could uncontrollably increase the salt concentrations, and that the vitrified layer could still be too thin and force molecules into unusual conformations due to the presence of the nearby interfaces (Gebe *et al.*, 1996).

A. Global Measures of DNA Curvature and Flexibility

A number of experimental techniques can measure observables dependent on the global conformation of DNA molecules and gather physicochemical data from them. For instance, in DNA cyclization experiments, the global conformation of the molecule causes an increased or reduced probability that the two ends of a molecule could come together properly to enable ring closure. In hydrodynamic techniques, the global shape of a molecule determines its behavior when subjected to a flow. For DNA, all these global characters are determined by the base sequence and consequently by the composition of the local conformations along the contour of a molecule.

The microscopy determinations of the global shape of DNA molecules have relied on the distance of the ends (*eed*) after the molecules have been adsorbed on a flat substrate (Bettini *et al.*, 1980). Such an observable is related to the global shape and condensation state of the population of molecules. The statistical physics of polymers can relate this observable not only to the solution state of DNA but also to other observables measurable with different techniques. The statistical values of the *eed* can be rationalized in terms of the mean intrinsic curvature or flexibility of DNA, and comparisons between the results of different sequences can single out the differential effect of short sequences or of other factors, such as self-avoiding effects. The shortcoming of this approach is that local effects are spread out over the entire length of the molecule, and the longer the molecule, the more difficult it is to pick up the origin of a "perturbation." Theoretical efforts have been devoted to squeezing out most of the information contained in a global parameter, such as the *eed,* and localize, for instance, single pronounced curvatures (Rivetti *et al.*, 1998). These types of determinations have also been performed with the SFM and have the advantage of being able to accurately control both the environmental conditions and the adhesion to the substrate to obtain thermodynamically relevant data (Rivetti *et al.*, 1996, 1998). Most of the interesting data on the flexibility and curvature of DNA have been obtained for particular DNA molecules with tracts of anomalous flexibility, due to the facility of studying the differential behavior with respect to a "standard" DNA.

B. A More Local Approach to DNA Curvature and Flexibility

If complex and multiple local contributions can compose, add, or cancel out to give the mean conformation of a DNA molecule, and the effect of a single tract of sequence is, at the very least, blurred over a long molecule, then the most natural direction for a study would be to make it local, i.e., exploit the high-resolution imaging capabilities of the electron and the probe microscopes. Using the transmission electron microscope, Muzard and collaborators mapped the local curvatures on the segments of pBR322 double-stranded DNA (Muzard *et al.*, 1990). What they found was in accordance with other well-established techniques that localized significant curvatures along that DNA sequence.

The SFM makes this type of experiment even easier to perform. The molecules are supposed to be in a more biologically relevant environment, and the specimen preparation is easier with SFM than with EM. If operating under stable conditions, the SFM can record images automatically, so that several hundred high-resolution images can be collected in a day.

The resolution of the micrographs is an issue to be considered for the conformational analyses. The length scale of DNA curvature is that of the double-helical turn, since the origin of curvature is frequently the phased composition of small base-step contributions. The best spatial resolution routinely attained with the SFM operating on DNA is in the order of nanometers, so it seems that one can effectively study the local curvature of DNA with the SFM. A DNA molecule is usually considered a very rigid polymer, but on the scale of the thermal fluctuations, a long DNA molecule is truly a quite flexible chain. A kilobase-sized DNA molecule enjoys a relatively high conformational freedom and can assume a surprising variety of shapes, apparently unrelated to the intrinsic curvature directed by its sequence (see Fig. 5), thanks to a high number of almost negligible angular deflections from the minimum energy states. It is only from the statistical analysis of the shapes that the intrinsic curvature of DNA can be obtained from a set of molecule conformation profiles derived from the micrographs. If the molecules are thermodynamically equilibrated on the surface, the conformational data should reliably give information about the intrinsic curvature.

The results shown in Figure 6 are from the analysis of the profiles of more than 2000 double-stranded DNA molecules (Zuccheri *et al.*, 2001). We constructed long palindromic dimers of DNA by joining two segments in either the head-to-head or the tail-to-tail configuration. We cut the 937-bp fragment between the EcoR V and the Pst I sites in pBR322 and dimerized it around either site to get an EcoR V–EcoR V and a

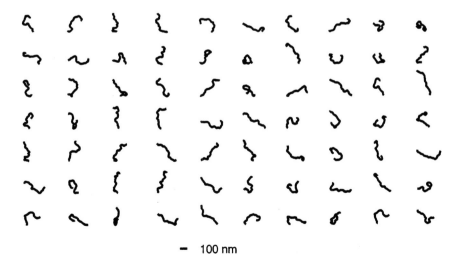

— 100 nm

Fig. 5 A sampling of the conformational variability of the 1874-bp dimeric molecules deposited on mica and imaged with the SFM after dehydration (EcoR V–Pst I–EcoR V dimers in this figure).

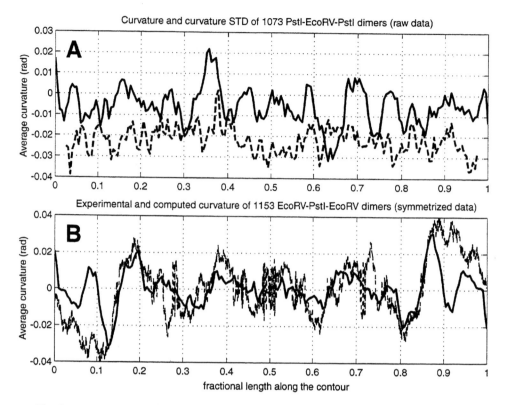

Fig. 6 Diagrams of the position-dependent curvature of the 1874-bp dimeric molecules. (A) Averaged experimental curvature profile (solid line) and experimental curvature standard deviation profile (off-scale, dashed
line) for more than 1000 Pst I–EcoR V–Pst I dimers. In this configuration, the significant curvatures are expected
and found in the central part of the molecules. The standard deviation profile shows the modulations of the
standard deviations that seem to correlate with the magnitude of the curvatures. (B) Symmetrized averaged experimental curvature profile (solid line) and computed theoretical curvature profile (off-scale, thin dashed line)
for more than 1000 EcoR V–Pst I–EcoR V dimers. The plot has been symmetrized to double the data on each
half of the molecule and to compare to the inherently symmetric computed curvature (thin dashed line). The
significant curvatures are found near the ends of the molecules as expected, and their location matches the
computational prediction very well.

Pst I–Pst I dimer. We have chosen this region since it comprises regions of both high
curvature and low curvature, as determined experimentally (Muzard *et al.,* 1990). The
EcoR V–EcoR V dimer should display significant curvatures near its ends, while in
the Pst I–Pst I dimer, the same curvatures should be located in the central portion. The
contour length distributions of the imaged molecules is shown in Fig. 7; their average
length is very close to that of the theoretical B-DNA and, as expected, it is very similar
between the two populations. From this result, we conclude that no significant B-to-A
transition seems to have occurred in our depositions. Analyses of the angular distributions on the populations [with the methods described also by Rivetti *et al.* (1996)]

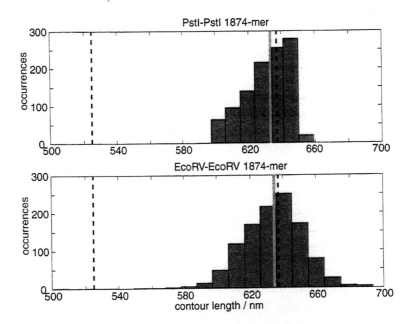

Fig. 7 Distributions of the contour lengths of the populations of dimeric molecules (whose curvature is reported in Fig. 6). The solid gray vertical lines are the average contour lengths, the dashed vertical lines are the theoretical contour lengths for a 1874-bp ds-DNA in the A form (the lower) or in the B form (the higher). Molecules longer than the B form might be the result of errors in the determination of the length or possibly of local overstretching of the DNA.

demonstrated that the data come from an equilibrated and complete population of adsorbed molecules.

Having thus verified the validity of the data set, we computed the curvature of the DNA molecules. All the traces of the molecular profiles were divided into an equal number of segments of equal length. The angles between adjacent segments were computed and averaged with the angles from the other molecules to give a plot of the position-dependent average curvature. The curvature value is algebraic, having defined as positive the angles between segments, where the second segment is rotated counterclockwise with respect to the first. The binary symmetry of the molecule helps to avoid a constant problem in microscopy of DNA: the orientational uncertainty. No one can tell the beginning from the end of a DNA molecule imaged by EM or SFM, just because they lack any distinctive feature. In the past, the beginning and the end of a sequence had been identified as a result of properly introduced tags. With a biotin–streptavidin bond, a bulky protein was bound to the only biotinylated end of a DNA molecule (Theveny and Revet, 1987). The problem with this approach is that the conformation of the tagged end could be altered by the presence of a bulky object. Furthermore, if we are interested in the dynamics of the DNA molecule, the different composition of one of the ends is very likely to impart a different mobility to that terminus, especially if the molecule is adsorbed on a surface. It sometimes occurs that proteins bind strongly to the surfaces: the tagged end could act as an anchor for the DNA molecule. As can be seen from the traces in Fig. 6, the curvature profile resulting from the averaging of approximately 1000 molecules is quite symmetric, as expected from the symmetry in the DNA sequence. It could be expected that the profile will turn out symmetric even if the sequence of the DNA, and so the physical properties, were not symmetric. The operator who records the profiles of the molecules (in a semiautomatic way (Rivetti et al., 1996)) cannot distinguish the ends of the molecules, so he/she chooses each terminus at the initial 50% of the time. Summing up the average curvature plots of two such subpopulations, a symmetric plot would result, even if the two plots were not symmetric. The difference from what we obtained is that neither the positions nor the magnitude of the curvature peaks would make any sense. The number of peaks would generally be doubled. As also shown in Fig. 6, the positions of the peaks we obtained match very well that which can be predicted by computational methods derived from experimental data (Anselmi et al., 1999; Shpigelman et al., 1993) and it is also is in agreement with experimental data on the same DNA (Muzard et al., 1990). The advantage of using the symmetric molecules is that no labels are necessary to distinguish the ends, and a direct correlation of the base sequence with the position on the molecule can be made. The binary symmetry practically doubles up the data about the structure of one of the dimerized fragments, so it is justifiable to make the plots symmetric by summing one-half with the other and using only this symmetrized curvature plot.

These curvature profiles demonstrate that the SFM can be used to directly evaluate the intrinsic curvature of a natural "random" sequence of DNA, without the need of any comparison with other DNA fragments of either known curvature or tracts of reference curvature. The surprising structural variability of the studied long DNA molecules is the result of their flexibility. The sequence-dependent flexibility of DNA is a very interesting subject on which there are many contradictory opinions (Crothers, 1998). The origin of curvature and its coupling with flexibility are not yet clear. We are proposing a new experimental technique to directly investigate the sequence-related DNA flexibility, by studying the local conformation of DNA. In a population of molecular shapes, the standard deviation of the local molecular curvature is a measure of flexibility. In Fig. 6, we also report the experimental averaged profile of the standard deviation of DNA curvature: the plot represents the local modulations of flexibility along a generally flexible chain. This plot is also symmetric. The correlation of the differential flexibility with curvature is such that highly curved parts seem to also be very flexible. This result might be in disagreement with some of the results obtained from X-ray diffraction, which seem to associate chain rigidity with the sequence elements which are also primarily responsible for DNA curvature (Crothers, 1998; and references therein.)

C. Ergodicity in SFM Studies of DNA

The ergodic principle also applies to the SFM imaging of DNA. The ensemble of the profiles of all the molecules with a given sequence at a given time is statistically identical to the ensemble of the profiles assumed with time by only one single molecule with the same sequence. We have seen that the sequence-directed mechanical properties of a DNA chain can be extracted from the profiles of a sample of molecules. Provided that this ensemble of molecules are sufficiently large and that they are well equilibrated on mica, these properties are equivalent to those we can obtain from the ensemble of profiles of only one molecule thermally fluctuating for a period of time long enough to allow it to sample all its conformational space. To reach the latter condition, the observation in fluid of a relatively well-adsorbed molecule must be protracted for a very long time. Alternatively, the molecule must be very mobile on the surface under the conditions of observation. An SFM experiment usually lies somewhere in between these two cases, for an experiment cannot be protracted for too long, and adsorption must be strong enough to enable the observation. Meaningful data can be still gathered if the ensemble is not ergodic (Diau *et al.*, 1998), for example, information about transient scarcely populated states coming from an incomplete statistical redistribution in the conformational space of the process under investigation. Natural selection may rely on nonergodic paths and single-molecule experiments are crucial for exploring these paths.

VII. An Interesting Issue: The Shape of Supercoiled DNA

In the early days of SFM of DNA, plasmids were chosen as specimens because they were easily recognizable in the images, and all the doubts about possible SFM artifacts could be erased with one image. Being circular is not the only virtue of plasmids: all the molecules have the same sequence and length, and they can be obtained in large quantities with simple and cheap protocols that can be followed by any normally equipped biology lab. However, there is more to plasmids than this. In their native state, plasmids are supercoiled, as all natural ds-DNA is. Supercoiling turns out to be a very important biological phenomenon and relevant for a number of processes (Bates and Maxwell, 1993); for example, supercoiling has a tremendous effect on the shape and the compaction of DNA molecules. Supercoiling, which originates from a topological constriction (generally from the underwinding of the two DNA chains), translates into geometric effects. The twisting and writhing of the chains in space are altered. Referring the readers to the many good sources of information about DNA supercoiling (Bates and Maxwell, 1993; Boles *et al.*, 1990; Calladine and Drew, 1992; Cozzarelli *et al.*, 1990; Vologodskii, 1992; White, 1989), we only need to state that free supercoiled DNA in solution is expected to have an interwound shape, named "plectonemic," that can be explained (and modeled) considering the DNA as an elastic chain.

A. Is All DNA Relaxed? Salt and Surface Effects

The physiologic supercoiling of DNA is constant in the cell. DNA has a native superhelical density of −0.03 to −0.09 (Vologodskii, 1992). This means that pBR322, a 4.4-kbp plasmid, could have a linking deficit of 25, of which at least 75% is expected to be allocated to the writhing of the helical axis. The helical axis should cross over itself approximately 18 times on average. When not complexed with proteins (like in nucleosomes), the supercoiled DNA should be in the interwound form, which is called "plectonemic."

What is almost constantly found in the SFM literature is the imaging of DNA plasmids that simply do not appear plectonemic, even if the authors describe them as supercoiled. Among other researchers, we have seen and reported plasmids that appeared too loosely coiled for their native superhelical density. A comparison between theoretical and computational evaluations would also suggest that plasmids with a native superhelical density should be significantly more coiled under the conditions of SFM imaging (Vologodskii, 1992; Vologodskii and Cozzarelli, 1994). The shape of supercoiled plasmids in SFM images is often very open, with only one or two apparent crossovers. Sometimes it is difficult to distinguish the loosely coiled plasmids from the relaxed plasmids that can have "random flops," especially if they are very long. At times, strange looking supercoiled plasmids have been presented. The superhelical density should be spread homogeneously along the entire length of the molecule, where no extended sections have a significantly looser appearance. On the contrary, many reports show plasmids with apparently highly coiled sections and other sections which are completely loose (Pope et al., 2000). It is our understanding that this could be due to local condensation conditions, which can drive some strand-to-strand contacts not only due to supercoiling. Under the conditions of DNA deposition and dehydration adopted for SFM imaging, some condensation could occur, especially if enabled by the superhelical tension. These conditions could even drive the molecules toward unnatural shapes. A seriously limited number of EM studies demonstrate nicely interwound plectonemic plasmids; among them, the cryo-EM studies demonstrate the nicest ones (Adrian et al., 1990). As mentioned earlier, the cryo-EM pictures generated much criticism (Gebe et al., 1996). Theoretical evaluations suggest that the motivation of the highly coiled forms is due to the very low temperature equilibration of the plasmids (Gebe et al., 1996). Evidence is building toward the idea that under the normal conditions of imaging the observed shapes are normally fairly loose. If the ionic strength of the solution to deposit is very high, then the deposited plasmids can appear as highly coiled. A good example is that of Lyubchenko and Shlyakhtenko (1997), who produced very nice images by depositing on AP-mica a plasmid solution with a very high concentration of NaCl. Supercoiled plasmids deposited on freshly cleaved mica are normally loosely coiled, unless uranyl-acetate is used in the dehydration phase to preserve the structure of the molecules (Cherny et al., 1998; F. Nagami et al., unpublished results). In this case (see Fig. 8), the plasmids can be highly coiled, and it is very easy to distinguish between relaxed and supercoiled plasmids in which case, the statistical evaluation of the two populations is very similar to the quantitative analysis by agarose gel electrophoresis. On the other hand, in a few cases, very coiled plasmids were imaged at low ionic strength on either freshly cleaved mica or oxidized silicon (Hansma et al., 1996; Yoshimura et al., 2000).

Under high ionic strength conditions, plasmids are extremely coiled and can be seen as coiled on the surfaces especially if the experimental conditions (AP-mica or uranyl acetate) can limit their mobility before the ionic conditions are changed (e.g., in a rinsing step). Uranyl acetate might have the dual role of limiting the mobility and increasing the ionic strength. (Not all salts would work in this way.)

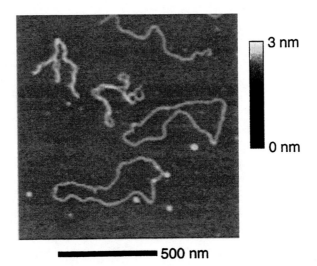

3 nm

0 nm

500 nm

Fig. 8 SFM image of plasmid DNA deposited on freshly cleaved mica and dehydrated from a solution containing uranyl acetate. Highly supercoiled, completely relaxed, and fragmented linear DNA molecules are very easy to distinguish from the images under these conditions. Courtesy of Fuji Nagami (F. Nagami, G. Zuccheri, B. Samorì, and R. Kuroda (2002). Time-lapse imaging of conformational changes in supercoiled DNA by scanning force microscopy. *Anal. Biochem.* **300,** 170–176).

Thus, why are the same plasmids so loosely coiled under conditions that should still preserve their coiling? We presently believe that the electrostatic interactions with the surface and the statistical effects of being confined in a very thin layer of solution strongly increase the intramolecular strand-to-strand repulsion. The electrostatic repulsion should keep the DNA strands apart, and thus uncoil them, except in the presence of a high concentration of salt that screens the charges. The natural partition of a linking deficit in DNA between writhing and twisting of the chain might be altered in favor of twisting under these conditions. We are not aware of computer simulations that consider these factors to determine the partition of supercoiling under strongly spatially confined conditions.

A partial B-to-A transition has been known to occur on DNA on the surface of mica (see Section III,E). Under these conditions, the superhelical tension should be greatly reduced (Krylov *et al.,* 1990) and might help to bring the supercoiled DNA to a low state of coiling. It appears to us that the shape of supercoiled DNA evidenced from SFM images could be the result of multiple factors and further evidences are awaited to completely clarify the issue.

Imaging supercoiled DNA in fluid at high ionic strength seems to be the only safe method to have nice plectonemic DNA on freshly cleaved mica under completely controlled conditions and without the aid of extraneous molecules. To image plectonemic DNA in air, strongly adhesive surfaces (like AP-mica) need to be coupled to the high-ionic-strength environment. Nicked plasmids appear open on freshly cleaved mica since they have the mobility to respond to the high electrostatic repulsion between strands. This same repulsion that opens them up like circles is also responsible for the self-avoiding that causes DNA strands to never overlap if imaged on mica under nontrapping conditions (Rivetti *et al.*, 1996). When imaged under trapping conditions, supercoiled DNA might appear more nicely plectonemic. Unfortunately, under the same conditions, nicked plasmids might display many chain crossovers, due to their random adsorption and trapping on the surface (such as for the many rosette-like shapes shown by Lyubchenko and Shlyakhlenko, 1997), and they will be more difficult to visualize and characterize.

VIII. Conclusions and Perspectives

After several years of SFM experiments on DNA, there are many questions that microscopists would still like to answer. Newer techniques and instruments are still emerging, often from the close collaboration of chemists and biologists with physicists. Newer technologies and interdisciplinary approaches will certainly expand the knowledge on the structure and behavior of DNA in the following years.

A. SFM Single–Molecule Stretching Experiments on DNA: Only a Brief Note on a Booming Issue

The last few years have seen the emergence of force spectroscopy as a new field of research. Using several kinds of force transducers, measured small forces can be applied to properly selected parts of a molecule in order to study its behavior and the forces that hold its structure together. The SFM cantilever is one of the most frequently used force transducers for force spectroscopy. Many groups are already working in SFM force spectroscopy of DNA, studying the forces that not only keep the double helix together but also might determine its behavior in the interaction with proteins or other molecules (Clausen-Schaumann *et al.*, 2000; Lee *et al.*, 1994; Noy *et al.*, 1997; Strunz *et al.*, 2000; Strunz *et al.*, 1999).

B. The Structure of ss–DNA

Relatively few SFM studies have been done on single-stranded nucleic acids. Some possible reasons are the variability, complexity, and fragility of their structures. Recent studies have shown that some useful information can also be gathered from single-stranded DNA (Zuccheri *et al.*, 2000). It can be easily predicted that some of the numerous interesting questions regarding single-stranded DNA and RNA will find answers as a result of SFM studies.

Acknowledgments

We are grateful to Pasquale De Santis (University La Sapienza, Rome, Italy) for the very helpful discussion on DNA curvature and flexibility, to Giuseppe Gargiulo (University of Bologna, Italy) for sharing his DNA constructs, and to Fuji Nagami (University of Tokyo, Japan) not only for helpful discussion but also for sharing his images of supercoiled plasmids. This work was supported by Programmi Biotecnologie Legge 95/95 (MURST 5%); MURST PRIN (Progetti Biologia Strutturale 1997–1999 and 1999–2001).

References

Adrian, M., Ten Heggeler-Bordier, B., Wahli, W., Stasiak, A. Z., Stasiak, A., and Dubochet, J. (1990). Direct visualization of supercoiled DNA molecules in solution. *EMBO J.* **9**(13), 4551–4554.

Aich, P., Labiuk, S. L., Tari, L. W., Delbaere, L. J., Roesler, W. J., Falk, K. J., Steer, R. P., and Lee, J. S. (1999). M-DNA: A complex between divalent metal ions and DNA which behaves as a molecular wire. *J. Mol. Biol.* **294**(2), 477–485.

Anselmi, C., Bocchinfuso, G., De Santis, P., Savino, M., and Scipioni, A. (1999). Dual role of DNA intrinsic curvature and flexibility in determining nucleosome stability. *J. Mol. Biol.* **286**(5), 1293–1301.

Argaman, M., Golan, R., Thomson, N. H., and Hansma, H. G. (1997). Phase imaging of moving DNA molecules and DNA molecules replicated in the atomic force microscope. *Nucl. Acids Res.* **25**(21), 4379–4384.

Bamdad, C. (1998). A DNA self-assembled monolayer for the specific attachment of unmodified double- or single-stranded DNA. *Biophys. J.* **75**(4), 1997–2003.

Bates, A. D., and Maxwell, A. (1993). "DNA Topology." Oxford, New York; IRL Press, Oxford University Press.

Baur, C., Gazen, B. C., Koel, B., Ramachandran, T. R., Requicha, A. A. G., and Zini, L. (1997). Robotic nanomanipulation with a scanning probe microscope in a networked computing environment. *J. Vacuum Sci. Technol. B (Microelectronics and Nanometer Structures)* **15**(4), 1577–1580.

Bednar, J., Furrer, P., Katritch, V., Stasiak, A. Z., Dubochet, J., and Stasiak, A. (1995). Determination of DNA persistence length by cryo-electron microscopy. Separation of the static and dynamic contributions to the apparent persistence length of DNA. *J. Mol. Biol.* **254**(4), 579–594.

Bettini, A., Pozzan, M., Valdevit, E., and Frontali, C. (1980). Microscopic persistence length of native DNA: its relation to average molecular dimensions. *Biopolymers* **19**, 1689–1694.

Bezanilla, M., Brake, B., Nudler, E., Kashlev, M., Hansma, P. K., and Hansma, H. G. (1994). Motion and enzymatic degradation of DNA in the atomic force microscope. *Biophys. J.* **67**(6), 2454–2459.

Bezanilla, M., Manne, S., Laney, D. E., Lyubchenko, Y. L., and Hansma, H. G. (1995). Adsorption of DNA to mica, silylated mica, and minerals—Characterization by atomic force microscopy. *Langmuir* **11**(2), 655–659.

Binnig, G. (1992). Force microscopy. *Ultramicroscopy* **42**(JUL), 7–15.

Binnig, G., Quate, C. F., and Gerber, C. (1986). Atomic force microscope. *Phys. Rev. Lett.* **56**(9), 930–933.

Binnig, G., and Rohrer, H. (1984). Scanning tunnelling microscopy. *In* "Trends in Physics" (J. Janta and J. Pantoflicek, eds.) Vol. 1, 38–48. European Physical Society, Prague.

Binnig, G., Rohrer, H., Gerber, C., and Weibel, E. (1982a). Surface studies by scanning tunneling microscopy. *Phys. Rev. Lett.* **49**, 57–61.

Binnig, G., Rohrer, H., Gerber, C., and Weibel, E. (1982b). Tunneling through a controllable vacuum gap. *Appl. Phys. Lett.* **40**(2), 178–180.

Boles, T. C., White, J. H., and Cozzarelli, N. R. (1990). Structure of plectonemically supercoiled DNA. *J. Mol. Biol.* **213**(4) 931–951.

Bustamante, C., Guthold, M., Zhu, X., and Yang, G. (1999). Facilitated target location on DNA by individual Escherichia coli RNA polymerase molecules observed with the scanning force microscope operating in liquid. *J. Biol. Chem.* **274**(24), 16,665–16,668.

Bustamante, C., and Rivetti, C. (1996). Visualizing protein-nucleic acid interactions on a large scale with the scanning force microscope. *Annu. Rev. Biophys. Biomol. Struct.* **25**, 395–429.

Bustamante, C., Vesenka, J., Tang, C. L., Rees, W., Guthold, M., and Keller, R. (1992). Circular DNA molecules imaged in air by scanning force microscopy. *Biochemistry* **31**(1), 22–26.

Bustamante, C., Zuccheri, G., Leuba, S. H., Yang, G., and Samorì, B. (1997). Visualization and analysis of chromatin by scanning force microscopy. *Methods—A Companion Methods Enzymol.* **12**(1), 73–83.

Calladine, C. R., and Drew, H. R. (1992). "Understanding DNA : The molecule and How it works." Academic Press, San Diego, London.

Cherny, D. I., Fourcade, A., Svinarchuk, F., Nielsen, P. E., Malvy, C., and Delain, E. (1998). Analysis of various sequence-specific triplexes by electron and atomic force microscopies. *Biophys. J.* **74**(2 Pt 1), 1015–1023.

Cheung, C. L., Hafner, J. H., and Lieber, C. M. (2000). Carbon nanotube atomic force microscopy tips: direct growth by chemical vapor deposition and application to high-resolution imaging. *Proc. Natl. Acad. Sci. U.S.A.* **97**(8), 3809–3813.

Clausen-Schaumann, H., Rief, M., Tolksdorf, C., and Gaub, H. E. (2000). Mechanical stability of single DNA molecules. *Biophys. J.* **78**(4), 1997–2007.

Coury, J. E., Anderson, J. R., McFail-Isom, L., Williams, L. D., and Bottomley, L. A. (1997). Scanning force microscopy of small ligand-nucleic acid complexes: Tris(o-phenanthroline)ruthenium(II) as a test for a new assay. *J. Am. Chem. Soc.* **119**, 3792–3796.

Coury, J. E., McFail-Isom, L., Williams, L. D., and Bottomley, L. A. (1996). A novel assay for drug-DNA binding mode, affinity, and exclusion number: scanning force microscopy. *Proc. Natl. Acad. Sci. U.S.A.* **93**(22), 12,283–12,286.

Cozzarelli, N. R., Boles, T. C., White, J. H., and Cozzarelli (1990). Primer on the topology and geometry of DNA supercoiling. *In* "DNA Topology and Its Biological Effects" (N. R. Cozzarelli and J. C. Wang, eds.), pp. 139–215. CSH Laboratory Press, Cold Spring Harbor, NY.

Crothers, D. M. (1998). DNA curvature and deformation in protein-DNA complexes: A step in the right direction [comment]. *Proc. Natl. Acad. Sci. U.S.A.* **95**(26), 15,163–15,165.

Crothers, D. M., Drak, J., Kahn, J. D., and Levene, S. D. (1992). DNA bending, flexibility, and helical repeat by cyclization kinetics. *Methods in Enzymology* **212**, 3–29.

Dai, H., Hafner, J. H., Rinzler, A. G., Colbert, D. T., and Smalley, R. E. (1996). Nanotubes as nanoprobes in scanning probe microscopy. *Nature* **384**, 147–150.

Dame, R. T., Wyman, C., and Goosen, N. (2000). H-NS mediated compaction of DNA visualized by atomic force microscopy. *Nucl. Acids Res.* **28**(18),

Davenport, R. J., Wuite, G. J., Landick, R., and Bustamante, C. (2000). Single-molecule study of transcriptional pausing and arrest by E. coli RNA polymerase [see comments]. *Science* **287**(5462), 2497–2500.

Diau, E. W. G., Herek, J. L., Kim, Z. H., and Zewail, A. H. (1998). Femtosecond activation of reactions and the concept of nonergodic molecules. *Science* **279**(5352), 847–851.

Dubochet, J., Adrian, M., Dustin, I., Furrer, P., and Stasiak, A. (1992). Cryoelectron microscopy of DNA molecules in solution. *Methods in Enzymology* **211**, 507–518.

Dustin, I., Furrer, P., Stasiak, A., Dubochet, J., Langowski, J., and Egelman, E. (1991). Spatial visualization of DNA in solution. *J. Struct. Biol.* **107**(1), 15–21.

Erie, D. A., Yang, G., Schultz, H. C., and Bustamante, C. (1994). DNA bending by Cro protein in specific and nonspecific complexes: implications for protein site recognition and specificity. *Science* **266**(5190), 1562–1566.

Fang, Y., and Hoh, J. H. (1998). Surface-directed DNA condensation in the absence of soluble multivalent cations. *Nucl. Acids Res.* **26**(2), 588–593.

Fang, Y., and Hoh, J. H. (1999). Cationic silanes stabilize intermediates in DNA condensation. *FEBS Lett.* **459**(2), 173–176.

Fang, Y., Spisz, T. S., and Hoh, J. H. (1999). Ethanol-induced structural transitions of DNA on mica. *Nucl. Acids Res.* **27**(8), 1943–1949.

Fang, Y., Spisz, T. S., Wiltshire, T., D'Costa, N. P., Bankman, I. N., Reeves, R. H., and Hoh, J. H. (1998). Solid-state DNA sizing by atomic force microscopy. *Anal. Chem.* **70**(10), 2123–2129.

Feng, X. Z., Bash, R., Balagurumoorthy, P., Lohr, D., Harrington, R. E., and Lindsay, S. M. (2000). Conformational transition in DNA on a cold surface. *Nucl. Acids Res.* **28**(2), 593–596.

Fritzsche, W., Schaper, A., and Jovin, T. M. (1995). Scanning force microscopy of chromatin fibers in air and in liquid. *Scanning* **17**(3), 148–155.

Frontali, C. (1988). Excluded-volume effect on the bidimensional conformation of DNA molecules adsorbed to protein films. *Biopolymers* **27**(8), 1329–1331.

Frontali, C., Dore, E., Ferrauto, A., Gratton, E., Bettini, A., Pozzan, M. R., and Valdevit, E. (1979). An absolute method for the determination of the persistence length of native DNA from electron micrographs. *Biopolymers* **18**(6), 1353–1373.

Furrer, P., Bednar, J., Stasiak, A. Z., Katritch, V., Michoud, D., Stasiak, A., and Dubochet, J. (1997). Opposite effect of counterions on the persistence length of nicked and non-nicked DNA. *J. Mol. Biol.* **266**(4), 711–721.

Gebe, J. A., Delrow, J. J., Heath, P. J., Fujimoto, B. S., Stewart, D. W., and Schurr, J. M. (1996). Effects of Na+ and Mg²+ on the structures of supercoiled DNAs: Comparison of simulations with experiments. *J. Mol. Biol.* **262**(2), 105–128.

Guthold, M., Zhu, X., Rivetti, C., Yang, G., Thomson, N. H., Kasas, S., Hansma, H. G., Smith, B., Hansma, P. K., and Bustamante, C. (1999). Direct observation of one-dimensional diffusion and transcription by Escherichia coli RNA polymerase. *Biophys. J.* **77**(4), 2284–2294.

Hafner, J. H., Cheung, C. L., and Lieber, C. M. (1999). Direct growth of single-walled carbon nanotube scanning probe microscopy tips. *J. Am. Chem. Soc.* **121**, 9750–9751.

Han, W., Dlakic, M., Zhu, Y. J., Lindsay, S. M., and Harrington, R. E. (1997). Strained DNA is kinked by low concentrations of Zn2+. *Proc. Natl. Acad. Sci. U.S.A.* **94**(20), 10,565–10,570.

Han, W., Lindsay, S. M., Dlakic, M., and Harrington, R. E. (1997). Kinked DNA [letter]. *Nature* **386**(6625), 563.

Hansma, H. G., Bezanilla, M., Zenhausern, F., Adrian, M., and Sinsheimer, R. L. (1993). Atomic force microscopy of DNA in aqueous solutions. *Nucl. Acids Res.* **21**(3), 505–512.

Hansma, P. K., Cleveland, J. P., Radmacher, M., Walters, D. A., Hillner, P. E., Bezanilla, M., Fritz, M., Vie, D., Hansma, H. G., Prater, C. B., Massie, J., Fukunaga, L., Gurley, J., and Elings, V. (1994). Tapping mode atomic force microscopy in liquids. *Appl. Phys. Lett.* **64**(13), 1738–1740.

Hansma, H., Golan, R., Hsieh, W., Daubendiek, S. L., and Kool, E. T. (1999). Polymerase Activities and RNA Structures in the Atomic Force Microscope. *J. Struct. Biol.* **127**, 240–247.

Hansma, H. G., and Laney, D. E. (1996). DNA binding to mica correlates with cationic radius: Assay by atomic force microscopy. *Biophys. J.* **70**(4), 1933–1939.

Hansma, H. G., Laney, D. E., Bezanilla, M., Sinsheimer, R. L., and Hansma, P. K. (1995). Applications for atomic force microscopy of DNA. *Biophys. J.* **68**(5), 1672–1677.

Hansma, H. G., Revenko, I., Kim, K., and Laney, D. E. (1996). Atomic force microscopy of long and short double-stranded, single-stranded and triple-stranded nucleic acids. *Nucl. Acids Res.* **24**(4), 713–720.

Hansma, H. G., Sinsheimer, R. L., Li, M. Q., and Hansma, P. K. (1992). Atomic force microscopy of single- and double-stranded DNA. *Nucl. Acids Res.* **20**(14), 3585–3590.

Hansma, H. G., Vesenka, J., Siegerist, C., Kelderman, G., Morrett, H., Sinsheimer, R. O., Elings, V., Bustamante, C., and Hansma, P. K. (1992). Reproducible imaging and dissection of plasmid DNA under liquid with the atomic force microscope. *Science* **256**(5060), 1180–1184.

Heckl, W. M. (1998). The combination of AFM Nanodissection with PCR. *BIOforum Int.* **2**, 133–138.

Hegner, M., Wagner, P., and Semenza, G. (1993). Immobilizing DNA on gold via thiol modification for atomic force microscopy imaging in buffer solutions. *FEBS Lett.* **336**(3), 452–456.

Henderson, E. (1992). Imaging and nanodissection of individual supercoiled plasmids by atomic force microscopy. *Nucl. Acids Res.* **20**(3), 445–447.

Israelachvili, J. N. (1992). "Intermolecular and Surface Forces.". Academic Press, London.

Jeltsch, A. (1998). Flexibility of DNA in complex with proteins deduced from the distribution of bending angles observed by scanning force microscopy. *Biophys. Chem.* **74**, 53–57.

Ji, X., Oh, J., Dunker, A. K., and Hipps, K. W. (1998). Effects of relative humidity and applied force on atomic force microscopy images of the filamentous phage fd. *Ultramicroscopy* **72**(3–4), 165–176.

Joanicot, M., and Revet, B. (1987). DNA conformational studies from electron microscopy. I. Excluded volume effect and structure dimensionality. *Biopolymers* **26**(2), 315–326.

Karrasch, S., Dolder, M., Schabert, F., Ramsden, J., and Engel, A. (1993). Covalent binding of biological samples to solid supports for scanning probe microscopy in buffer solution. *Biophys. J.* **65**(6), 2437–2446.

Kasas, S., Thomson, N. H., Smith, B. L., Hansma, H. G., Zhu, X., Guthold, M., Bustamante, C., Kool, E. T., Kashlev, M., and Hansma, P. K. (1997). Escherichia coli RNA polymerase activity observed using atomic force microscopy. *Biochemistry* **36**(3), 461–468.

Keller, D. J., and Chih-Chung, C. (1992). Imaging steep, high structures by scanning force microscopy with electron beam deposited tips. *Surf. Sci.* **268**(1–3), 333–339.

Krylov, D., Makarov, V. L., and Ivanov, V. I. (1990). The B-A transition in superhelical DNA. *Nucl. Acids Res.* **18**(4), 759–761.

Kumar, A., Larsson, O., Parodi, D., and Liang, Z. (2000). Silanized nucleic acids: a general platform for DNA immobilization. *Nucl. Acids Res.* **28**(14), e71.

Lee, G. U., Chrisey, L. A., and Colton, R. J. (1994). Direct measurement of the forces between complementary strands of DNA. *Science* **266**(5186), 771–773.

Lindsay, S. M., Tao, N. J., DeRose, J. A., Oden, P. I., Lyubchenko, Y. L., Harrington, R. E., and Shlyakhtenko, L. (1992). Potentiostatic deposition of DNA for scanning probe microscopy. *Biophys. J.* **61**(6), 1570–1584.

Lyubchenko, Y. L., Gall, A. A., Shlyakhtenko, L. S., Harrington, R. E., Jacobs, B. L., Oden, P. I., and Lindsay, S. M. (1992). Atomic force microscopy imaging of double stranded DNA and RNA. *J. Biomol. Struct. Dyn.* **10**(3), 589–606.

Lyubchenko, Y. L., Jacobs, B. L., and Lindsay, S. M. (1992). Atomic force microscopy of reovirus dsRNA: A routine technique for length measurements. *Nucl. Acids Res.* **20**(15), 3983–3986.

Lyubchenko, Y. L., and Shlyakhtenko, L. S. (1997). Visualization of supercoiled DNA with atomic force microscopy in situ. *Proc. Natl. Acad. Sci. U.S.A.* **94**(2), 496–501.

Mou, J., Czajkowsky, D. M., Zhang, Y., and Shao, Z. (1995). High-resolution atomic-force microscopy of DNA: the pitch of the double helix. *FEBS Lett.* **371**(3), 279–282.

Murakami, M., Hirokawa, H., and Hayata, I. (2000). Analysis of radiation damage of DNA by atomic force microscopy in comparison with agarose gel electrophoresis. *J. Biochem. Biophys. Methods* **44**, 31–40.

Murray, M. N., Hansma, H. G., Bezanilla, M., Sano, T., Ogletree, D. F., Kolbe, W., Smith, C. L., Cantor, C. R., Spengler, S., Hansma, P. K., and Salmeron, M. (1993). Atomic force microscopy of biochemically tagged DNA. *Proc. Natl. Acad. Sci. U.S.A.* **90**(9), 3811–3814.

Muzard, G., Theveny, B., and Revet, B. (1990). Electron microscopy mapping of pBR322 DNA curvature. Comparison with theoretical models. *EMBO J.* **9**(4), 1289–1298.

Muzzalupo, I., Nigro, C., Zuccheri, G., Samorì, B., Quagliariello, C., and Buttinelli, M. (1995). Deposition on mica and scanning force microscopy imaging of DNA molecules whose original B structure is retained. *J. Vacuum Sci. Technol. A (Vacuum, Surfaces, and Films)* **13**(3, pt. 2), 1752–1754.

Nishimura, S., Biggs, S., Scales, P. J., Healy, T. W., Tsunematsu, K., and Tateyama, T. (1994). Molecular-scale structure of the cation modified muscovite mica masal plane. *Langmuir* **10**(12), 4554–4559.

Nishimura, S., Scales, P. J., Tateyama, H., Tsunematsu, K., and Healy, T. W. (1995). Cationic modification of muscovite mica - an electrokinetic study. *Langmuir* **11**(1), 291–295.

Noy, A., Vezenov, D. V., Kayyem, J. F., Meade, T. J., and Lieber, C. M. (1997). Stretching and breaking duplex DNA by chemical force microscopy. *Chem. Biol.* **4**(7), 519–527.

Oussatcheva, E. A., Shlyakhtenko, L. S., Glass, R., Sinden, R. R., Lyubchenko, Y. L., and Potaman, V. N. (1999). Structure of branched DNA molecules: gel retardation and atomic force microscopy studies. *J. Mol. Biol.* **292**(1), 75–86.

Paige, C. R., Kornicker, W. A., Hileman, O. E., and Snodgrass, W. J. (1992). Kinetics of desorption of ions from quartz and mica surfaces. *J. Radioanal. Nucl. Chem. Art.* **159**, 37–46.

Pietrasanta, L. I., Schaper, A., and Jovin, T. M. (1994). Probing specific molecular conformations with the scanning force. *Nucl. Acids Res.* **22**(16), 3288–3292.

Pope, L. H., Davies, M. C., Laughton, C. A., Roberts, C. J., Tendler, S. J., and Williams, P. M. (1999). Intercalation-induced changes in DNA supercoiling observed in real-time by atomic force microscopy. *Anal. Chim. Acta* **400**, 27–32.

Pope, L. H., Davies, M. C., Laughton, C. A., Roberts, C. J., Tendler, S. J., and Williams, P. M. (2000). Atomic force microscopy studies of intercalation-induced changes in plasmid DNA tertiary structure [In Process Citation]. *J. Microsc.* **199**(Pt 1), 68–78.

Putman, C. A. J., van der Werf, K. O., de Grooth, B. G., and van Hulst, N. F. (1994). Tapping mode atomic force microscopy in liquid. *Appl. Phys. Lett.* **64**(18), 2454–2456.

Rabke, C. E., Wenzler, L. A., and Beebe, T. P. Jr. (1994). Electron spectroscopy and atomic force microscopy studies of DNA adsorption on mica. *Scan. Microsc.* **8**(3), 471–480.

Ramachandran, T. R., Madhukar, A., Chen, P., and Koel, B. E. (1998). Imaging and direct manipulation of nanoscale three-dimensional features using the noncontact atomic force microscope. *J. Vacuum Sci. Technol. A (Vacuum, Surfaces, and Films)* **16**(3), 1425–1429.

Rippe, K., Guthold, M., von Hippel, P. H., and Bustamante, C. (1997). Transcriptional activation via DNA-looping: visualization of intermediates in the activation pathway of E. coli RNA polymerase × sigma 54 holoenzyme by scanning force microscopy. *J. Mol. Biol.* **270**(2), 125–138.

Rippe, K., Mücke, N., and Langowski, J. (1997). Superhelix dimensions of a 1868 base pair plasmid determined by scanning force microscopy in air and in acqueous solution. *Nucl. Acids Res.* **25**(9), 1736–1744.

Rivetti, C., and Codeluppi, S. (2001). Accurate length determination of DNA molecules visualized by atomic force microscopy: Evidence for a partial B- to A-form transition on mica. *Ultramicroscopy* **87**(1–2), 55–66.

Rivetti, C., Guthold, M., and Bustamante, C. (1996). Scanning force microscopy of DNA deposited onto mica: equilibration versus kinetic trapping studied by statistical polymer chain analysis. *J. Mol. Biol.* **264**(5), 919–932.

Rivetti, C., Guthold, M., and Bustamante, C. (1999). Wrapping of DNA around the E. coli RNA polymerase open promoter complex. *EMBO J.* **18**(16), 4464–4475.

Rivetti, C., Walker, C., and Bustamante, C. (1998). Polymer chain statistics and conformational analysis of DNA molecules with bends or sections of different flexibility. *J. Mol. Biol.* **280**(1), 41–59.

Samorì, B. (1998). Stretching, tearing, and dissecting single molecules of DNA. *Angew. Chem. Int. Ed.* **37**(16), 2198–2200.

Samorì, B., Muzzalupo, I., and Zuccheri, G. (1996). Deposition of supercoiled DNA on mica for scanning force microscopy imaging. *Scan. Microsc.* **10**(4), 953–962.

Samorì, B., Nigro, C., Armentano, V., Cimieri, S., Zuccheri, G., and Quagliariello, C. (1993). DNA supercoiling imaged in 3 Dimensions by scanning force microscopy. *Angew. Chem. Int. Ed.* **32**(10), 1461–1463.

Schaper, A., Pietrasanta, L. I., and Jovin, T. M. (1993). Scanning force microscopy of circular and linear plasmid DNA spread on mica with a quaternary ammonium salt. *Nucl. Acids Res.* **21**, 6004–6009.

Schepartz, A. (1995). Nonspecific DNA bending and the specificity of protein-DNA interactions. *Science* **269**(5226), 989–990.

Shlyakhtenko, L. S., Gall, A. A., Weimer, J. J., Hawn, D. D., and Lyubchenko, Y. L. (1999). Atomic force microscopy imaging of DNA covalently immobilized on a functionalized mica surface. *Biophys. J.* **77**, 568–576.

Shlyakhtenko, L. S., Potaman, V. N., Sinden, R. R., Gall, A. A., and Lyubchenko, Y. L. (2000). Structure and dynamics of three-way DNA junctions: atomic force microscopy studies. *Nucl. Acids Res.* **28**(18), 34–72.

Shpigelman, E. S., Trifonov, E. N., and Bolshoy, A. (1993). Curvature—Software for the analysis of curved DNA. *CABIOS* **9**(4), 435–440.

Spisz, T. S., Fang, Y., Reeves, R. H., Seymour, C. K., Bankman, I. N., and Hoh, J. H. (1998). Automated sizing of DNA fragments in atomic force microscope images. *Med. Biol. Eng. Comput.* **36**(6), 667–672.

Steinberg, I. Z. (1994). Brownian motion of the end-to-end distance in oligopeptide molecules: Numerical solution of the diffusion equations as coupled first order linear differential equations. *J. Theor. Biol.* **166**(2), 173–187.

Strother, T., Hamers, R. J., and Smith, L. M. (2000). Covalent attachment of oligodeoxyribonucleotides to amine-modified Si (001) surfaces. *Nucl. Acids Res.* **28**(18), 3535–3541.

Strunz, T., Hegner, M., Oroszlan, K., Schumakovic, I., and Güntherdt, H.-J. (2000). Force spectroscopy and dissociation kinetics of single molecules under an applied force. *Single Mol.* **1**(2), 175.

Strunz, T., Oroszlan, K., Schafer, R., and Guntherodt, H. J. (1999). Dynamic force spectroscopy of single DNA molecules. *Proc. Natl. Acad. Sci. U.S.A.* **96**(20), 11,277–11,282.

Thalhammer, S., Stark, R. W., Muller, S., Wienberg, J., and Heckl, W. M. (1997). The atomic force microscope as a new microdissecting tool for the generation of genetic probes. *J. Struct. Biol.* **119**(2), 232–237.

Theveny, B., and Revet, B. (1987). DNA orientation using specific avidin-ferritin biotin end labelling. *Nucl. Acids Res.* **15**(3), 947–958.

Thundat, T., Allison, D. P., Warmack, R. J., and Ferrell, T. L. (1992). Imaging isolated strands of DNA molecules by atomic force microscopy. *Ultramicroscopy* **42**(44, pt.B), 1101–1106.

Thundat, T., Warmack, R. J., Allison, D. P., Bottomley, L. A., Lourenco, A. J., and Ferrell, T. L. (1992). Atomic force microscopy of deoxyribonucleic acid strands adsorbed on mica: The effect of humidity on apparent width and image contrast. *J. Vacuum Sci. Technol. A (Vacuum, Surfaces, and Films)* **10**(4, pt. 1), 630–635.

Thundat, T., Zheng, X. Y., Chen, G. Y., and Warmack, R. J. (1993). Role of relative humidity in atomic force microscopy imaging. *Surf. Sci.* **294**(1,2), L939–L943.

van Noort, S. J., van Der Werf, K. O., de Grooth, B. G., and Greve, J. (1999). High speed atomic force microscopy of biomolecules by image tracking. *Biophys. J.* **77**(4), 2295–2303.

van Noort, S. J., van der Werf, K. O., Eker, A. P., Wyman, C., de Grooth, B. G., van Hulst, N. F., and Greve, J. (1998). Direct visualization of dynamic protein-DNA interactions with a dedicated atomic force microscope [see comments]. *Biophys. J.* **74**(6), 2840–2849.

Vesenka, J., Guthold, M., Tang, C. L., Keller, D., Delaine, E., and Bustamante, C. (1992). Substrate preparation for reliable imaging of DNA molecules with the scanning force microscope. *Ultramicroscopy* **42**(44, pt. B), 1243–1249.

Vesenka, J., Manne, S., Yang, G., Bustamante, C. J., and Henderson, E. (1993). Humidity effects on atomic force microscopy of gold-labeled DNA on mica. *Scan. Microsc.* **7**(3), 781–788.

Vologodskii, A. V. (1992). "Topology and Physics of Circular DNA." CRC Press, Boca Raton, FL.

Vologodskii, A. V., and Cozzarelli, N. R. (1994). Conformational and thermodynamic properties of supercoiled DNA. *Annu. Rev. Biophys. Biomol. Struct.* **23**, 609–643.

Wagner, P., Nock, S., Spudich, J. A., Volkmuth, W. D., Chu, S., Cicero, R. L., Wade, C. P., Linford, M. R., and Chidsey, C. E. (1997). Bioreactive self-assembled monolayers on hydrogen-passivated Si(111) as a new class of atomically flat substrates for biological scanning probe microscopy. *J. Struct. Biol.* **119**(2), 189–201.

Weisenhorn, A. L., Egger, M., Ohnesorge, F., Gould, S. A. C., Heyn, S.-P., Hansma, H. G., Sinsheimer, R. L., Gaub, H. E., and Hansma, P. K. (1991). Molecular-resolution images of Langmuir-Blodgett and DNA by atomic force microscopy. *Langmuir* **7**, 8–12.

Weisenhorn, A. L., Gaub, H. E., Hansma, H. G., Sinsheimer, R. L., Kelderman, G. L., and Hansma, P. K. (1990). Imaging single-stranded DNA, antigen-antibody reaction and polymerized langmuir-blodgett films with an atomic force microscope. *Scan. Microsc.* **4**(3), 511–516.

White, J. H. (1989). An introduction to the geometry and topology of DNA structure. *In* "Mathematical Methods for DNA Sequences," pp. 225–253. CRC Press, Boca Raton, FL.

Wong, S. S., Harper, J. D., Lansbury, P. T., and Lieber, C. M. (1998). Carbon nanotube tips: High-resolution probes for imaging biological systems. *J. Am. Chem. Soc.* **120**(3), 603–604.

Wong, S. S., Joselevich, E., Woolley, A. T., Cheung, C. L., and Lieber, C. M. (1998). Covalently functionalized nanotubes as nanometre-sized probes in chemistry and biology. *Nature* **394**(6688), 52–55.

Wyman, C., Rombel, I., North, A. K., Bustamante, C., and Kustu, S. (1997). Unusual oligomerization required for activity of NtrC, a bacterial enhancer-binding protein [see comments]. *Science* **275**(5306), 1658–1661.

Xu, Y. C., and Bremer, H. (1997). Winding of the DNA helix by divalent metal ions. *Nucl. Acids Res.* **25**(20), 4067–4071.

Xu, X. M., and Ikai, A. (1998). Retrieval and amplification of single-copy genomic DNA from a nanometer region of chromosomes: a new and potential application of atomic force microscopy in genomic research. *Biochem. Biophys. Res. Commun.* **248**(3), 744–748.

Yang, J., and Shao, Z. (1993). Effect of probe force on the resolution of atomic force microscopy of DNA. *Ultramicroscopy* **50**(2), 157–170.

Yang, J., Takeyasu, K., and Shao, Z. (1992). Atomic force microscopy of DNA molecules. *FEBS Lett.* **301**(2), 173–176.

Yang, G., Vesenka, J. P., and Bustamante, C. J. (1996). Effects of tip-sample forces and humidity on the imaging of DNA with a scanning force microscope. *Scanning* **18**(5), 344–350.

Yoshimura, S. H., Ohniwa, R. L., Sato, M. H., Matsunaga, F., Kobayashi, G., Uga, H., Wada, C., and Takeyasu, K. (2000). DNA phase transition promoted by replication initiator. *Biochemistry* **39**(31), 9139–9145.

Zenhausern, F., Adrian, M., Ten Heggeler-Bordier, B., Emch, R., Jobin, M., Taborelli, M., and Descouts, P. (1992). Imaging of DNA by scanning force microscopy. *J. Struct. Biol.* **108**(1), 69–73.

Zenhausern, F., Adrian, M., Ten Heggeler-Bordier, B., Eng, L. M., and Descouts, P. (1992). DNA and RNA polymerase/DNA complex imaged by scanning force microscopy: influence of molecular-scale friction. *Scanning* **14**(4), 212–217.

Zenhausern, F., Eng, L. M., Adrian, M., Kasas, S., Weisenhorn, A. L., and Descouts, P. (1992). A scanning force microscopy investigation of DNA adsorbed on mica. *Helv. Phys. Acta* **65**(6), 820–821.

Zuccheri, G., Bergia, A., Gallinella, G., Musiani, M., and Samorì, B. (2000). Scanning force microscopy study on a single-stranded DNA: the genome of Parvovirus B19. *Chem. Bio. Chem.* **2**(3), 199–204.

Zuccheri, G., Dame, R. T., Aquila, M., Muzzalupo, I., and Samorì, B. (1998). Conformational fluctuations of supercoiled DNA molecules observed in real time with a scanning force microscope. *Appl. Phys. A* **66**(suppl., pt. 1–2), S585–S589.

Zuccheri, G., Scipioni, A., Cavaliere, V., Gargiulo, G., De Santis, P., and Samorì, B. (2001). Mapping the instrinsic curvature and flexibility along the DNA chain. *Proc. Natl. Acad. Sci U.S.A.* **98**(6), 3074–3079.

INDEX

VOLUMES IN SERIES

Founding Series Editor
DAVID M. PRESCOTT

Volume 1 (1964)
Methods in Cell Physiology
Edited by David M. Prescott

Volume 2 (1966)
Methods in Cell Physiology
Edited by David M. Prescott

Volume 3 (1968)
Methods in Cell Physiology
Edited by David M. Prescott

Volume 4 (1970)
Methods in Cell Physiology
Edited by David M. Prescott

Volume 5 (1972)
Methods in Cell Physiology
Edited by David M. Prescott

Volume 6 (1973)
Methods in Cell Physiology
Edited by David M. Prescott

Volume 7 (1973)
Methods in Cell Biology
Edited by David M. Prescott

Volume 8 (1974)
Methods in Cell Biology
Edited by David M. Prescott

Volume 9 (1975)
Methods in Cell Biology
Edited by David M. Prescott

Volume 10 (1975)
Methods in Cell Biology
Edited by David M. Prescott

Volume 11 (1975)
Yeast Cells
Edited by David M. Prescott

Volume 12 (1975)
Yeast Cells
Edited by David M. Prescott

Volume 13 (1976)
Methods in Cell Biology
Edited by David M. Prescott

Volume 14 (1976)
Methods in Cell Biology
Edited by David M. Prescott

Volume 15 (1977)
Methods in Cell Biology
Edited by David M. Prescott

Volume 16 (1977)
Chromatin and Chromosomal Protein Research I
Edited by Gary Stein, Janet Stein, and
Lewis J. Kleinsmith

Volume 17 (1978)
Chromatin and Chromosomal Protein Research II
Edited by Gary Stein, Janet Stein, and
Lewis J. Kleinsmith

Volume 18 (1978)
Chromatin and Chromosomal Protein Research III
Edited by Gary Stein, Janet Stein, and
Lewis J. Kleinsmith

Volume 19 (1978)
Chromatin and Chromosomal Protein Research IV
Edited by Gary Stein, Janet Stein, and
Lewis J. Kleinsmith

Volume 20 (1978)
Methods in Cell Biology
Edited by David M. Prescott

Advisory Board Chairman
KEITH R. PORTER

Volume 21A (1980)
Normal Human Tissue and Cell Culture, Part A: Respiratory, Cardiovascular, and Integumentary Systems
Edited by Curtis C. Harris, Benjamin F. Trump, and Gary D. Stoner

Volume 21B (1980)
Normal Human Tissue and Cell Culture, Part B: Endocrine, Urogenital, and Gastrointestinal Systems
Edited by Curtis C. Harris, Benjamin F. Trump, and Gray D. Stoner

Volume 22 (1981)
Three-Dimensional Ultrastructure in Biology
Edited by James N. Turner

Volume 23 (1981)
Basic Mechanisms of Cellular Secretion
Edited by Arthur R. Hand and Constance Oliver

Volume 24 (1982)
The Cytoskeleton, Part A: Cytoskeletal Proteins, Isolation and Characterization
Edited by Leslie Wilson

Volume 25 (1982)
The Cytoskeleton, Part B: Biological Systems and *in Vitro* Models
Edited by Leslie Wilson

Volume 26 (1982)
Prenatal Diagnosis: Cell Biological Approaches
Edited by Samuel A. Latt and Gretchen J. Darlington

Series Editor
LESLIE WILSON

Volume 27 (1986)
Echinoderm Gametes and Embryos
Edited by Thomas E. Schroeder

Volume 28 (1987)
***Dictyostelium discoideum:* Molecular Approaches to Cell Biology**
Edited by James A. Spudich

Volume 29 (1989)
Fluorescence Microscopy of Living Cells in Culture, Part A: Fluorescent Analogs, Labeling Cells, and Basic Microscopy
Edited by Yu-Li Wang and D. Lansing Taylor

Volume 30 (1989)
Fluorescence Microscopy of Living Cells in Culture, Part B: Quantitative Fluorescence Microscopy—Imaging and Spectroscopy
Edited by D. Lansing Taylor and Yu-Li Wang

Volume 31 (1989)
Vesicular Transport, Part A
Edited by Alan M. Tartakoff

Volume 32 (1989)
Vesicular Transport, Part B
Edited by Alan M. Tartakoff

Volume 33 (1990)
Flow Cytometry
Edited by Zbigniew Darzynkiewicz and Harry A. Crissman

Volume 34 (1991)
Vectorial Transport of Proteins into and across Membranes
Edited by Alan M. Tartakoff

Selected from Volumes 31, 32, and 34 (1991)
Laboratory Methods for Vesicular and Vectorial Transport
Edited by Alan M. Tartakoff

Volume 35 (1991)
Functional Organization of the Nucleus: A Laboratory Guide
Edited by Barbara A. Hamkalo and Sarah C. R. Elgin

Volume 36 (1991)
***Xenopus laevis:* Practical Uses in Cell and Molecular Biology**
Edited by Brian K. Kay and H. Benjamin Peng

Series Editors
LESLIE WILSON AND PAUL MATSUDAIRA

Volume 37 (1993)
Antibodies in Cell Biology
Edited by David J. Asai

Volume 50 (1995)
Methods in Plant Cell Biology, Part B
Edited by David W. Galbraith, Don P. Bourque, and Hans J. Bohnert

Volume 51 (1996)
Methods in Avian Embryology
Edited by Marianne Bronner-Fraser

Volume 52 (1997)
Methods in Muscle Biology
Edited by Charles P. Emerson, Jr. and H. Lee Sweeney

Volume 53 (1997)
Nuclear Structure and Function
Edited by Miguel Berrios

Volume 54 (1997)
Cumulative Index

Volume 55 (1997)
Laser Tweezers in Cell Biology
Edited by Michael P. Sheez

Volume 56 (1998)
Video Microscopy
Edited by Greenfield Sluder and David E. Wolf

Volume 57 (1998)
Animal Cell Culture Methods
Edited by Jennie P. Mather and David Barnes

Volume 58 (1998)
Green Fluorescent Protein
Edited by Kevin F. Sullivan and Steve A. Kay

Volume 59 (1998)
The Zebrafish: Biology
Edited by H. William Detrich III, Monte Westerfield, and Leonard I. Zon

Volume 60 (1998)
The Zebrafish: Genetics and Genomics
Edited by H. William Detrich III, Monte Westerfield, and Leonard I. Zon

Volume 61 (1998)
Mitosis and Meiosis
Edited by Conly L. Rieder

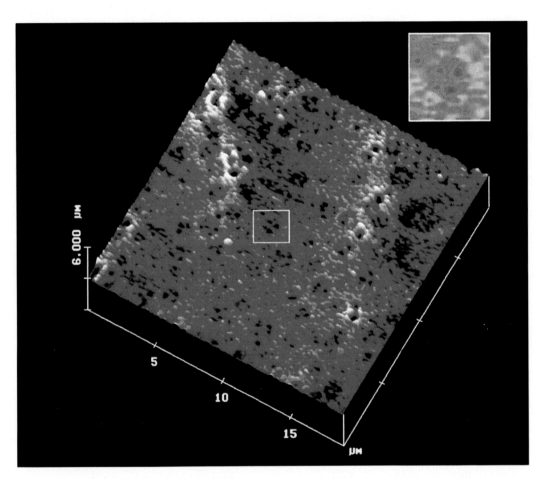

Fig. 2.3 Topology of the apical cell surface of isolated pancreatic acini, observed using AFM. Scattered pits are seen at the apical plasma membrane. One pit (inset) with four depressions is shown. Reproduced with permission from Schneider, S. W., Sritharin, K. C., Geibel, J. P., Oberleithner, H., and Jena, B. P. (1997). Surface dynamics in living acinar cells imaged by atomic force microscopy: Identification of plasma membrane structures involved in exocytosis. *Proc. Nat Acad. Sci. U.S.A.* **94,** 316-321. Copyright (1997) *National Academy of Sciences, U.S.A.*

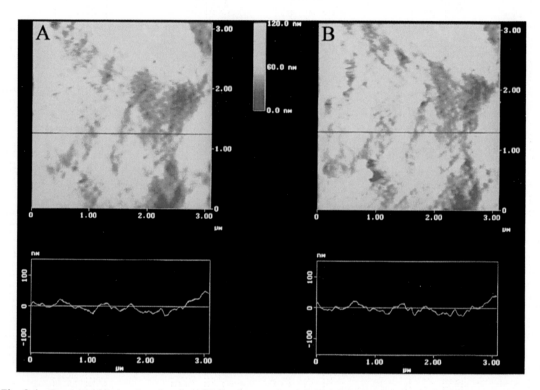

Fig. 2.4 Topology of the basolateral surface of isolated pancreatic acini, observed using the AFM (A). The scan line depicts the relative height and depth of the surface imaged. Following stimulation of secretion, no major change in the basolateral surface profile is observed (B).

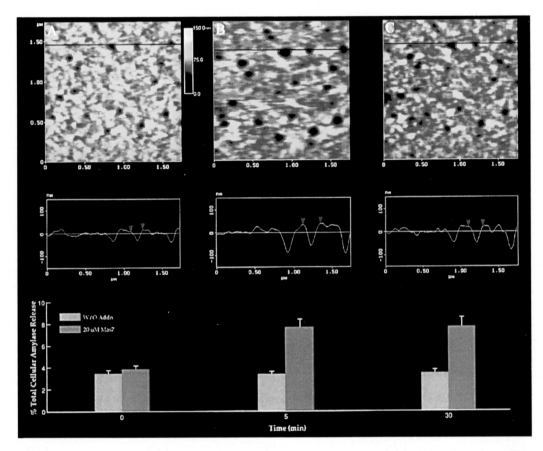

Fig. 2.5 Dynamics of depressions following stimulation of secretion. (A) Several depressions within a pit are shown. The scan line across three depressions in the top panel is represented graphically in the middle panel and defines the diameter and relative depth of the depressions. The middle depression is represented by red arrowheads. The bottom panel represents the percentage of total cellular amylase release in the presence and absence of the secretagogue Mas 7. (B) Notice an increase in the diameter and depth of depressions, correlating with an increase in total cellular amylase release at 5 min after stimulation of secretion. (C) At 30 min after stimulation of secretion, there is a decrease in diameter and depth of depressions, with no further increase in amylase release over the 5-min time point. No significant increases in amylase secretion or depression diameter were observed in either resting acini or those exposed to the nonstimulatory mastoparan analog Mas 17. Reproduced with permission from Schneider, S.W., Sritharin, K. C., Geibel, J. P., Oberleithner, H., and Jena, B. P. (1997). Surface dynamics in living acinar cells imaged by atomic force microscopy: Identification of plasma membrane structures involved in exocytosis. *Proc. Nat Acad. Sci. U.S.A.* **94**, 316-321. Copyright (1997) *National Academy of Sciences, U.S.A.*

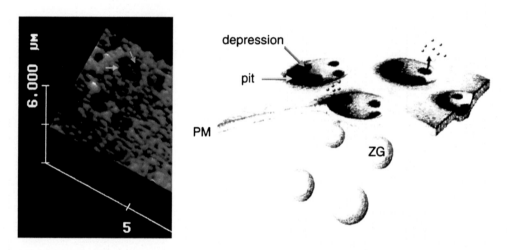

Fig. 2.8 Atomic force micrograph (left) of the apical plasma membrane (PM) in live acinar cells. Large pits containing small depressions (inset) are seen, where secretory vesicles may dock and fuse transiently to release their contents to the extracellular space. A diagram of PM pits containing small depressions, where vesicles fuse to release their contents to the outside (right).

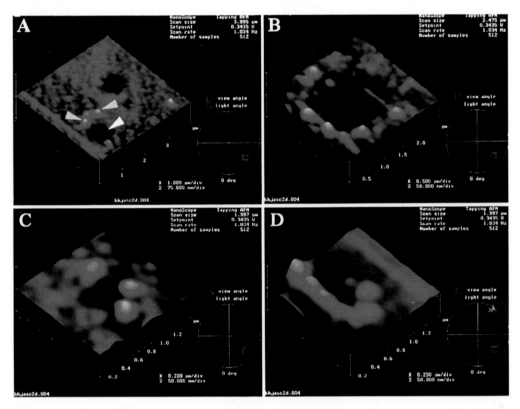

Fig. 2.9 Localization of amylase at the "pit." Atomic force micrography of the apical plasma membrane of fixed pancreatic aci-nar cells following exposure to amylase-gold during stimulation of secretion. Note gold clusters decorating primarily the edges of "pits."

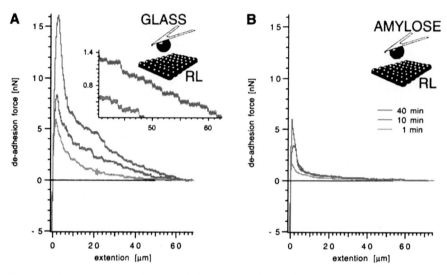

Fig. 5.5 Two sets of deadhesion force traces recorded with a plain glass surface on the sensor (A) and an amylose-passivated surface (B) after contacts of 1, 10, and 40 min at 5 nN on a confluent monolayer of epithelial cells (RL95-2). The contact area was about 500 μm². Inset (A) zooms into the traces where indicated by the circle revealing single rupture steps (Thie *et al.*, 1988).

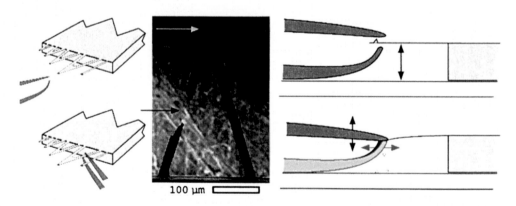

Fig. 5.8 Scheme of the "surgery" on a cantilever with tweezers and a SEM image of such a sensor.

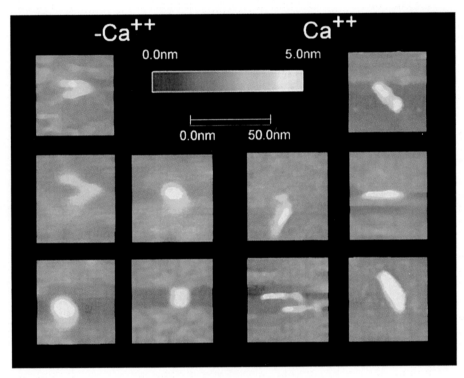

Fig. 6.4 AFM images of VE-cadherin-Fc dimers. The topography images of the proteins adsorbed to mica were recorded with dynamic force microscopy in isotonic buffer. Cadherin dimers show elongated rod-like structure in the presence of Ca^{2+} and globular to V-shaped morphology in the absence of Ca^{2+}. Reproduced with permission from Baumgartner, W., Hinterdorfer, P., Ness, W., Raab, A., Vestweber, D., Schindler, H., and Drenckhahn, D. (2000). Cadherin interaction probed by atomic force microscopy. *Proc. Natl. Acad. Sci. U.S.A.* **8**, 4005-4010. Copyright (2000) *National Academy of Sciences, U.S.A.*

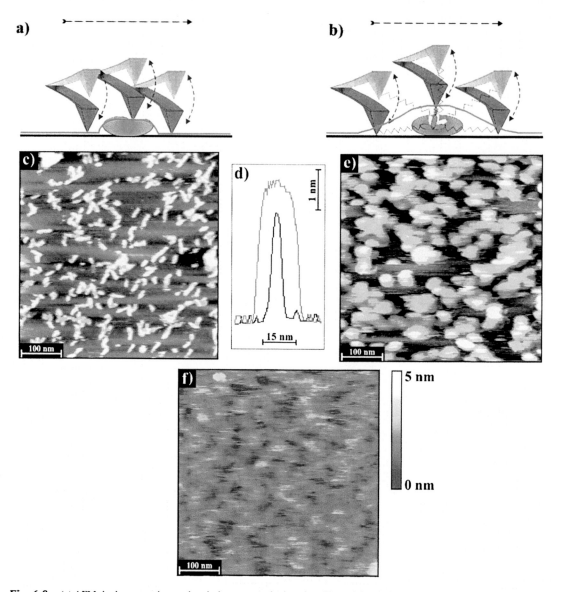

Fig. 6.9 (a) AFM tip–lysozyme interaction during topography imaging. The red line indicates the height profile obtained from a single lysozyme molecule (shown in green) with the AFM using a bare tip. (b) AFM tip–lysozyme interaction during recognition imaging. Half-antibodies (shown in red) are bound to the AFM tip via a flexible tether (jagged line) for the recognition of lysozyme (shown in green) on the surface. Imaging results in a height profile as indicated (red line). (c) Topography image. Single lysozyme molecules can be clearly resolved. Sometimes small lysozyme aggregates are observed. Image size was 500 nm. False color bar for heights from 0 (dark) to 5 (bright) nm. (d) Height profiles. Cross-section profiles of single lysozyme molecules obtained from the topography (black line) and the recognition (red line) image. (e) Recognition image. The bright dots represent recognition profiles of single lysozyme molecules. Imaging was performed using an AFM tip carrying one half-antibody with access to the antigens on the surface. Conditions were exactly the same as those in (c). (f) Block image. The image was obtained under the same conditions as those in the recognition image (e) in the presence of free antibody in solution. Recognition is blocked as apparent from the lack of recognition profiles. Reproduced with permission from Raab, A., Han, W., Badt, D., Smith-Gill, S. J., Lindsay, S. M., Schindler, H., and Hinterdorfer, P. (1999). Antibody recognition imaging by force miscroscopy. *Nature Biotechnol.* **17**, 902-905.

5 µm

Fig. 7.11 Principle of simultaneous AFM and patch-clamp recording. This image displays an arrangement of three separate SEM images with identical scale. The center represents a part of the organ of Corti with V-shaped hair bundles on top. The upper right side displays a small part of the cantilever beam with the pyramidal-shaped tip scanning the top of a V-shaped hair bundle of an OHC. The arrow indicates the scan direction. The lower left side shows a glass pipette in contact with the lipid membrane of an OHC allowing the transmembrane currents to be recorded as the transduction current.

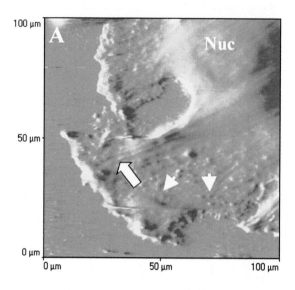

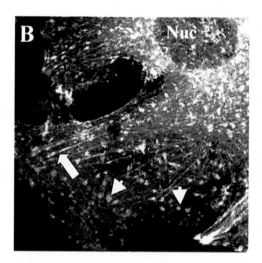

Fig. 8.6 Combined topography AFM imaging with confocal fluorescence microscopy using cells expressing recombinant green fluorescent protein (GFP)-actin. (A) is a topographic AFM image of a melanoma cell showing intracellular cytoskeletal elements (arrow), demonstrated by their height profile in contact mode scanning under different applied forces. (B) is a fluorescence image of the same cellular region as (A) showing the distribution of fluorescent GFP-actin (i.e., F-actin fibers) analyzed using the FITC channel of the linked confocal microscope. A representative region rich in actin fibers is arrowed to show the coincident distribution by both imaging techniques; actin-rich patches ("focal adhesion complexes") are also seen and examples are marked in both images (arrowheads). The nucleus of the cell (Nuc) is identifiable in both images. (The images are sized at $100 \times 100 \ \mu$m.) Adapted from Horton *et al.* (2000), with permission.

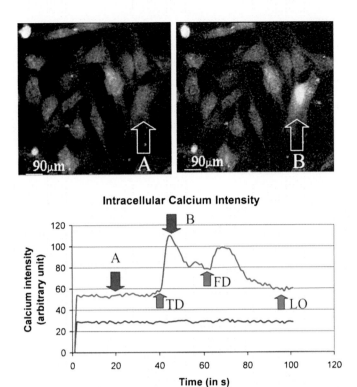

Fig. 8.7 (A) Osteoblasts loaded with Fluo-3 prior to indentation. The cell about to be indented is indicated by the white arrow. (B) Osteoblasts after indentation. The indented cell (indicated by the white arrow) has increased its intracellular calcium concentration. Time course of the calcium intensity within the indented cell. TD (touch down) indicates the time when the AFM cantilever contacts the cell. FD (force–distance) indicates the time when a force–distance curve is taken on the cell. LO (lift off) indicates the time when the AFM cantilever is lifted from contact with the cell surface. Adapted from Charras *et al.* (2000), with permission.

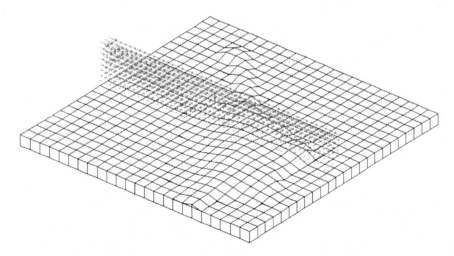

Fig. 8.8 Computational fluid dynamics (CFD) model of a primary osteoblast submitted to laminar flow. The cellular profile was acquired, extracted from its substrate, digitally plated on a flat surface, and then transformed into a mesh suitable for CFD. The flow speed and direction are represented by the red arrows for a region around the center of the cell.

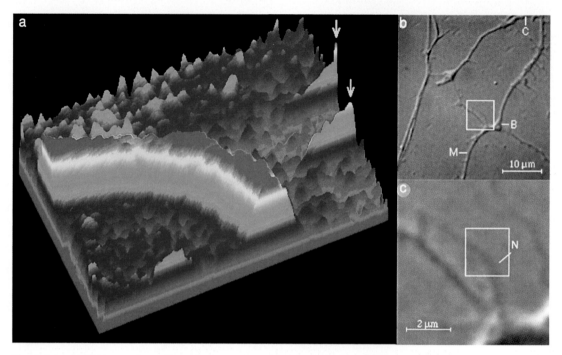

Fig. 9.1 (a) A PFM scan of a small neurite (N) branching (B) from a major neurite (M) of a growing hippocampal neuron. (b, c) Different scale DIC overviews including the scan area. The PFM scan measures the neurite to be 400 nm high and 300 nm wide. Adapted with permission from Florin et al. (1997). *J. Struc. Biol.* **119**, 202–211.

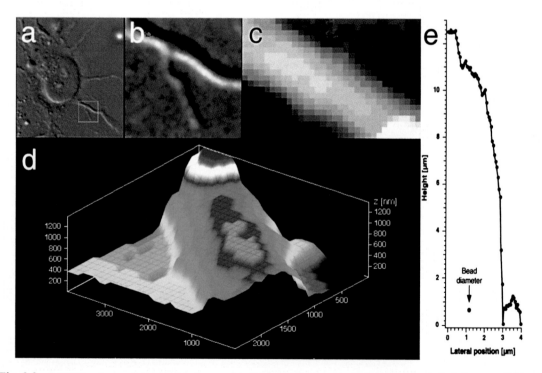

Fig. 9.2 A PFM tapping mode image of another neurite of a hippocampal neuron is shown in (c) and (d). Being up to 1200 nm high, these structures could only be resolved in the PFM tapping mode. Different scale DIC overviews including the scan area are presented in (a) and (b). Part (e) displays a line scan taken from an extreme example of the tapping mode in which the probe climbed down the side of a fibroblast cell near the nucleus.

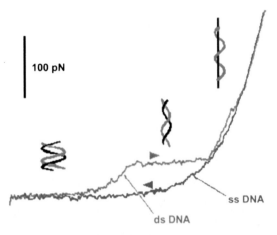

100 pN

ss DNA

ds DNA

Fig. 10.1 A single molecule of overstretched DNA. This graph shows a force measurement of a single tethered molecule of Lambda Digest DNA showing the B–S and the melting transition. Arrowheads indicate pulling direction as follows: DNA stretch is ▶ and DNA relaxation is ◀. During the extension of the molecule (red trace), the DNA first goes through the B–S transition (the plateau) and then melts to single-stranded DNA (ss-DNA) at a higher force. During relaxation of the molecule (blue trace), the DNA does not reanneal, so the curve is a simple freely jointed chain, indicative of ssDNA. The traces were made at a pulling speed of 1 mm/s. Data courtesy of H. Clausen-Schaumann and R. Krautbauer, Gaub Lab, LMU-München. Data were obtained with a cantilever from Park Scientific Microlevers on a Molecular Force Probe from Asylum Research (http://www.asylumresearch.com).

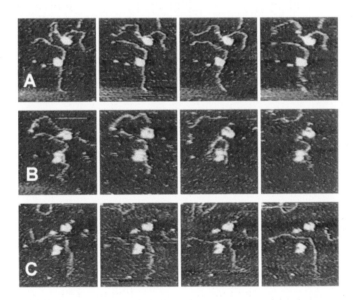

Fig. 10.2 Two active complexes of DNA with RNAP under fluid in an AFM. *E. coli* RNAP transcription complexes were prepared with a 1047-bp DNA template (Guthold *et al.*, 1999; Kasas *et al.*, 1997). (A), (B), and (C) each show a series of four consecutive images at 42-s intervals. (A) DNA strands move near the surface in Zn(II) buffer. (B) 3.5–6 min after the last image in (A). RNAP transcribes and/or detaches from DNA strands after NTPs are introduced. (C) 6–8 min after the last image in (B); Zn(II) buffer is reintroduced. Note that the image quality deteriorates in Zn(II)-free buffer and improves as Zn(II) buffer is reintroduced [see Hansma and Laney (1996)]. DNA Images are nm × 330 nm.

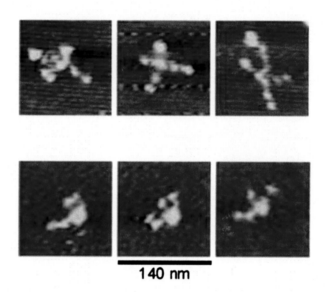

Fig. 10.3 AFM imaging laminin molecules in air shows submolecular structure in the laminin arms (top row). In the sequential images, a single laminin molecule in aqueous solution waves its arms (bottom row).

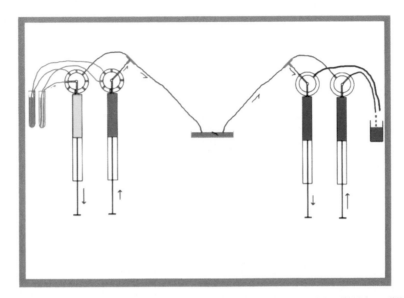

Fig. 10.4 The setup of the fluid-handling system. On the left are the pump-modules that inject fluid from different source solutions. On the right are the additional pump-modules sucking solution from an open fluid cell at the same rate. In the center is the fluid chamber around the sample with the cantilever above the sample.

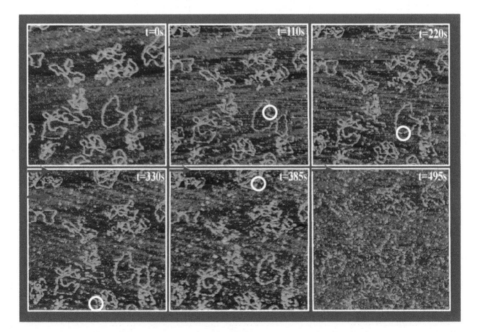

Fig. 10.5 Enzymatic degradation of single DNA molecules in the AFM. A field of DNA molecules (0.5 μg/mL of BlueScript plasmid DNA) in a buffer containing 20 mM Hepes, 5 mM MnCl$_2$, pH 7.6, continuously pumped at 5 μL/s. After the injection of DNase I into the same buffer, the degradation of the molecules can be observed; arrows indicate frame and position in frame where the 10-μL injections occurred. The circles highlight new cuts in DNA molecules. The scan size is 1 μm × 1 μm; the Z range is 7 nm. All imaging was done on a Nanoscope III Multimode-AFM (Digital Instruments). The microscope was operated in fluid tapping mode using cantilever oscillation frequencies between 10 and 20 kHz.

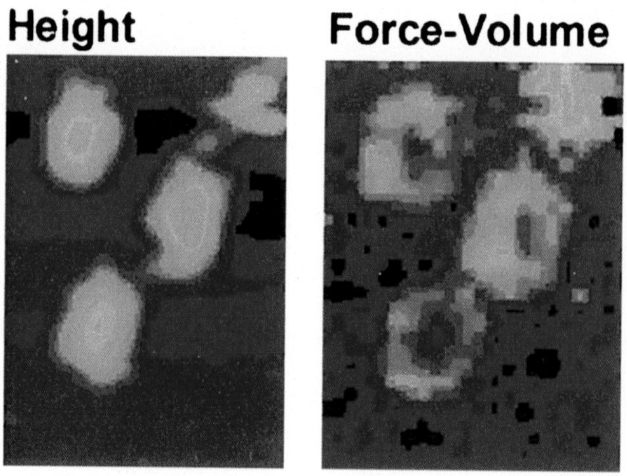

Height   **Force-Volume**

Fig. 10.6 Three cholinergic synaptic vesicles. Height image (left) and force–volume (FV) image (right) of three synaptic vesicles from the electric organ of *Torpedo*. The centers of the vesicles are harder or stiffer than the edges of the vesicles (see Laney *et al.*, 1997).

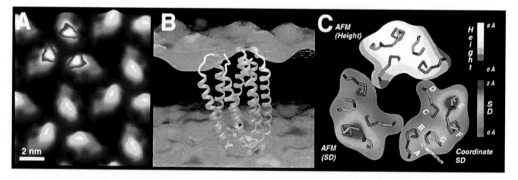

Fig. 13.13 Quantitative analyses of the native cytoplasmic bacteriorhodopsin surface. (A) Correlation average of the AFM topograph recorded at an applied force of 100 pN (Müller, Sass, *et al.*, 1999). Regions with enhanced flexibility are derived from SD maps and superimposed in red to white shades. The vertical brightness range of topograph corresponds to 1 nm. The raw image is displayed as a relief tilted by 5°. (B) Mapping the structural variance of bacteriorhodopsin on the atomic model and the AFM envelope. The atomic model is an average of six models derived from electron and X-ray crystallography, with the coordinate variance mapped from blue (low variance) to red (high variance). The surfaces are derived from the AFM height images, with the SD mapped onto each surface from blue (low SD) to red (high SD). The minimum separation between the surfaces is ~4 nm. Calculations are as given in Heymann *et al.*, 1999. (C) Cytoplasmic surface with each bacteriorhodopsin monomer displaying a different surface property. The surface loops are shown as backbone tracings colored according to the backbone coordinate root-mean-square deviation (SD) calculated after merging five different atomic models of bacteriorhodopsin (Heymann *et al.*, 1999). The gray scale image shows the height map determined by AFM (Fig. 2); the prominent protrusion is the EF loop. The colored monomers represent the coordinate variation (SD) between the atomic models and the SD of the height measured by AFM, respectively. Height and SD maps determined by AFM correlate amazingly well to corresponding data from X-ray and electron crystallography.

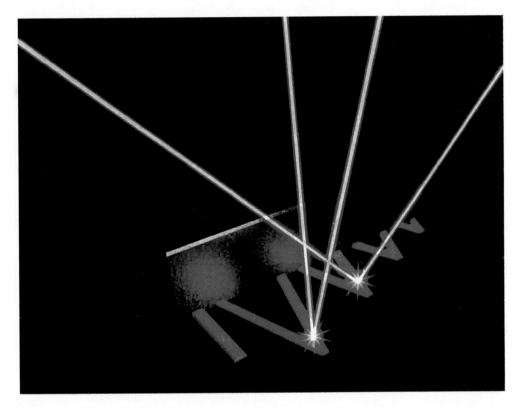

Fig. 15.14 The two optical detection units are used to simultaneously detect the deflection signals from two levers on the same substrate independently.

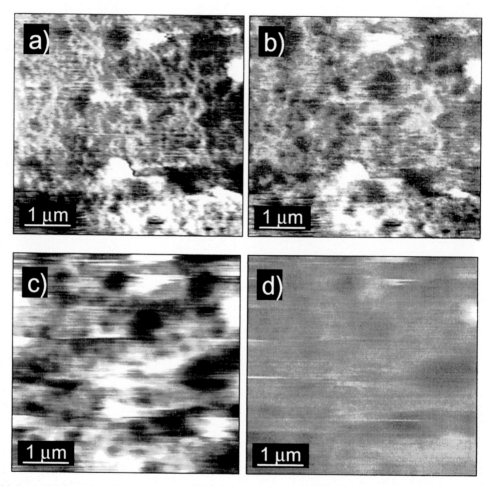

Fig. 16.4 A 30% isotactic polystyrene (iPS)/decahydronapthalene (dekalin) gel under dekalin, imaged with (a and b) and without (c and d) Q control. Topography (a and c) is displayed with a z range of 80 nm and phase contrast (b and d) with a range of 5°.

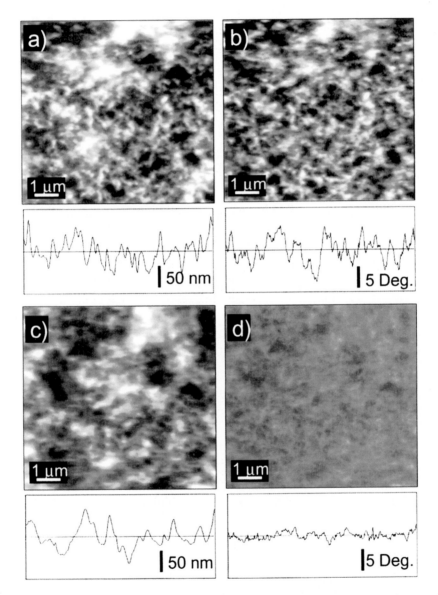

Fig. 16.5 A 1% agarose gel under water, imaged with an effective quality factor of 300 (a and b) and with conventional tapping mode (c and d). Topography (a and c) is displayed with a z range of 150 nm and phase contrast (b and d) with a range of 5°.

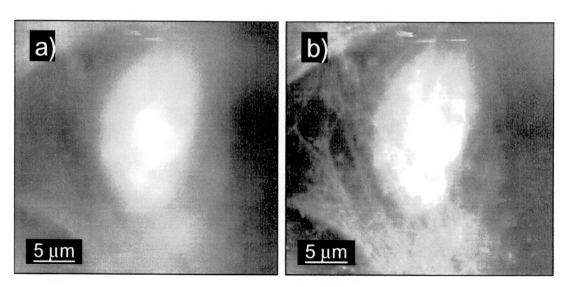

Fig. 16.6 A living rat kidney cell imaged in buffer with an effective quality factor of 300, height (a) z range 2.5 mm and phase (b) range 60°.

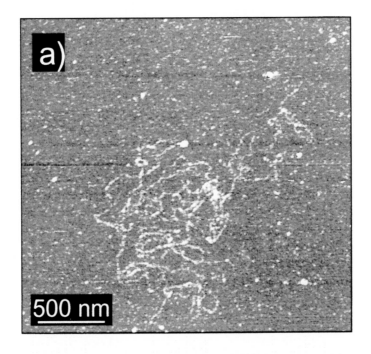

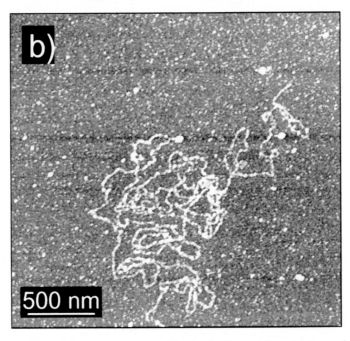

Fig. 16.7 DNA imaged under butanol with (b) and without (a) Q control, z range 3 nm.

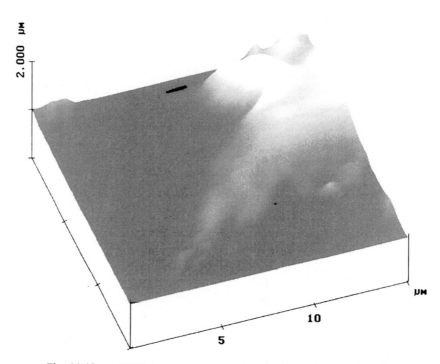

Fig. 16.13 A TDFM image of a living rat kidney cell imaged in growth medium.